BAYESIAN DISEASE MAPPING
HIERARCHICAL MODELING in SPATIAL EPIDEMIOLOGY

CHAPMAN & HALL/CRC
Interdisciplinary Statistics Series

Series editors: N. Keiding, B.J.T. Morgan, C.K. Wikle, P. van der Heijden

Published titles

AN INVARIANT APPROACH TO STATISTICAL ANALYSIS OF SHAPES	S. Lele and J. Richtsmeier
ASTROSTATISTICS	G. Babu and E. Feigelson
BAYESIAN DISEASE MAPPING: HIERARCHICAL MODELING IN SPATIAL EPIDEMIOLOGY	Andrew B. Lawson
BIOEQUIVALENCE AND STATISTICS IN CLINICAL PHARMACOLOGY	S. Patterson and B. Jones
CLINICAL TRIALS IN ONCOLOGY SECOND EDITION	J. Crowley, S. Green, and J. Benedetti
CORRESPONDENCE ANALYSIS IN PRACTICE, SECOND EDITION	M. Greenacre
DESIGN AND ANALYSIS OF QUALITY OF LIFE STUDIES IN CLINICAL TRIALS	D.L. Fairclough
DYNAMICAL SEARCH	L. Pronzato, H. Wynn, and A. Zhigljavsky
GENERALIZED LATENT VARIABLE MODELING: MULTILEVEL, LONGITUDINAL, AND STRUCTURAL EQUATION MODELS	A. Skrondal and S. Rabe-Hesketh
GRAPHICAL ANALYSIS OF MULTI-RESPONSE DATA	K. Basford and J. Tukey
INTRODUCTION TO COMPUTATIONAL BIOLOGY: MAPS, SEQUENCES, AND GENOMES	M. Waterman
MARKOV CHAIN MONTE CARLO IN PRACTICE	W. Gilks, S. Richardson, and D. Spiegelhalter
MEASUREMENT ERROR AND MISCLASSIFICATION IN STATISTICS AND EPIDEMIOLOGY: IMPACTS AND BAYESIAN ADJUSTMENTS	P. Gustafson

Published titles

META-ANALYSIS OF BINARY DATA USING PROFILE LIKELIHOOD   D. Böhning, R. Kuhnert, and S. Rattanasiri

STATISTICAL ANALYSIS OF GENE EXPRESSION MICROARRAY DATA   T. Speed

STATISTICAL AND COMPUTATIONAL PHARMACOGENOMICS   R. Wu and M. Lin

STATISTICS IN MUSICOLOGY   J. Beran

STATISTICAL CONCEPTS AND APPLICATIONS IN CLINICAL MEDICINE   J. Aitchison, J.W. Kay, and I.J. Lauder

STATISTICAL AND PROBABILISTIC METHODS IN ACTUARIAL SCIENCE   P.J. Boland

STATISTICS FOR ENVIRONMENTAL BIOLOGY AND TOXICOLOGY   A. Bailer and W. Piegorsch

STATISTICS FOR FISSION TRACK ANALYSIS   R.F. Galbraith

Interdisciplinary Statistics

BAYESIAN DISEASE MAPPING
HIERARCHICAL MODELING in SPATIAL EPIDEMIOLOGY

Andrew B. Lawson
Medical University of South Carolina (MUSC)
Charleston, U.S.A.

CRC Press
Taylor & Francis Group
Boca Raton London New York

CRC Press is an imprint of the
Taylor & Francis Group, an **informa** business

A CHAPMAN & HALL BOOK

Chapman & Hall/CRC
Taylor & Francis Group
6000 Broken Sound Parkway NW, Suite 300
Boca Raton, FL 33487-2742

© 2009 by Taylor & Francis Group, LLC
Chapman & Hall/CRC is an imprint of Taylor & Francis Group, an Informa business

No claim to original U.S. Government works
Printed in the United States of America on acid-free paper
10 9 8 7 6 5 4 3 2 1

International Standard Book Number-13: 978-1-58488-840-6 (Hardcover)

This book contains information obtained from authentic and highly regarded sources. Reasonable efforts have been made to publish reliable data and information, but the author and publisher cannot assume responsibility for the validity of all materials or the consequences of their use. The authors and publishers have attempted to trace the copyright holders of all material reproduced in this publication and apologize to copyright holders if permission to publish in this form has not been obtained. If any copyright material has not been acknowledged please write and let us know so we may rectify in any future reprint.

Except as permitted under U.S. Copyright Law, no part of this book may be reprinted, reproduced, transmitted, or utilized in any form by any electronic, mechanical, or other means, now known or hereafter invented, including photocopying, microfilming, and recording, or in any information storage or retrieval system, without written permission from the publishers.

For permission to photocopy or use material electronically from this work, please access www.copyright.com (http://www.copyright.com/) or contact the Copyright Clearance Center, Inc. (CCC), 222 Rosewood Drive, Danvers, MA 01923, 978-750-8400. CCC is a not-for-profit organization that provides licenses and registration for a variety of users. For organizations that have been granted a photocopy license by the CCC, a separate system of payment has been arranged.

Trademark Notice: Product or corporate names may be trademarks or registered trademarks, and are used only for identification and explanation without intent to infringe.

Library of Congress Cataloging-in-Publication Data

Lawson, Andrew (Andrew B.)
  Bayesian disease mapping : hierarchical modeling in spatial epidemiology / Andrew B. Lawson.
     p. ; cm. -- (Chapman & Hall/CRC interdisciplinary statistics series)
  Includes bibliographical references and index.
  ISBN 978-1-58488-840-6 (hardcover : alk. paper)
  1. Medical mapping. 2. Epidemiology--Statistical methods. 3. Bayesian statistical decision theory. I. Title. II. Series: Interdisciplinary statistics.
  [DNLM: 1. Epidemiologic Methods. 2. Bayes Theorem. 3. Statistics as Topic. 4. Topography, Medical--methods. WA 950 L425b 2008]

RA792.5.L387 2008
614.4'2--dc22   2008022718

Visit the Taylor & Francis Web site at
http://www.taylorandfrancis.com

and the CRC Press Web site at
http://www.crcpress.com

Contents

List of Tables   xiii

Preface   xv

Author   xvii

I   Background   1

1 Introduction   **3**
   1.1 Datasets 5

2 Bayesian Inference and Modeling   **19**
   2.1 Likelihood Models 19
       2.1.1 Spatial Correlation 20
   2.2 Prior Distributions 22
       2.2.1 Propriety 22
       2.2.2 Noninformative Priors 22
   2.3 Posterior Distributions 24
       2.3.1 Conjugacy 25
       2.3.2 Prior Choice 25
   2.4 Predictive Distributions 26
       2.4.1 Poisson–Gamma Example 26
   2.5 Bayesian Hierarchical Modeling 27
   2.6 Hierarchical Models 27
   2.7 Posterior Inference 28
       2.7.1 A Bernoulli and Binomial Example 30
   2.8 Exercises 34

3 Computational Issues   **35**
   3.1 Posterior Sampling 35
   3.2 Markov Chain Monte Carlo Methods 36
   3.3 Metropolis and Metropolis–Hastings Algorithms 37
       3.3.1 Metropolis Updates 38
       3.3.2 Metropolis–Hastings Updates 38
       3.3.3 Gibbs Updates 38
       3.3.4 M–H versus Gibbs Algorithms 39
       3.3.5 Special Methods 40

	3.3.6	Convergence	40
	3.3.7	Subsampling and Thinning	45
3.4	Perfect Sampling		47
3.5	Posterior and Likelihood Approximations		48
	3.5.1	Pseudolikelihood and Other Forms	48
	3.5.2	Asymptotic Approximations	50
3.6	Exercises		53

4 Residuals and Goodness-of-Fit 55
4.1	Model GOF Measures		55
	4.1.1	Deviance Information Criterion	56
	4.1.2	Posterior Predictive Loss	57
4.2	General Residuals		59
4.3	Bayesian Residuals		61
4.4	Predictive Residuals and the Bootstrap		62
	4.4.1	Conditional Predictive Ordinates	63
4.5	Interpretation of Residuals in a Bayesian Setting		64
4.6	Exceedence Probabilities		64
4.7	Exercises		67

II Themes 71

5 Disease Map Reconstruction and Relative Risk Estimation 73
5.1	Introduction to Case Event and Count Likelihoods		73
	5.1.1	Poisson Process Model	73
	5.1.2	Conditional Logistic Model	75
	5.1.3	Binomial Model for Count Data	76
	5.1.4	Poisson Model for Count Data	76
5.2	Specification of the Predictor in Case Event and Count Models		77
	5.2.1	Bayesian Linear Model	79
5.3	Simple Case and Count Data Models with Uncorrelated Random Effects		80
	5.3.1	Gamma and Beta Models	82
	5.3.2	Log-Normal/Logistic-Normal Models	84
5.4	Correlated Heterogeneity Models		84
	5.4.1	Conditional Autoregressive (CAR) Models	86
	5.4.2	Fully-Specified Covariance Models	90
5.5	Convolution Models		91
5.6	Model Comparison and Goodness-of-Fit Diagnostics		92
	5.6.1	Residual Spatial Autocorrelation	94
5.7	Alternative Risk Models		96
	5.7.1	Autologistic Models	96
	5.7.2	Spline-Based Models	101
	5.7.3	Zip Regression Models	102
	5.7.4	Ordered and Unordered Multicategory Data	107

Contents ix

        5.7.5   Latent Structure Models 108
  5.8 Edge Effects 111
        5.8.1   Edge Weighting Schemes and McMC Methods 113
        5.8.2   Discussion and Extension to Space–Time 115
  5.9 Exercises 116
        5.9.1   Maximum Likelihood 116
        5.9.2   Poisson–Gamma Model: Posterior and Predictive
                    Inference 117
        5.9.3   Poisson-Gamma Model: Empirical Bayes 117

6 Disease Cluster Detection   119

  6.1 Cluster Definitions 119
        6.1.1   Hot Spot Clustering 121
        6.1.2   Clusters as Objects or Groupings 121
        6.1.3   Clusters Defined as Residuals 121
  6.2 Cluster Detection using Residuals 122
        6.2.1   Case Event Data 122
        6.2.2   Count Data 126
  6.3 Cluster Detection Using Posterior Measures 130
  6.4 Cluster Models 133
        6.4.1   Case Event Data 133
        6.4.2   Count Data 143
        6.4.3   Markov Connected Component Field (MCCF) Models 148
  6.5 Edge Detection and Wombling 149

7 Ecological Analysis   151

  7.1 General Case of Regression 151
  7.2 Biases and Misclassification Error 158
        7.2.1   Ecological Biases 158
  7.3 Putative Hazard Models 165
        7.3.1   Case Event Data 166
        7.3.2   Aggregated Count Data 172
        7.3.3   Spatiotemporal Effects 176

8 Multiple Scale Analysis   185

  8.1 Modifiable Areal Unit Problem (MAUP) 185
        8.1.1   Scaling Up 185
        8.1.2   Scaling Down 187
        8.1.3   Multiscale Analysis 187
  8.2 Misaligned Data Problem(MIDP) 190
        8.2.1   Predictor Misalignment 191
        8.2.2   Outcome Misalignment 198
        8.2.3   Misalignment and Edge Effects 200

9 Multivariate Disease Analysis — 201

- 9.1 Notation for Multivariate Analysis 201
  - 9.1.1 Case Event Data 201
  - 9.1.2 Count Data 202
- 9.2 Two Diseases 202
  - 9.2.1 Case Event Data 202
  - 9.2.2 Count Data 204
  - 9.2.3 Georgia County Level Example (3 Diseases) 206
- 9.3 Multiple Diseases 207
  - 9.3.1 Case Event Data 209
  - 9.3.2 Count Data 216
  - 9.3.3 Multivariate Spatial Correlation and MCAR Models . 219
  - 9.3.4 Georgia Chronic Ambulatory Care-Sensitive Example 222

10 Spatial Survival and Longitudinal Analysis — 227

- 10.1 General Issues 227
- 10.2 Spatial Survival Analysis 228
  - 10.2.1 Endpoint Distributions 228
  - 10.2.2 Censoring 229
  - 10.2.3 Random Effect Specification 230
  - 10.2.4 General Hazard Model 232
  - 10.2.5 Cox Model 232
  - 10.2.6 Extensions 233
- 10.3 Spatial Longitudinal Analysis 234
  - 10.3.1 General Model 237
  - 10.3.2 Seizure Data Example 237
  - 10.3.3 Missing Data 241
- 10.4 Extensions to Repeated Events 243
  - 10.4.1 Simple Repeated Events 243
  - 10.4.2 More Complex Repeated Events ... 244
  - 10.4.3 Fixed Time Periods 247

11 Spatiotemporal Disease Mapping — 255

- 11.1 Case Event Data 255
- 11.2 Count Data 257
  - 11.2.1 Georgia Low Birth Weight Example 262
- 11.3 Alternative Models 266
  - 11.3.1 Autologistic Models 266
  - 11.3.2 Latent Structure ST Models 268
- 11.4 Infectious Diseases 271
  - 11.4.1 Case Event Data 272
  - 11.4.2 Count Data 273
  - 11.4.3 Special Case: Veterinary Disease Mapping 276

A Basic R and WinBUGS — 283
- A.1 Basic R Usage ... 283
    - A.1.1 Data ... 283
    - A.1.2 Graphics ... 284
- A.2 Use of R in Bayesian Modeling ... 287
- A.3 WinBUGS ... 290
    - A.3.1 Simulation ... 291
    - A.3.2 Model Code ... 291
- A.4 R2WinBUGS Function ... 298
- A.5 BRugs ... 302
- A.6 Maps on R and GeoBUGS ... 305

B Selected WinBUGS Code — 307
- B.1 Code for the Convolution Model (Chapter 5) ... 307
- B.2 Code for Spatial Spline Model (Chapter 5) ... 308
- B.3 Code for the Spatial Autologistic Model (Chapter 6) ... 308
- B.4 Code for Logistic Spatial Case Control Model (Chapter 6) ... 309
- B.5 Code for PP Residual Model (Chapter 6) ... 309
    - B.5.1 Same Model with Uncorrelated Random Effect ... 310
- B.6 Code for the Logistic Spatial Case-Control Model (Chapter 6) ... 310
- B.7 Code for Poisson Residual Clustering Example (Chapter 6) ... 312
- B.8 Code for the Proper CAR Model (Chapter 7) ... 312
- B.9 Code for the Multiscale Model for PH and County Level Data (Chapter 8) ... 313
- B.10 Code for the Shared Component Model for Georgia Asthma and COPD (Chapter 9) ... 314
- B.11 Code for the Seizure Example with Spatial Effect (Chapter 10) ... 315
- B.12 Code for the Knorr-Held Model for Space–Time Relative Risk Estimation (Chapter 11) ... 316
- B.13 Code for the Space–Time Autologistic Model (Chapter 11) ... 316

C R Code for Thematic Mapping — 319

References — 321

Index — 339

List of Tables

5.1 Comparison of convolution and uncorrelataed heterogeneity models for the Georgia oral cancer dataset 93
5.2 DICs for three models: autologistic with no randon effects; autlogistic with UH component; convolution model with UH and CH 98

7.1 Model fitting results for a variety of models for the South Carolina congenital mortality data, (See text for details). 154
7.2 Results for a variety of models fitted to 1988 respiratory cancer incident counts for counties of Ohio 175
7.3 Ohio respiratory cancer (1979–1988): putative source model fits 181

8.1 Goodness of fit results for separate and joint models for Georgia oral cancer PH-county level data 190
8.2 Model fit results for the point misalignment example 197

9.1 Model comparisons for the three disease examples: joint, common, and shared 207
9.2 Correlation of spatial random effects under the MCAR model for the Georgia chronic disease example: upper triangle: correlation of the spatially structured effects; lower triangle: correlation of the relative risks 225

10.1 Comparison of four models for the seizure data: basic and Models 1–3 239
10.2 Posterior average estimates of the model parameters under the four different seizure models 240
10.3 Parameter estimates for the basic model compared to the best fitting model (Model 1) 240

11.1 Space–time models for the Georgia oral cancer dataset; models are explained in text. 264
11.2 Autologistic space–time models: models 1–4; convolution model (Model 5) 267

Preface

Bayesian approaches to biostatistical problems have become commonplace in epidemiological, medical, and public health applications. Indeed the use of Bayesian methodology has seen great advances since the introduction of, first, BUGS, and then WinBUGS. WinBUGS is a free software package that allows the development and fitting of relatively complex hierarchical Bayesian models. The introduction of fast algorithms for sampling posterior distributions in the 1990s has meant that relatively complex Bayesian models can be fitted in a straightforward manner. This has led to a great increase in the use of Bayesian approaches not only to medical research problems but also in the field of public heath. One area of important practical concern is the analysis of the geographical distribution of health data found commonly in both public health databases and in clinical settings. Often population level data are available via government data sources such as online community health systems (e.g., for the US state of South Carolina this is http://scangis.dhec.sc.gov/scan/, while in the US state of Georgia it is http://oasis.state.ga.us/) or via centrally organized data registries where individual patient records are held. Cancer registry data (such as SEER in the United States) usually include individual diagnosis type and date as well as demographic information and so is at a finer level of resolution.

Most government sources hold publicly accessible *aggregated* health data due to confidentiality requirements. The resulting count data, usually available at county or postal/census region level, can yield important insights into the general spatial variation of disease in terms of incidence or prevalence. It can also be analyzed with respect to health inequalities or disparities related to health service provision. While this form of data and its analysis are relatively well documented, there are other areas of novel application of spatial methodology that are less well recognized currently. For example, one source of *individual level* data are disease registries where notification of a disease case leads to registering the individual and their demographic details. In addition, some diagnostic information is usually held. This is typically found on cancer registries, but other diseases have similar registration processes. In clinical trials, or community-based behavioral intervention trials, individual patient information is often held and disease progression is noted over the duration of the trial. In two ways, it may be important or relevant to consider spatial information in such applications. First, the recruitment or dropout process for trials may have a spatial component. Second, there may be unobserved confounding variables that have a spatial expression over the course of the trial. These issues may lead to the consideration of longitudinal or survival analyses where geo-referencing is admitted as a confounding factor. In general, the focus area of this work is, in effect, *spatial biostatistics*, as the inclusion of clinical and registry-level analysis, as well as population level analyses, lies within the range of applications for the methods covered.

In this work I have tried to provide an overview of the main areas of Bayesian hierarchical modeling in its application to geographical analysis of disease. I have tried to orient the coverage to both deal with population level analyses and also individual level analyses resulting from cancer registry data and also the possibility of the use of data on health service utilization (disease progression via health practitioner visits, etc.), and designed studies (clinical or otherwise). To this end, as well as including chapters on more conventional topics such as relative risk estimation and clustering, I have included coverage of spatial survival and longitudinal analysis, with a section on repeated event analysis.

There are many people that have helped in the production of this work. In particular, I would like to recognize sources of encouragement from Andrew Cliff, Sudipto Banerjee, Emmanuel Lesaffre, Peter Rogerson, and Allan Clark. In addition, I have to thank a range of postdoctoral fellows and graduate students who have provided help at various times: Hae-Ryoung Song, Ji-in Kim, Huafeng Zhou, Kun Huang, Junlong Wu, Yuan Liu, and Bo Ma. I must also thank those at CRC press for great help in finalizing the work. In particular, Rob Calver for general production support and Shashi Kumar for Latex help.

Finally I would like to acknowledge the continual and patient support of my family, and, in particular, Pat for her understanding during the sometimes fraught activity of book writing.

Andrew Lawson
Charleston, United States
2008

Author

Andrew B. Lawson is professor of biostatistics in the Department of Biostatistics, Bioinformatics, and Epidemiology, Medical University of South Carolina. Previously, he was professor of biostatistics in the Department of Epidemiology and Biostatistics, University of South Carolina. Prior to USC, he was reader in statistics in the Mathematical Sciences Department, University of Aberdeen, United Kingdom. He received a PhD in spatial statistics from the University of St Andrews, United Kingdom. He has over 70 journal papers on the subject of spatial epidemiology, spatial statistics, and related areas. In addition to a number of book chapters, he is the author of 7 books in areas related to spatial epidemiology. In addition to associate editorships on a variety of journals, he is an advisor in disease mapping and risk assessment for the World Health Organization (WHO).

Dr. Lawson's research interests currently focus on analysis of clustered disease maps and in the broad area of spatial and spatiotemporal disease surveillance. He is currently involved in several National Institutes of Health (NIH) funded projects in the area of surveillance and missing data and cluster method evaluation. He also works in the area of nutritional measurement error and Bayesian modeling.

Part I

Background

1

Introduction

Some basic ideas and history concerning Bayesian methods

Bayesian methods have become commonplace in modern statistical applications. The acceptance of these methods is a relatively recent phenomenon however. This acceptance has been facilitated in large measure by the development of fast computational algorithms that were simply not commonly available or accessible as recently as the late 1980s. The widespread adoption of Markov chain Monte Carlo (McMC) methods for posterior distribution sampling has led to a large increase in Bayesian applications. Most recently Bayesian methods have become commonplace in epidemiology, and the pharmaceutical industry, and they are becoming more widely accepted in Public Health practice. As early as 1993, review articles appeared extolling the virtues of McMC in medical applications (Gilks et al., 1993). This increase in use has been facilitated by the implementation of software which provides a platform for the posterior distribution sampling which is necessary when relatively complex Bayesian models are employed. The development of the package BUGS (Bayesian inference Using Gibbs Sampling) and its Windows incarnation WinBUGS (Spiegelhalter et al., 2007) have had a huge effect on the dissemination and acceptance of these methods. To quote Cowles (2004): "A brief search for recently published papers referencing WinBUGS turned up applications in food safety, forestry, mental health policy, AIDS clinical trials, population genetics, pharmacokinetics, pediatric neurology, and other diverse fields, indicating that Bayesian methods with WinBUGS indeed are finding widespread use."

Basic ideas in Bayesian modeling stem from the extension of the likelihood paradigm to allow parameters within the likelihood model to have distributions. These distributions are called prior distributions. Thus parameters are allowed to be stochastic. By making this allowance, in turn, parameters in the prior distributions of the likelihood parameters can also be stochastic. Hence a natural parameter hierarchy is established. These hierarchical models form the basis of inference under the Bayesian paradigm. By combining the likelihood (data) model with suitable prior distributions for the parameters, a so-called posterior distribution is formed which describes the behavior of the parameters after having seen the data. There is a natural sequence underlying this approach that allows it to well describe the progression of scientific advance. The prior distribution is the current idea about the variation in the parameter set, data is collected (and modeled within the likelihood) and this updates our understanding of the parameter set variation via a posterior distribution. This posterior distribution can become the prior distribution for the parameter set before the next data experiment.

Various reviews of the different aspects of the Bayesian paradigm and modeling are now available. Among these a seminal work on Bayesian Theory has been provided by Bernardo and Smith (1994). Recent general reviews of Bayesian methods appear in Leonard and Hsu (1999), Carlin and Louis (2000), and Gelman et al. (2004). Overviews of Bayesian modeling are also provided in Congdon (2003), Congdon (2005), Gelman and Hill (2007), and Congdon (2007). For overview of McMC methods, Gamerman and Lopes (2006) and Marin and Robert (2007) are useful starting points. For fuller coverage then Gilks et al. (1996) and Robert and Casella (2005) are useful resources.

Some basic ideas and history concerning disease mapping

Disease mapping goes under a variety of names, some of which are: spatial epidemiology, environmental epidemiology, disease mapping, small area health studies. However at the center of these different names are two characteristics. First a spatial or geographical distribution is the focus and so the relative location of events is important. This brings the world of geographical information systems into play, while also including spatial statistics as a key component. The second ingredient is that of disease and it is the spatial distribution of disease that is the focus. Hence the fundamental issue is how to analyze disease incidence or prevalence when we have geographical information. Sometimes this is called *geo-referenced* disease data, specifying the labeling of outcomes with spatial tags.

It is apparent that none of the names listed above include the term "statistics." This is unfortunate as it is often the case that statistical methodology (especially methodology from spatial statistics) is involved in the analysis of maps of disease. A more appropriate description of the area of focus of this work is *Spatial Biostatistics* as this emphasizes the broad nature of the focus. In later chapters, I focus on both population level analysis and analysis of clinical studies where longitudinal and survival data arise and so more conventional biostatistical applications are also stressed here.

The area of disease mapping has had a long but checkered history. Some of the first epidemiological studies were geographic in nature. For example, the study of the spatial distribution of cholera victims around the Broad

Street pump by John Snow (Snow, 1854) was one of the earliest epidemiological studies and it was innately geographical. The use of geo-referenced data in observational studies was overtaken and subordinated to more rigorous clinical studies in medicine and often the geo-referencing is assumed to be irrelevant. In more recent decades, the development of fast computational platforms and with them geographical information system (GIS) capabilities has allowed a much greater sophistication in the handling of geo-referenced data. This coupled with advances in computational algorithms has allowed many spatial problems to be addressed effectively with accessible software. The recent rise on open source (free) software has provided wide access to students and professionals. The existence of free software, such as **GRASS**, **R**, and **WinBUGS**, enhances this access.

Within the area of spatial biostatistics the major advances have been relatively recent. In the area of risk estimation and modeling the development of Bayesian models with random effects fitted via McMC was first proposed by Besag et al. (1991). Since that time there has been a large increase in the use of such methods. The use of scanning methods for disease cluster detection was also developed quite recently by Kulldorff and Nagarwalla (1995). There is now widespread use of scanning methods in cluster detection and surveillance. The widespread use of Bayesian methods in most areas of disease mapping is now well established and there is a need to review and summarize these disparate strands in one place. The focus of this is on the use of Bayesian models and computational methods in application to studies in spatial biostatistics.

Recent reviews of the general area of application of statistics in disease mapping can be found in Lawson et al. (1999), Elliott et al. (2000), Waller and Gotway (2004), Lawson (2006b), and Lawson and Banerjee (2008). For a more epidemiologic slant, the edited work by Elliott et al. (1992) is useful. For a GIS slant on health both human and veterinary see for example Maheswaran and Craglia (2004), Durr and Gatrell (2004), or Pfeiffer et al. (2008).

1.1 Datasets

In the following chapters, a range of data sets are analyzed. Most of these are available publicly and can be downloaded from public domain web sites. In a few cases the data are confidential and cannot be accessed widely without approval. These latter datasets are not made available here. All other datasets are available (along with a selection of relevant programs) at the Web site http://www.sph.sc.edu/alawson/default.htm.

The datasets listed here are in order of appearance in the book, and are those where it is possible to make available the data. Some datasets are well

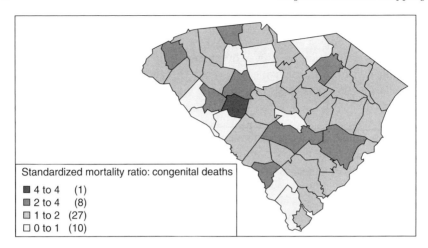

FIGURE 1.1
South Carolina congenital deaths by county 1990: standardized mortality ratio.

known and are not displayed as they are viewable elsewhere.

1. South Carolina (SC) county level congenital anomaly deaths 1990. This dataset consists of counts of deaths from congenital anomalies within the year 1990 in 46 counties of South Carolina, United States, along with expected death rates computed from the statewide rate, without age–gender standardization, and applied to the county total population. The data is available from the SCAN system at South Carolina Department of Health and Environmental Control (SCDHEC): (http://scangis.dhec.sc.gov/scan/). Both maps and tabulations of this data are available online from that source. Figure 1.1 displays the standardized mortality ratios for this example computed as the ratio of the count of disease to the expected rate within each county. The expected rate is calculated from the standardized rate from the SC statewide incidence rate of the condition.

2. Georgia oral cancer mortality 2004. This dataset consists of counts of oral cancer deaths within the 159 counties of the state of Georgia, United States. It also includes expected rates computed from the statewide overall rate for 2004, and applied to the total county population. The data is available from the OASIS online system of the Georgia DHR Division of Public Health (http://oasis.state.ga.us/). Both maps and tabulations of this data are available online from that source. Figure 1.2 displays the standardized mortality ratios for this example computed as the ratio of the count of disease to the expected rate within each county. Expected rates are computed from the statewide incidence rate.

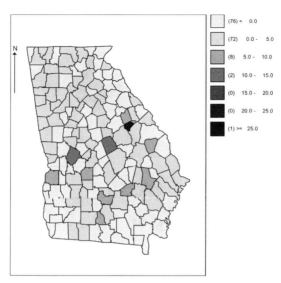

FIGURE 1.2
Georgia, United States, oral cancer standardized mortality ratio by county for 2004, using statewide rate.

3. Ohio respiratory cancer mortality count data set. The full dataset covers 1968–1988 and consists of count data broken by age, gender, and race for the state of Ohio, United States. This is available from http://www.stat.uni-muenchen.de/service/datenarchiv/ohio/ohio_e.html. This full data set has been analyzed many times (see e.g., Knorr-Held and Besag, 1998; Carlin and Louis, 2000). Subsets of the data set using total counts in counties, or functions of counts, are used here: for example 1968, 1979 in Chapter 5 and 1979–1988 in Chapter 7. The data are accompanied by expected rates computed for each county from the statewide rate stratified by age–gender groups and applied to these groups in the county population and summed.

4. Ohio county data for the autologistic model: counts of first order and second order neighbors of each county and their totalled binary outcome after thresholding by exceedence for the state of Ohio, United States. These data are used in autologistic modeling when binary outcomes are observed.

5. Georgia asthma mortality 2000. This dataset consists of counts of asthma deaths within the 159 counties of the state of Georgia, USA. It also includes expected rates computed from the statewide overall rate for 2000, and applied to the total county population. The data is available from the OASIS online system of the Georgia Department of Human

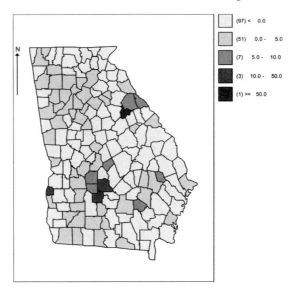

FIGURE 1.3
Georgia county level asthma mortality for year 2000: standardized mortality ratios based on the statewide standard population rate.

Resources (DHR) Division of Public Health (http://oasis.state.ga.us/). Both maps and tabulations of this data are available online from that source. Figure 1.3 displays the standardized mortality ratios for this example computed as the ratio of the count of deaths to the expected death rate within each county. Expected rates are computed from the statewide incidence rate.

6. Larynx cancer incidence, Lancashire NW England (1973–1984). This dataset was made available by Peter Diggle. Variants of the dataset have appeared at different times. The dataset consists of the residential addresses of cases of larynx cancer (58) reported for the period 1973–1984 in the Charnock Richards area of Lancashire NW England, United Kingdom. Within the map area is an incinerator (location: easting 35450, northing 41400) and the data was originally collected to help in the analysis of larynx cancer incidence around this location. Besides the case address locations there are 978 control disease addresses (respiratory cancer incident cases) within the same study region.

7. South Carolina congenital anomaly deaths 1990: additional covariate information. For each county, the percentage poverty listed under the US census of 1990 and also the average household income for the same census is given.

Introduction 9

8. Georgia oral cancer 2004 multi-level data: This dataset consists of counts of oral cancer and expected rates for the 159 counties of Georgia as well as the counts and expected rates for the 18 public health districts of Georgia for the same period. The public health districts are groupings of counties and are an aggregation of the county level data. The expected rates are computed from statewide rates and applied to the local unit population (district or county). Figure 1.4 displays the geographies for the 18 public health districts and 159 counties of Georgia, United States. In Chapter 8 these geographies are used with associated count data to examine multiple scale models.

9. Anonymized binary outcome (misalignment example): this dataset consists of 140 binary indicators (0: control, 1: case) and their address locations and the measured soil chemical concentration of arsenic (As) found in a network of 119 sampling sites. The soil chemical values must be interpolated to the sites of the binary outcome variable.

10. Georgia chronic multiple disease example. For the state of Georgia, United States, for the year of 2005, this dataset consists of 3 ambulatory care sensitive chronic diseases: asthma, chronic obstructive pulmonary disease (COPD), and angina. These diseases could be affected by poor air quality and so could have common patterning or correlation. The specific data used was counts of disease at county level in Georgia for the year 2005 for all age and gender groups. This data is publicly available from the OASIS online system of the Georgia DHR Division of Public Health (http://oasis.state.ga.us/). Both maps and tabulations of this data are available online from that source. Figure 1.5 displays the standardized incidence ratios for the three diseases. The expected rates are computed from the statewide rate for each disease.

11. UK industrial town multiple disease example. The data set consists of residential locations of death certificates for respiratory disease (bronchitis) and air-way cancers (respiratory, gastric, and oesophageal) for the period 1966–1976. These diseases were chosen as a set of diseases potentially related to adverse air pollution. A control disease (lower body cancers composite control) was also obtained. Another control (coronary heart disease) was available, but it has a confounding with smoking). The data consist of 630 coordinates of the residential locations of the composite control and the three other diseases. As an example of the data that is available in this study, Figure 1.6 displays three plots of the spatial distribution of the case diseases of interest: gastric and oesophageal cancer, respiratory cancer, and bronchitis.

12. Seizure data example. This dataset consists of seizure counts on 59 participants in a clinical trial of an anti-convulsive therapy for epilepsy.

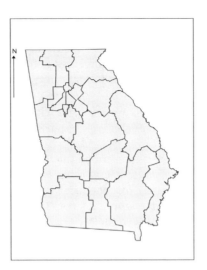

FIGURE 1.4
Georgia, United States, public health district geographies (top panel) and county geographies (bottom panel).

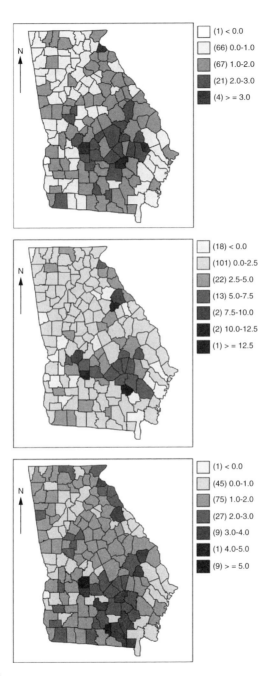

FIGURE 1.5
Georgia, United States, county level standardized incidence ratios for three diseases: asthma (top left), COPD (bottom), angina (top right).

Each participant has available: a group indicator (0/1: control, treatment), seizure count at four time points, baseline seizure count, and age in years. The dataset has been analyzed before by Breslow and Clayton (1993), and is discussed in detail by Diggle et al. (2002). I have added a randomly assigned spatial county indicator for South Carolina. The full data set, for each individual, consists of variables: seizure count, county indicator, baseline count, age, and group.

13. Burkitt's lymphoma dataset: This dataset appears in the splancs package in R. It consists of locations of Burkitt's lymphoma cases in the Western Nile district of Uganda for the period 1960–1975. In the dataset the location of cases and a diagnosis date (days from January 1st 1960) are given as well as the age of the case. There is no background population information in this example. There are a total of 188 cases.

14. Georgia's very low birthweight ST example: this dataset consists of counts of very low birthweight births for the counties of Georgia, United States for the sequence of 11 years 1994–2004. The total birth count for the same period and county is also available. The data is available from the OASIS online system of the Georgia DHR Division of Public Health (http://oasis.state.ga.us/). Both maps and tabulations of this data are available online from that source. Figure 1.7 displays the rate ratios for the 11 years of very low birth weight births in relation to the county birth rate over the 11 years over all counties.

15. Ohio respiratory cancer dataset 21 years: This dataset is, as for dataset 4, except that it is for 21 years (1968–1988) with a binary outcome created by threshold exceedence. Both a one time unit lag and a 1st and 2nd order spatial neighborhood are available as covariates.

16. Georgia asthma ST dataset: this dataset is from Georgia, United States and consists of ambulatory sensitive asthma case counts at county level for 8 years for <1 year age group (1999–2006). Expected rates are also available calculated from the statewide rate over the eight year period. The data is available from the OASIS online system of the Georgia DHR Division of Public Health (http://oasis.state.ga.us/). Both maps and tabulations of this data are available online from that source. The expected rates were computed from the overall rate (= total rate/total population <1year) times the local population count <1 year. Figures 1.8 and 1.9 display the standardized incidence ratios for the ambulatory asthma dataset for 1999–2004 and 2005–2006, respectively.

17. South Carolina flu season C+ notifications data: this consists of count data for laboratory C+ notifications for the 46 counties of South Carolina, United States by 13 time periods (biweekly recording) for the 2004–2005 flu season. This data is publicly available from the SCAN system at South Carolina Department of Health and Environmental Control (SCDHEC): (http://scangis.dhec.sc.gov/scan/). Both maps and

Introduction 13

tabulations of this data are available online from that source. Figure 1.10 displays the count of notifications for four selected counties of South Carolina for the flu season over 13 bi-weekly periods.

18. Foot and Mouth dataset: this dataset consists of parish level counts of foot and mouth disease (FMD) in Cumbria, NW England during the FMD outbreak in the year 2001. The time period is biweekly, starting February 2nd and ending June 1st (8 biweekly periods). This dataset was made available by Dr. Mark Stevenson, Massey University, New Zealand. Figure 1.11 displays the standardized incidence ratios for FMD by parish.

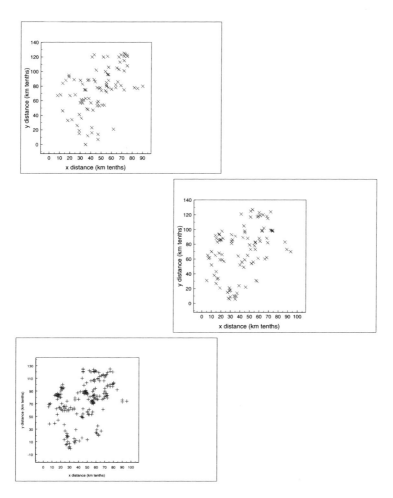

FIGURE 1.6
UK industrial town mortality study: gastric and oesophageal cancer (top); respiratory cancer (middle); bronchitis (bottom).

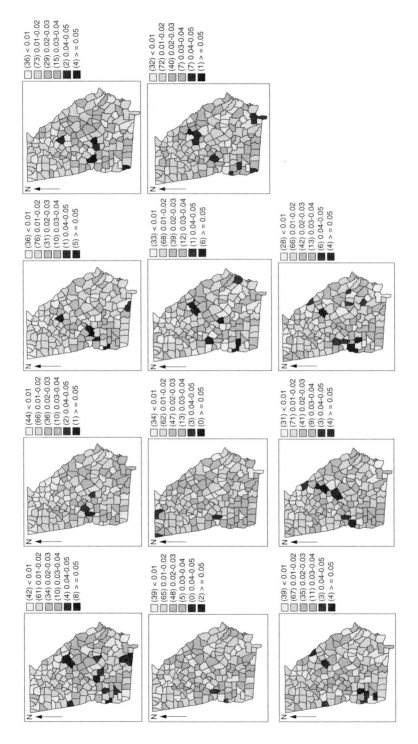

FIGURE 1.7
Georgia county level very low birth weight (VLBW) risk ratios

Introduction

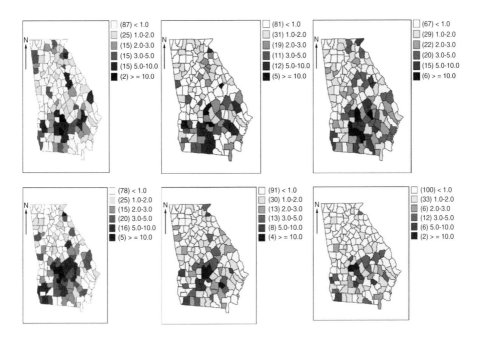

FIGURE 1.8
Standardized incidence ratios for Georgia ambulatory asthma for 1999–2000.

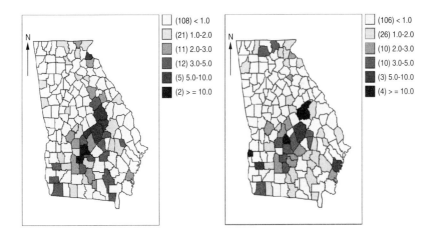

FIGURE 1.9
Standarized incidence ratios for Georgia ambulatory asthma for 2005–2006.

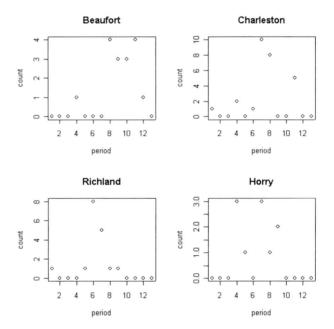

FIGURE 1.10
South Carolina influenza C+ notifications for the 2004–2005 flu season: counts for 13 biweekly time periods.

Introduction

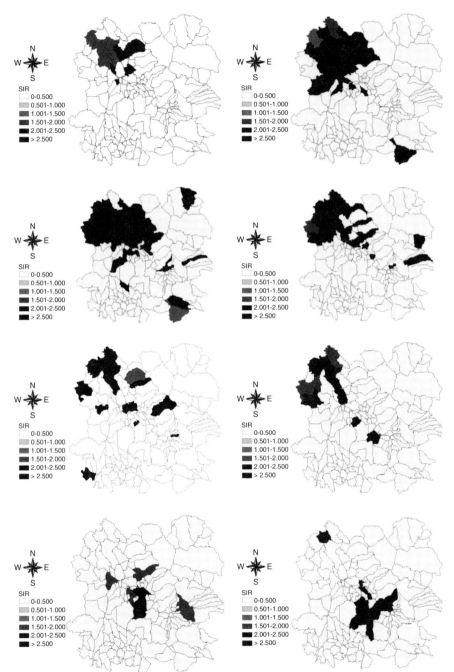

FIGURE 1.11
Northwest England Foot-and-Mouth disease (FMD) during the 2001 epidemic: parish level standardized incidence ratios for 8 biweekly periods.

2

Bayesian Inference and Modeling

The development of Bayesian inference has as its kernel the data likelihood. The likelihood is the joint distribution of the data evaluated at the sample values. It can also be regarded as a function describing the dependence of a parameter or parameters on sample values. Hence there can be two interpretations of this function. In Bayesian inference it is this latter interpretation that is of prime importance. In fact, the *likelihood principle,* by which observations come into play through the likelihood function, and only through the likelihood function, is a fundamental part of the Bayesian paradigm (Bernardo and Smith (1994) Section 5.1.4). This implies that the information content of the data is entirely expressed by the likelihood function. Furthermore, the likelihood principle implies that any event that did not happen has no effect on an inference, since if an unrealized event does affect an inference then there is some information not contained in the likelihood function.

2.1 Likelihood Models

The likelihood for data $\{y_i\}, i = 1, ..., m$, is defined as

$$L(\mathbf{y}|\boldsymbol{\theta}) = \prod_{i=1}^{m} f(y_i|\boldsymbol{\theta}) \qquad (2.1)$$

where $\boldsymbol{\theta}$ is a p length vector $\boldsymbol{\theta} : \{\theta_1, \theta_2, ..., \theta_p\}$ and $f(.|.)$ is a probability density (or mass) function. The assumption is made here that the 'sample' values of $\mathbf{y}$ given the parameters are independent, and hence it is possible to take the product of individual contributions in (2.1). Hence the data are assumed to be conditionally independent. Note that in many spatial applications the data would not be unconditionally independent and would in fact be correlated. This conditional independence is an important assumption fundamental to many disease mapping applications. The logarithm of the likelihood is also useful in model development and is defined as

$$l(\mathbf{y}|\boldsymbol{\theta}) = \sum_{i=1}^{m} \log f(y_i|\boldsymbol{\theta}). \qquad (2.2)$$

2.1.1 Spatial Correlation

Within spatial applications it is often found that correlation will exist between spatial units. This correlation is geographical and relates to the basic idea that locations close together in space often have similar values of outcome variables while locations far apart are often different. This spatial correlation (or autocorrelation as it's sometimes called) must be allowed for in spatial analyses. This may have an impact on the structure and form of likelihood models that are assumed for spatial data. The assumption made in the construction of conventional likelihoods is that the individual contribution to the likelihood is independent and this independence allows the likelihood to be derived as a product of probabilities. However, if this independence criterion is not met, then a different approach would be required.

2.1.1.1 Conditional independence

In some circumstances it is possible to consider *conditional* independence of the data given parameters at a higher level of the hierarchy. For instance in count data examples y_i from the i th area might be thought to be independent of other outcomes given knowledge of the model parameters. In the simple case, of dependence on a parameter vector $\boldsymbol{\theta}$, then conditioning on the parameters can allow [$y_i|\boldsymbol{\theta}$] to be assumed to be an independent contribution. This simply states that dependence only exists unconditionally (i.e., unobserved effects can induce dependence). This is often true in disease mapping examples where confounders that have spatial expression may or may not be measured in a study and their exclusion may leave residual correlation in the data. Note that this approach to correlation does not completely account for spatial effects as there can be residual correlation effects after inclusion of confounders. These effects could be due to unobserved or unknown confounders. Alternatively they could be due to intrinsic correlation in the process. Hence the assumption of conditional independence may only be valid if correlation is accounted for somewhere within the model.

The idea of inclusion of spatial correlation at a hierarchical level *above* the likelihood is a fundamental assumption often made in Bayesian small area health modeling. This means that the correlation appears in prior distributions rather than in the likelihood itself. Often parameters are given such priors and it is assumed that conditional independence applies in the likelihood. This is valid for many situations and will be the focus of most of this book.

2.1.1.2 Joint densities with correlation

Situations exist where spatial correlation can be incorporated within a joint distribution of the data. For example if a continuous spatial process is observed at measurement sites (such as air pollutants, soil chemical concentration, water quality) then often a spatial Gaussian process (SGP) will be assumed

(Ripley, 1981). This process assumes that any realisation of the process is multivariate normal with spatially-defined covariance, within its specification. Hence, if these data were observed outcome data, then the joint density would include spatial correlation (see Section 5.4.2).

Alternatively, it is possible to consider discrete outcome data where correlation is explicitly modeled. The autologistic and auto-Poisson models were developed for lattice data with spatial correlation included via dependence on a spatial neighborhood (Besag and Tantrum, 2003). In this approach, the normalization of the likelihood is computationally prohibitive and resort is often made to likelihood approximation (see Section 2.1.1.3).

2.1.1.3 Pseudolikelihood approximation

Pseudolikelihood has been proposed as an option to exact likelihood analysis when correlation exists. It has a number of variants (composite, local, pairwise: Lindsay (1988), Tibshirani and Hastie (1987), Kauermann and Opsomer (2003), Nott and Rydén (1999), Varin et al. (2005)). Pseudolikelihood has been used for autologistic models both in space and time (most recently by Besag and Tantrum (2003)). In space, the likelihood is given by

$$L_p(\mathbf{y}|\boldsymbol{\theta}) = \prod_{i=1}^{m} f(y_i|y_{j\neq i}, \boldsymbol{\theta}).$$

For the autologistic model, with binary outcome y_i, a simple version could be

$$f(y_i|y_{j\neq i}) = \frac{\exp[m(\beta, \{y_j\}_{j\in\delta_i})]}{1 + \exp[m(\beta, \{y_j\}_{j\in\delta_i})]}$$

where δ_i is a neighbourhood set of the i th location/area, and $m(.)$ is a specified function (such as mean or median) and β is a parameter controlling the spatial smoothing or degree of correlation. For nonlattice data the neighborhood can be defined by adjacency (for count data this could be adjacent regions and for case event data this could be tesselation neighbours). It is known that pseudolikelihood is least biased when relatively low spatial correlation exists (see e.g., Diggle et al., 1994). While the autologistic model has seen some application, the auto-Poisson model is limited by its awkward negative correlation structure. An autobinomial model is also available for the situation where y_i is a count of disease out of a finite local population n_i (see e.g., Cressie (1993), 431).

2.2 Prior Distributions

All parameters within Bayesian models are stochastic and are assigned appropriate probability distributions. Hence a single parameter value is simply one possible realization of the possible values of the parameter, the probability of which is defined by the prior distribution. The prior distribution is a distribution assigned to the parameter before seeing the data. Note also that one interpretation of prior distributions are that they provide additional 'data' for a problem and so they can be used to improve estimation or identification of parameters. For a single parameter, θ, the prior distribution can be denoted $\mathbf{g}(\theta)$, while for a parameter vector, $\boldsymbol{\theta}$, the joint prior distribution is $\mathbf{g}(\boldsymbol{\theta})$.

2.2.1 Propriety

It is possible that a prior distribution can be *improper*. Inpropriety is defined as the condition that integration of the prior distribution of the random variable θ over its range (Ω) is not finite:

$$\int_\Omega g(\theta)d\theta = \infty.$$

A prior distribution is improper if its normalizing constant is infinite. While impropriety is a limitation of any prior distribution, it is not necessarily the case that an improper prior will lead to impropriety in the posterior distribution. The posterior distribution can often be proper even with an improper prior specification.

2.2.2 Noninformative Priors

Often prior distributions are assumed that do not make strong preferences over values of the variables. These are sometimes known as *vague*, or *reference* or *flat* or *noninformative* prior distributions. Usually, they have a relatively flat form yielding close-to-uniform preference for different values of the variables. This tends to mean that in any posterior analysis (see Section 2.3) that the prior distribution(s) will have little impact compared to the likelihood of the data. *Jeffrey's* priors were developed in an attempt to find such reference priors for given distributions. They are based on the Fisher information matrix. For example, for the binomial data likelihood with common parameter p, then the Jeffrey's prior distribution is $p \sim Beta(0.5, 0.5)$. This is a proper prior distribution. However it is not completely noninformative as it has asymptotes close to 0 and 1. Jeffrey's prior for the Poisson data likelihood with common mean θ is given by $g(\theta) \propto \theta^{-\frac{1}{2}}$ which is *improper*. This also is not particularly noninformative. The Jeffrey's prior is locally uniform, but can often be improper.

Bayesian Inference and Modeling

Choice of noninformative priors can often be made with some general understanding of the range and behavior of the variable. For example, variance parameters must have prior distributions on the positive real line. Noninformative distributions in this range are often in the gamma, inverse gamma, or uniform families. For example, $\tau \sim G(0.001, 0.001)$ will have a small mean (1) but a very large variance (1000) and hence will be relatively flat over a large range. Another specification chosen is $\tau \sim G(0.1, 0.1)$ with variance 10 for a more restricted range. On the other hand, a uniform distribution on a large range has been advocated for the standard deviation (Gelman, 2006), $\sqrt{\tau} \sim U(0, 1000)$. For parameters on an infinite range, such as regression parameters, then a distribution centered on zero with a large variance will usually suffice. The zero-mean Gaussian or Laplace distribution could be assumed. For example,

$$\beta \sim N(0, \tau_\beta)$$
$$\tau_\beta = 100000.$$

is typically assumed in applications. The Laplace distribution is favoured in large scale Bayesian regression to encourage removal of covariates (Balakrishnan and Madigan (2006)).

Of course sometimes it is important to be informative with prior distributions. Identifiability is an issue relating to the ability to distinguish between parameters within a parametric model (see e.g., Bernardo and Smith (1994, 239). In particular, if a restricted range must be assumed to allow a number of variables to be *identified*, then it may be important to specify distributions that will provide such support. Ultimately, if the likelihood has little or no information about the separation of parameters then separation or identification can only come from prior specification. In general, if proper prior distributions are assumed for parameters then they are identified in the posterior distribution. However, how far they are identified may depend on the assumed variability. An example of identification which arises in disease mapping is where a linear predictor is defined to have two random effect components:

$$\log \theta_i = v_i + u_i,$$

and the components have different normal prior distributions with variances (say, τ_v, τ_u). These variances can have gamma prior distributions such as:

$$\tau_v \sim G(0.001, 0.001)$$
$$\tau_u \sim G(0.1, 0.1).$$

The difference in the variability of the second prior distribution allows there to be some degree of identification. Note that this means that a priori τ_v will be allowed greater variability in the variance of v_i than that found in u_i.

2.3 Posterior Distributions

Prior distributions and likelihood provide two sources of information about any problem. The likelihood informs about the parameter via the data, while the prior distributions inform via prior beliefs or assumptions. When there are large amounts of data, i.e., the sample size is large, the likelihood will contribute more to the relative risk estimation. When the example is data poor, then the prior distributions will dominate the analysis.

The product of the likelihood and the prior distributions is called the posterior distribution. This distribution describes the behavior of the parameters after the data are observed and prior assumptions are made. The posterior distribution is defined as

$$p(\boldsymbol{\theta}|\mathbf{y}) = L(\mathbf{y}|\boldsymbol{\theta})\mathbf{g}(\boldsymbol{\theta})/C \qquad (2.3)$$
$$\text{where } C = \int_p L(\mathbf{y}|\boldsymbol{\theta})\mathbf{g}(\boldsymbol{\theta})d\boldsymbol{\theta}.$$

where $\mathbf{g}(\boldsymbol{\theta})$ is the joint distribution of the $\boldsymbol{\theta}$ vector. Alternatively this distribution can be specified as a proportionality: $p(\boldsymbol{\theta}|\mathbf{y}) \propto L(\mathbf{y}|\boldsymbol{\theta})\mathbf{g}(\boldsymbol{\theta})$.

A simple example of this type of model in disease mapping is where the data likelihood is Poisson and there is a common relative risk parameter with a single gamma prior distribution:

$$p(\boldsymbol{\theta}|\mathbf{y}) \propto L(\mathbf{y}|\boldsymbol{\theta})g(\boldsymbol{\theta})$$

where $g(\theta)$ is a gamma distribution with parameters α, β i.e., $G(\alpha, \beta)$, and $L(\mathbf{y}|\theta) = \prod_{i=1}^m \{(e_i\theta)^{y_i} \exp(-e_i\theta)\}$ bar a constant only dependent in the data. A compact notation for this model is

$$y_i|\theta \sim Pois(e_i\theta)$$
$$\theta \sim G(\alpha, \beta).$$

This leads to a posterior distribution for fixed α, β of:

$$[\theta|\{y_i\}, \alpha, \beta] = L(\mathbf{y}|\theta, \alpha, \beta).p(\theta)/C$$
$$\text{where } C = \int L(\mathbf{y}|\theta, \alpha, \beta).p(\theta)d\theta.$$

In this case the constant C can be calculated directly and it leads to another gamma distribution:

$[\theta|\mathbf{y}, \alpha, \beta] = \frac{\beta^{*\alpha^*}}{\Gamma(\alpha^*)}\theta^{\alpha^*-1}\exp(-\theta\beta^*)$ where $\alpha^* = \sum y_i + \alpha$, $\beta^* = \sum e_i + \beta$.

2.3.1 Conjugacy

Certain combinations of prior distributions and likelihoods lead to the same distribution family in the posterior as for the prior distribution. This can lead to advantages in inference as the posterior form will follow from the prior specification. For instance, for the Poisson likelihood with mean parameter θ then with a gamma prior distribution for θ, the posterior distribution of θ is also gamma. Similar results hold for binomial likelihood and beta prior distribution and for a normal data likelihood with a normal prior distribution for the mean. The table below gives a small selection of results of this conjugacy. Conjugacy can often be found by examining the kernel of the prior-likelihood product. The unnormalized kernel should have a recognizable form related to the conjugate distribution. For example, a beta form has unnormalised kernel $\theta^{\alpha-1}(1-\theta)^{\beta-1}$. Conjugacy always guarantees a proper posterior distribution. Note that conjugacy may not be possible within a large parameter hierarchy but conditional conjugacy could be useful to exploit when examining model adequacy. It is also the case that for the sophisticated hierarchical models found in disease mapping, simple conjugacy is less likely to be available.

Likelihood	Prior	Posterior
$\mathbf{y} \sim \text{Poisson}(\theta)$	$\theta \sim G(\alpha, \beta)$	$\theta\|\mathbf{y} \sim G(\sum y_i + \alpha, m + \beta)$
$\mathbf{y} \sim \text{binomial}(\mathbf{p},1)$	$\mathbf{p} \sim Beta(\alpha_1, \alpha_2)$	$\mathbf{p}\|\mathbf{y} \sim Beta(\sum y_i + \alpha_1, m - \sum y_i + \alpha_2)$
$\mathbf{y} \sim \text{normal}(\mu, \tau)$, τ fixed	$\mu \sim N(\alpha_0, \tau_0)$	$\mu\|\mathbf{y} \sim N(\frac{\tau_0 \sum y_i + \alpha_0 \tau}{m\tau_0 + \tau}, \frac{\tau_0 \tau}{m\tau_0 + \tau})$
$\mathbf{y} \sim \text{gamma}(1, \beta)$	$\beta \sim G(\alpha_0, \beta_0)$	$\beta\|\mathbf{y} \sim G(1 + \alpha_0, \beta_0 + \sum y_i)$

2.3.2 Prior Choice

Choice of prior distributions is very important as it can be the case that the prior distributions of parameters can affect the posterior significantly. The balance between prior and posterior evidence is related to the dominance of the likelihood and is a sample size issue. For example, with large samples the likelihood usually dominates the prior distributions. This effectively means that current data are given priority in their weight of evidence. Prior distributions that dominate the likelihood are informative, but have less influence as simple size increases. Hence, with additional data, the data speak more. Of course when parameters are not identified within a likelihood then additional data are unlikely to change the importance of informative priors in identification. Propriety of posterior distributions is important as only under propriety can the absolute statements about probability of posterior parameter values be made.

2.4 Predictive Distributions

The posterior distribution summarizes our understanding about the parameters given observed data and plays a fundamental role in Bayesian modeling. However we can also examine other related distributions that are often useful when prediction of new data (or future data) is required. Define a new observation of y as y^*. We can determine the predictive distribution of y^* in two ways. In general the predictive distribution is defined as

$$p(y^*|\mathbf{y}) = \int L(y^*|\boldsymbol{\theta})p(\boldsymbol{\theta}|\mathbf{y})d\boldsymbol{\theta}. \tag{2.4}$$

Here the prediction is based on marginalizing over the parameters in the likelihood of the new data ($L(y^*|\boldsymbol{\theta})$) using the posterior distribution $p(\boldsymbol{\theta}|\mathbf{y})$ to define the contribution of the observed data to the prediction. This is termed the posterior predictive distribution. A variant of this definition uses the prior distribution instead of the posterior distribution:

$$p(y^*|\mathbf{y}) = \int L(y^*|\boldsymbol{\theta})p(\boldsymbol{\theta})d\boldsymbol{\theta}. \tag{2.5}$$

This emphasizes the prediction based only on the prior distribution (before seeing any data). Note that this distribution (2.5) is just the marginal distribution of y^*.

2.4.1 Poisson–Gamma Example

A classic example of a predictive distribution that arises in disease mapping is the negative binomial distribution. Let y_i, $i = 1, ..., n$ be counts of disease in arbitrary small areas (e.g., census tracts, zip codes, districts). Also define, for the same areas, expected rates $\{e_i\}$ and relative risks $\{\theta_i\}$. We assume that independently $y_i \sim Poisson(e_i\theta_i)$ given θ_i. Assume that $\theta_i = \theta$ $\forall i$ and that the prior distribution of θ, $p(\theta)$, is $\theta \sim Gamma(\alpha, \beta)$ where $E(\theta) = \alpha/\beta$, and $var(\theta) = \alpha/\beta^2$. The posterior distribution of θ is

$[\theta|\mathbf{y}, \alpha, \beta] = \frac{\beta^{*\alpha^*}}{\Gamma(\alpha^*)}\theta^{\alpha^*-1}\exp(-\theta\beta^*)$ where $\alpha^* = \sum y_i + \alpha$, $\beta^* = \sum e_i + \beta$.

It follows that the (prior) predictive distribution is

$$[\mathbf{y}^*|\mathbf{y},a,b] = \int f(\mathbf{y}^*|\boldsymbol{\theta})f(\boldsymbol{\theta}|a,b)d\boldsymbol{\theta} \qquad (2.6)$$

$$= \prod_{i=1}^{m}\left[\frac{b^a}{\Gamma(a)}\frac{\Gamma(y_i^*+a)}{(e_i+b)^{(y_i^*+a)}}\right].$$

2.5 Bayesian Hierarchical Modeling

In Bayesian modeling the parameters have distributions. These distributions control the form of the parameters and are specified by the investigator based, usually, on their prior belief concerning their behavior. These distributions are prior distributions and I will denote such a distribution by $g(\theta)$. In the disease mapping context a commonly assumed prior distribution for θ in a Poisson likelihood model is the gamma distribution and the resulting model is the Poisson–gamma model.

2.6 Hierarchical Models

A simple example of a hierarchical model that is commonly found in disease mapping is where the data likelihood is Poisson and there is a common relative risk parameter with a single gamma prior distribution:

$$p(\boldsymbol{\theta}|\mathbf{y}) \propto L(\mathbf{y}|\boldsymbol{\theta})g(\boldsymbol{\theta})$$

where $g(\theta)$ is a gamma distribution with parameters α, β i.e., $G(\alpha, \beta)$, and $L(\mathbf{y}|\theta) = \prod_{i=1}^{m}\{(e_i\theta)^{y_i}\exp(e_i\theta)\}$ bar a constant only dependent in the data. A compact notation for this model is:

$$y_i|\theta \sim Pois(e_i\theta)$$
$$\theta \sim G(\alpha, \beta).$$

In the previous section a simple example of a likelihood and prior distribution was given. In that example the prior distribution for the parameter also had parameters controlling its form. These parameters (α, β) can have assumed values, but more usually an investigator will not have a strong belief in the prior parameters values. The investigator may want to estimate these parameters from the data. Alternatively and more formally, as parameters within models are regarded as stochastic (and thereby have probability distributions governing their behavior), then these parameters must also have distributions. These distributions are known as hyperprior distributions, and the parameters are known as hyperparameters.

The idea that the values of parameters could arise from distributions is a fundamental feature of Bayesian methodology and leads naturally to the use of models where parameters arise within hierarchies. In the Poisson–gamma example there is a two level hierarchy: θ has a $G(\alpha, \beta)$ distribution at the first level of the hierarchy and α will have a hyperprior distribution (h_α) as will β (h_β), at the second level of the hierarchy. This can be written as:

$$y_i|\theta \sim Pois(e_i\theta)$$
$$\theta|\alpha, \beta \sim G(\alpha, \beta)$$
$$\alpha|\nu \sim h_\alpha(\nu)$$
$$\beta|\rho \sim h_\beta(\rho).$$

For these types of models it is also possible to use a graphical tool to display the linkages in the hierarchy. This is known as a directed acyclic graph or DAG for short. On such a graph lines connect the levels of the hierarchy and parameters are nodes at the ends of the lines. Clearly it is important to terminate a hierarchy at an appropriate place, otherwise one could always assume an infinite hierarchy of parameters. Usually the cutoff point is chosen to lie where further variation in parameters will not affect the lowest level model. At this point the parameters are assumed to be fixed. For example, in the Poisson–gamma model if you assume α and β were fixed then the Gamma prior would be fixed and the choice of α and β would be uninformed. The data would not inform about the distribution at all. However by allowing a higher level of variation i.e., hyperpriors for α, β, then we can fix the values of ν and ρ without heavily influencing the lower level variation. Figure 2.1 displays the DAG for the simple two level Poisson–gamma model just described.

2.7 Posterior Inference

When a simple likelihood model is employed, often maximum likelihood is used to provide a point estimate and associated variability for parameters. This is true for simple disease mapping models. For example, in the model $y_i|\theta \sim Pois(e_i\theta)$ the maximum likelihood estimate of θ is the the overall rate for the study region i.e., $\sum y_i / \sum e_i$. On the other hand, the SMR is the maximum likelihood estimate for the model $y_i|\theta_i \sim Pois(e_i\theta_i)$.

When a Bayesian hierarchical model is employed it is no longer possible to provide a simple point estimate for any of the $\theta_i s$. This is because the parameter is no longer assumed to be fixed but to arise from a distribution of possible values. Given the observed data, the parameter or parameters of interest will

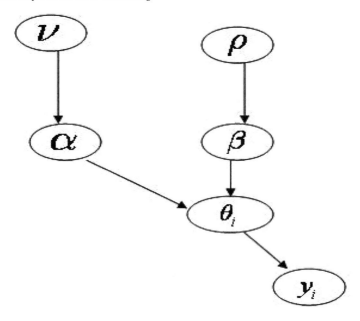

FIGURE 2.1
Directed acyclic graph for the Possion–gamma hiararchical model.

be described by the posterior distribution, and hence this distribution must be found and examined. It is possible to examine the expected value (mean) or the mode of the posterior distribution to give a point estimate for a parameter or parameters: e.g., for a single parameter θ, say, then $E(\theta|\mathbf{y}) = \int \theta\, p(\theta|\mathbf{y})d\theta$, and $\arg\max_{\theta} p(\boldsymbol{\theta}|\mathbf{y})$. Just as the maximum likelihood estimate is the mode of the likelihood, then the *maximum a posteriori* estimate is that value of the parameter or parameters at the mode of the posterior distribution. More commonly the expected value of the parameter or parameters is used. This is known as the posterior mean (or Bayes estimate). For simple unimodal symmetrical distributions, the modal and mean estimates coincide.

For some simple posterior distributions it is possible to find the exact form of the posterior distribution and to find explicit forms for the posterior mean or mode. However, it is commonly the case that for reasonably realistic models within disease mapping, it is not possible to obtain a closed form for the posterior distribution. Hence it is often not possible to derive simple estimators for parameters such as the relative risk. In this situation resort must be made to posterior sampling i.e., using simulation methods to obtain samples from the posterior distribution which then can be summarized to yield estimates of relevant quantities. In the next section we discuss the use of sampling algorithms for this purpose.

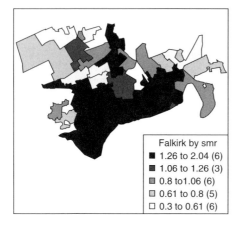
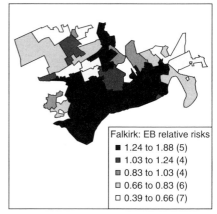

FIGURE 2.2
26 census enumeration districts (tracts) in Falkirk, Scotland: respirartory cancer mortality counts 1978–1983. Left panel is the standardised mortality ratio map using external age × sex standardised expected rates and the right panel is the Poisson–gamma estimates of relative risk using the empirical Bayes approach of Clayton and Kaldor (1987).

An exception to this situation where a closed form posterior distribution can be obtained is the Poisson–gamma model where α, β are fixed. In that case, the relative risks have posterior distribution given by:

$$\theta_i | y_i, e_i, \alpha, \beta \sim G(y_i + \alpha, e_i + \beta)$$

and the posterior expectation of θ_i is $(y_i + \alpha)/(e_i + \beta)$. The posterior variance is also available: $(y_i + \alpha)/(e_i + \beta)^2$, as is the modal value which is

$$\arg\max_\theta p(\theta|\mathbf{y}) = \begin{cases} [(y_i + \alpha) - 1]/(e_i + \beta) & \text{if } (y_i + \alpha) \geq 1 \\ 0 & \text{if } (y_i + \alpha) < 1 \end{cases}$$

Of course, if α and β are not fixed and have hyperprior distributions then the posterior distribution is more complex. Clayton and Kaldor (1987) use an approximation procedure to obtain estimates of α and β from a marginal likelihood apparently on the assumption that α and β had uniform hyperprior distributions. These estimates are those displayed in Figure 2.2. Note that these are not the full posterior expected estimates of the parameters from within a two level model hierarchy.

2.7.1 A Bernoulli and Binomial Example

Another example of a model hierarchy that arises commonly is the small area health data where a finite population exists within an area and within that population binary outcomes are observed. A fuller discussion of these models is given in Section 5.1.3. In the case event example, define the case events as $s_i : i = 1, ..., m$ and the control events as $s_i : i = m+1, ..., N$ where $N = m + n$ the total number of events in the study area. Associated with each location is a binary variable (y_i) which labels the event either as a case $(y_i = 1)$ or a control $(y_i = 0)$. A conditional Bernoulli model is assumed for the binary outcome where p_i is the probability of an individual being a case, given the location of the individual. Hence we can specify that $y_i \sim Bern(p_i)$. Here the probability will usually have either a prior distribution associated with it, or will be linked to other parameters and covariate or random effects, possibly via a linear predictor. Assume that a logistic link is appropriate for the probability and that two covariates are available for the individual: x_1 : age, x_2 : exposure level (of a health hazard). Hence,

$$p_i = \frac{\exp(\alpha_0 + \alpha_1 x_{1i} + \alpha_2 x_{2i})}{1 + \exp(\alpha_0 + \alpha_1 x_{1i} + \alpha_2 x_{2i})}$$

is a valid logistic model for this data with three parameters $(\alpha_0, \alpha_1, \alpha_2)$. Assume that the regression parameters will have independent zero-mean Gaussian prior distributions. The hierarchical model is specified in this case as:

$$y_i | p_i \sim Bern(p_i)$$
$$logit(p_i) = \mathbf{x}_i' \boldsymbol{\alpha}$$
$$\alpha_j | \tau_j \sim N(0, \tau_j)$$
$$\tau_j \sim G(\psi_1, \psi_2).$$

In this case, $\mathbf{x}_i'$ is the i th row of the design matrix (including an intercept term), $\boldsymbol{\alpha}$ is the (3×1) parameter vector, τ_j is the variance for the j th parameter, and ψ_1 and ψ_2 are fixed scale and shape parameters. Figure 2.3 displays the hierarchy for this model.

In the binomial case we would have a collection of small areas within which we observe events. Define the number of small areas as m and the total population as n_i. Within the population of each area individuals have a binary label which denotes the case status of the individual. The number of cases are denoted as y_i and it is often assumed that the cases follow an independent binomial distribution, conditional on the probability that an individual is a case, defined as p_i: $y_i \sim Bin(p_i, n_i)$.

The likelihood is given by $L(y_i | p_i, n_i) = \prod_{i=1}^{m} \binom{n_i}{y_i} p_i^{y_i} (1-p_i)^{(n_i - y_i)}$. Here the probability will usually have either a prior distribution associated with it, or will be linked to other parameters and covariate or random effects, possibly via a linear predictor such as $logit(p_i) = \mathbf{x}_i' \boldsymbol{\alpha} + \mathbf{z}_i' \boldsymbol{\gamma}$. In this general case, the $\mathbf{z}_i'$ are a vector of individual level random effects and the $\boldsymbol{\gamma}$ is a unit vector. Assume that a logistic link is appropriate for the probability and that a random effect

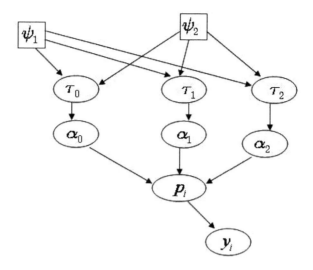

FIGURE 2.3
The Bernoulli hierarchical model where the logit link to a linear predication is assumed to a linear predictor. There are three parameters for intercept and two covariates. These parameters' distributions have variances with a gamma hyperprior with fixed and common parameters.

at the individual level is to be included: v_i. Hence,

$$p_i = \frac{\exp(\alpha_0 + v_i)}{1 + \exp(\alpha_0 + v_i)}$$

would represent a basic model with intercept to capture the overall rate and prior distribution for the intercept and the random effect could be assumed to be $\alpha_0 \sim N(0, \tau_{\alpha_0})$, and $v_i \sim N(0, \tau_v)$. The hyperprior distribution for the variance parameters could be a distribution on the positive real line such as the gamma, inverse gamma, or uniform. The uniform distribution has been proposed for the standard deviation ($\sqrt{\tau_*}$) by Gelman (2006). Here for illustration, I define a gamma distribution:

$$y_i \sim Bin(p_i, n_i)$$
$$logit(p_i) = \alpha_0 + v_i$$
$$\alpha_0 \sim N(0, \tau_{\alpha_0})$$
$$v_i \sim N(0, \tau_v)$$
$$\tau_{\alpha_0} \sim G(\psi_1, \psi_2)$$
$$\tau_v \sim G(\phi_1, \phi_2)$$

Bayesian Inference and Modeling

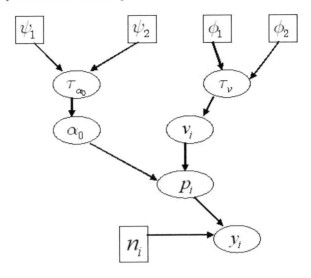

FIGURE 2.4
The hierarchical model for the binomial example with a logit link to a single intercept term and an individual level random effect. It is assumed that the hyperparameters ψ_* and ϕ_* are fixed and that the total population n_i is also fixed.

The hierarchy for this case would be as displayed in Figure 2.4.

An alternative approach to the Bernoulli or binomial distribution at the second level of the hierarchy is to assume a distribution directly for the case probability p_i. This might be appropriate when limited information about p_i is available. This is akin to the assumption of a gamma distribution as prior distribution for the Poisson relative risk parameter. Here one choice for the prior distribution could be a beta distribution:

$$p_i \sim Beta(\alpha_1, \alpha_2).$$

In general, the parameters α_1 and α_2 could be assigned hyperprior distributions on the positive real line, such as gamma or exponential. However if a uniform prior distribution for p_i is favored then $\alpha_1 = \alpha_2 = 1$ can be chosen. The hierarchy for this last situation with a Bernoulli model is displayed in Figure 2.5.

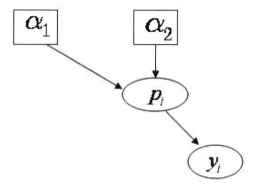

FIGURE 2.5
The hierarchical model for the beta Bernoulli hierarchy with fixed α_1, α_2 parameters.

2.8 Exercises

1) Derive the posterior distribution for θ where $L(\mathbf{y}|\theta) = \prod_{i=1}^{m}\{(e_i\theta)^{y_i}\exp(-e_i\theta)\}$ and the prior distribution for θ is $Exp(\beta)$, where $Exp(\beta)$ denotes an exponential distribution with mean β.

2) For the Poisson–gamma distribution in Section 2.4.1 derive the prior predictive distribution (negative binomial).

3) Show that the posterior predictive distribution is also negative binomial (hint: use gamma-gamma conjugacy).

4) Observed data is given as counts of birth abnormalities in m small areas: $\{y_i\}$ $i = 1, ..., m$. The total births within the same areas are $\{n_i\}$ and are assumed fixed. The probability of an abnormal birth in the i th area is ψ_i. For the following hierarchical model define the directed acyclic graph (DAG) assuming that $\tau_{\alpha_0}, \tau_{\alpha_1}$ are fixed:

$$[\{y_i\}|n_i, \psi_i] \sim Bin(n_i, \psi_i)$$
$$\log it(\psi_i) = \alpha_0 + \alpha_{1i}$$
$$\alpha_0 \sim N(0, \tau_{\alpha_0})$$
$$\alpha_{1i} \sim N(0, \tau_{\alpha_1}),$$

where $N(0, \tau)$ denotes a Gaussian distribution with zero mean and variance τ.

3
Computational Issues

3.1 Posterior Sampling

Once a posterior distribution has been derived, from the product of likelihood and prior distributions, it is important to assess how the form of the posterior distribution is to be evaluated. If single summary measures are needed then it is sometimes possible to obtain these directly from the posterior distribution either by direct maximization (mode: maximum a posteriori estimation) or analytically in simple cases (mean or variance for example)(see Section 2.3). If a variety of features of the posterior distribution are to be examined then often it will be important to be able to access the distribution via posterior sampling. Posterior sampling is a fundamental tool for exploration of posterior distributions and can provide a wide range of information about their form. Define a posterior distribution for data $\mathbf{y}$ and parameter vector $\boldsymbol{\theta}$ as $p(\boldsymbol{\theta}|\mathbf{y})$. We wish to represent features of this distribution by taking a sample from $p(\boldsymbol{\theta}|\mathbf{y})$. The sample can be used to estimate a variety of posterior quantities of interest. Define the sample size as m_p. For analytically tractable posterior distributions may be available to directly simulate the distribution. For example the Poisson–gamma model with α, β known, in Section 2.7, leads to the gamma posterior distribution: $\theta_i \sim G(y_i + \alpha, e_i + \beta)$. This can either be simulated directly (on R: rgamma) or sample estimation can be avoided by direct computation from known formulas. For example, in this instance, the moments of a gamma distribution are known: $E(\theta_i) = (y_i + \alpha)/(e_i + \beta)$ etc.

Define the sample values generated as: θ^*_{ij}, $j = 1, ..., m_p$. As long as a sample of reasonable size has been taken then it is possible to approximate the various functionals of the posterior distribution from these sample values. For example, an estimate of the posterior mean would be $\widehat{E}(\theta_i) = \widehat{\theta}_i = \sum_{j=1}^{m_p} \theta^*_{ij}/m_p$, while the posterior variance could be estimated as $\widehat{var}(\theta_i) = \frac{1}{m_p-1} \sum_{j=1}^{m_p} (\theta^*_{ij} - \widehat{\theta}_i)^2$, the sample variance. In general, any real function of the j th parameter $\gamma_j = t(\theta_j)$ can also be estimated in this way.

For example, the mean of γ_j is given by $\widehat{E}(\gamma_j) = \widehat{\gamma}_j = \sum_{j=1}^{m_p} t(\theta_{ij}^*)/m_p$. Note that credibility intervals can also be found for parameters by estimating the respective sample quantiles. For example if $m_p = 1000$ then 25th and 975th largest values would yield an equal tail 95% credible interval for γ_j. The median is also available as the 50th percentile of the sample, as are other percentiles.

The empirical distribution of the sample values can also provide an estimate of the marginal posterior density of θ_i. Denote this density as $\pi(\theta_i)$. A smoothed estimate of this marginal density can be obtained from the histogram of sample values of θ_i. Improved estimators can be obtained by using conditional distributions. A Monte Carlo estimator of $\pi(\theta_i)$ is given by

$$\widehat{\pi}(\theta_i) = \frac{1}{n} \sum_{j=1}^{n} \pi(\theta_i | \theta_{j,-i})$$

where the $\theta_{j,-i}$ $j = 1, ..., n$ are a sample from the marginal distribution $\pi(\theta_{-i})$.

Often m_p is chosen to be ≥ 500, more often 1000 or 10,000. If computation is not expensive then large samples such as these are easily obtained. The larger the sample size the closer the posterior sample estimate of the functional will be.

Generally, the complete sample output from the distribution is used to estimate functionals. This is certainly true in the case when independent sample values are available (such as when the distribution is analytically tractable and can be sampled from directly, such as in the Poisson–gamma case). In other cases, where iterative sampling must be used, it is sometimes necessary to sub-sample the output sample. In the next section, this is discussed more fully.

3.2 Markov Chain Monte Carlo Methods

Often in disease mapping, realistic models for maps have two or more levels and the resulting complexity of the posterior distribution of the parameters requires the use of sampling algorithms. In addition, the flexible modeling of disease could require switching between a variety of relatively complex models. In this case, it is convenient to have an efficient and flexible posterior sampling method which could be applied across a variety of models. Efficient algorithms for this purpose were developed within the fields of physics and image processing to handle large scale problems in estimation. In the late 1980s and early 1990s these methods were developed further particularly for dealing with Bayesian posterior sampling for more general classes of problems

Computational Issues

(Gilks et al., 1993; Gilks et al., 1996). Now posterior sampling is commonplace and a variety of packages (including WinBUGS, MlwiN, R) have incorporated these methods. For general reviews of this area the reader is referred to Cassella and George (1992), Robert and Casella (2005). Markov chain Monte Carlo (McMC) methods are a set of methods which use iterative simulation of parameter values within a Markov chain. The convergence of this chain to a stationary distribution, which is assumed to be the posterior distribution, must be assessed.

Prior distributions for the p components of $\boldsymbol{\theta}$ are defined as $g_i(\theta_i)$ for $i = 1, ..., p$. The posterior distribution of $\boldsymbol{\theta}$ and $\mathbf{y}$ is defined as

$$P(\boldsymbol{\theta}|\mathbf{y}) \propto L(\mathbf{y}|\boldsymbol{\theta}) \prod_i g_i(\theta_i). \tag{3.1}$$

The aim is to generate a sample from the posterior distribution $P(\boldsymbol{\theta}|\mathbf{y})$. Suppose we can construct a Markov chain with state space $\boldsymbol{\theta}_c$, where $\boldsymbol{\theta} \in \boldsymbol{\theta}_c \subset \Re^k$. The chain is constructed so that the equilibrium distribution is $P(\boldsymbol{\theta}|\mathbf{y})$, and the chain should be easy to simulate from. If the chain is run over a long period, then it should be possible to reconstruct features of $P(\boldsymbol{\theta}|\mathbf{y})$ from the realized chain values. This forms the basis of the McMC method, and algorithms are required for the construction of such chains. A selection of recent literature on this area is found in Ripley (1987), Gelman and Rubin (1992), Smith and Roberts (1993), Besag and Green (1993), Cressie (1993), Smith and Gelfand (1992), Tanner (1996), Robert and Casella (2005).

The basic algorithms used for this construction are

1. The Metropolis and its extension Metropolis–Hastings algorithm

2. The Gibbs Sampler algorithm

3.3 Metropolis and Metropolis–Hastings Algorithms

In all McMC algorithms, it is important to be able to construct the correct *transition probabilities* for a chain which has $P(\boldsymbol{\theta}|\mathbf{y})$ as its equilibrium distribution. A Markov chain consisting of $\boldsymbol{\theta}^1, \boldsymbol{\theta}^2, ...\boldsymbol{\theta}^t$ with state space Θ and equilibrium distribution $P(\boldsymbol{\theta}|\mathbf{y})$ has transitions defined as follows.

Define $q(\boldsymbol{\theta}, \boldsymbol{\theta}')$ as a transition probability function, such that, if $\boldsymbol{\theta}^t = \boldsymbol{\theta}$, the vector $\boldsymbol{\theta}^t$ drawn from $q(\boldsymbol{\theta}, \boldsymbol{\theta}')$ is regarded as a proposed possible value for $\boldsymbol{\theta}^{t+1}$.

3.3.1 Metropolis Updates

In this case choose a symmetric proposal $q(\boldsymbol{\theta},\boldsymbol{\theta}')$ and define the transition probability as

$$p(\boldsymbol{\theta},\boldsymbol{\theta}') = \begin{cases} \alpha(\boldsymbol{\theta},\boldsymbol{\theta}')q(\boldsymbol{\theta},\boldsymbol{\theta}') & \text{if } \boldsymbol{\theta}' \neq \boldsymbol{\theta} \\ 1 - \sum_{\boldsymbol{\theta}''} q(\boldsymbol{\theta},\boldsymbol{\theta}'')\alpha(\boldsymbol{\theta},\boldsymbol{\theta}'') & \text{if } \boldsymbol{\theta}' = \boldsymbol{\theta} \end{cases}.$$

where $\alpha(\boldsymbol{\theta},\boldsymbol{\theta}') = \min\left\{1, \frac{P(\boldsymbol{\theta}'|\mathbf{y})}{P(\boldsymbol{\theta}|\mathbf{y})}\right\}$.

In this algorithm a proposal is generated from $q(\boldsymbol{\theta},\boldsymbol{\theta}')$ and is accepted with probability $\alpha(\boldsymbol{\theta},\boldsymbol{\theta}')$. The acceptance probability is a simple function of the ratio of posterior distributions as a function of the ratio of posterior distributions as a function of $\boldsymbol{\theta}$ values. The proposal function $q(\boldsymbol{\theta},\boldsymbol{\theta}')$ can be defined to have a variety of forms but must be an irreducible and aperiodic transition function. Specific choices of $q(\boldsymbol{\theta},\boldsymbol{\theta}')$ lead to specific algorithms.

3.3.2 Metropolis–Hastings Updates

In this extension to the Metropolis algorithm the proposal function is not confined to symmetry and

$$\alpha(\boldsymbol{\theta},\boldsymbol{\theta}') = \min\left\{1, \frac{P(\boldsymbol{\theta}'|\mathbf{y})q(\boldsymbol{\theta}',\boldsymbol{\theta})}{P(\boldsymbol{\theta}|\mathbf{y})q(\boldsymbol{\theta},\boldsymbol{\theta}')}\right\}.$$

Some special cases of chains are found when $q(\boldsymbol{\theta},\boldsymbol{\theta}')$ has special forms. For example, if $q(\boldsymbol{\theta},\boldsymbol{\theta}') = q(\boldsymbol{\theta}',\boldsymbol{\theta})$ then the original Metropolis method arises and further, with $q(\boldsymbol{\theta},\boldsymbol{\theta}') = q(\boldsymbol{\theta}')$, (i.e., when no dependence on the previous value is assumed) then

$$\alpha(\boldsymbol{\theta},\boldsymbol{\theta}') = \min\left\{1, \frac{w(\boldsymbol{\theta}')}{w(\boldsymbol{\theta})}\right\}$$

where $w(\boldsymbol{\theta}) = P(\boldsymbol{\theta}|\mathbf{y})/q(\boldsymbol{\theta})$ and $w(.)$ are importance weights. One simple example of the method is $q(\boldsymbol{\theta}') \sim \text{Uniform}(\boldsymbol{\theta}_a,\boldsymbol{\theta}_b)$ and $g_i(\theta_i) \sim \text{Uniform}(\theta_{ia},\theta_{ib})$ $\forall i$, this leads to an acceptance criterion based on a likelihood ratio. Hence the original Metropolis algorithm with uniform proposals and prior distributions leads to a stochastic exploration of a likelihood surface. This, in effect, leads to the use of prior distributions as proposals. However, in general, when the $g_i(\theta_i)$ are not uniform this leads to inefficient sampling. The definition of $q(\boldsymbol{\theta},\boldsymbol{\theta}')$ can be quite general in this algorithm and, in addition, the posterior distribution only appears within a ratio as a function of $\boldsymbol{\theta}$ and $\boldsymbol{\theta}'$. Hence, the distribution is only required to be known up to proportionality.

3.3.3 Gibbs Updates

The Gibbs Sampler has gained considerable popularity, particularly in applications in medicine, where hierarchical Bayesian models are commonly applied

Computational Issues

(see, e.g., Gilks et al., 1993). This popularity is mirrored in the availability of software which allows its application in a variety of problems (e.g., WinBUGS, MLwiN, BACC). This sampler is a special case of the Metropolis–Hastings algorithm where the proposal is generated from the conditional distribution of θ_i given all other $\boldsymbol{\theta}$'s, and the resulting proposal value is accepted with probability 1.

More formally, define

$$q(\theta_j, \theta'_j) = \begin{cases} p(\theta^*_j | \theta^{t-1}_{-j}) & if\ \theta^*_{-j} = \theta^{t-1}_{-j} \\ 0 & \text{otherwise} \end{cases}$$

where $p(\theta^*_j | \theta^{t-1}_{-j})$ is the conditional distribution of θ_j given all other $\boldsymbol{\theta}$ values $(\boldsymbol{\theta}_{-j})$ at time $t-1$. Using this definition it is straightforward to show that

$$\frac{q(\boldsymbol{\theta}, \boldsymbol{\theta}')}{q(\boldsymbol{\theta}', \boldsymbol{\theta})} = \frac{P(\boldsymbol{\theta}'|\mathbf{y})}{P(\boldsymbol{\theta}|\mathbf{y})}$$

and hence $\alpha(\boldsymbol{\theta}, \boldsymbol{\theta}') = 1$.

3.3.4 M–H versus Gibbs Algorithms

There are advantages and disadvantages to M–H and Gibbs methods. The Gibbs Sampler provides a *single* new value for each $\boldsymbol{\theta}$ at each iteration, but requires the evaluation of a conditional distribution. On the other hand the M–H step does not require evaluation of a conditional distribution but does not guarantee the acceptance of a new value. In addition, block updates of parameters are available in M–H, but not usually in Gibbs steps (unless joint conditional distributions are available). If conditional distributions are difficult to obtain or computationally expensive, then M–H can be used and is usually available.

In summary, the Gibbs Sampler may provide faster convergence of the chain if the computation of the conditional distributions at each iteration are not time consuming. The M–H step will usually be faster at each iteration, but will not necessarily guarantee exploration. In straightforward hierarchical models where conditional distributions are easily obtained and simulated from, then the Gibbs Sampler is likely to be favoured. In more complex problems, such as many arising in spatial statistics, resort may be required to the M–H algorithm.

A simple M–H example: Assume that for m regions, the count n_i $i = 1, ..., m$ is observed. In addition, the expected count in the i th region, e_i is also observed. Assume also that the counts are independently distributed and have a Poisson distribution with $E(n_i) = \theta.e_i$, where θ is a constant parameter describing the relative risk over the whole study window. The likelihood in this case, bar a constant, is given by

$$L(\theta) = \exp(-\theta \sum_{i=1}^{m} e_i) . \prod_{i=1}^{m} (\theta e_i)^{n_i}. \tag{3.2}$$

Assuming a flat prior distribution for θ, then the M–H sampler for this problem reduces to a stochastic exploration of the likelihood surface. Hence the following sampler criterion is found for the θ parameter in this case:

$$\frac{L(\theta')}{L(\theta)} = \exp\{s_e(\theta - \theta')\} \cdot \left(\frac{\theta'}{\theta}\right)^{s_n}$$

$$where \ s_e = \sum_{i=1}^{m} e_i \ and \ s_n = \sum_{i=1}^{m} n_i.$$

3.3.5 Special Methods

Alternative methods exist for posterior sampling when the basic Gibbs or M–H updates are not feasible or appropriate. For example, if the range of the parameters are restricted then slice sampling can be used (Robert and Casella, 2005, Ch. 7; Neal, 2003). When exact conditional distributions are not available but the posterior is log-concave then adaptive rejection sampling algorithms can be used. The most general of these algorithms (ARS algorithm; Robert and Casella, 2005, 57–59) has wide applicability for continuous distributions, although may not be efficient for specific cases. Block updating can also be used to effect in some situations. When generalized linear model components are included then block updating of the covariate parameters can be effected via multivariate updating.

3.3.6 Convergence

McMC methods require the use of diagnostics to assess whether the iterative simulations have reached the equilibrium distribution of the Markov chain. Sampled chains require to be run for an initial burn-in period until they can be assumed to provide approximately correct samples from the posterior distribution of interest. This burn-in period can vary considerably between different problems. In addition, it is important to ensure that the chain manages to explore the parameter space properly so that the sampler does not 'stick' in local maxima of the surface of the distribution. Hence, it is crucial to ensure that a burn-in period is adequate for the problem considered. Judging convergence has been the subject of much debate and can still be regarded as art rather than science: a qualitative judgement has to be made at some stage as to whether the burn-in period is long enough.

There are a wide variety of methods now available to assess convergence of chains within McMC. Robert and Casella (2005) and Liu (2001) provide recent reviews. The available methods are largely based on checking the distributional properties of samples from the chains. In general define an output stream for a parameter vector $\boldsymbol{\theta}$ as $\{\boldsymbol{\theta}^1, \boldsymbol{\theta}^2, ...\boldsymbol{\theta}^m, \boldsymbol{\theta}^{m+1}....\boldsymbol{\theta}^{m+m_p}\}$. Here the m th value is the end of the burn-in period and a (converged) sample of size m_p is taken. Hence the converged sample is $\{\boldsymbol{\theta}^{m+1}....\boldsymbol{\theta}^{m+m_p}\}$. Define a function of the output stream as $\gamma = t(\boldsymbol{\theta})$ so that $\gamma^1 = t(\boldsymbol{\theta}^1)$.

3.3.6.1 Single chain methods

First, global methods for assessing convergence have been proposed which involve monitoring functions of the posterior output at each iteration. Globally this output could be the log posterior value (log $p(\widehat{\boldsymbol{\theta}}|\mathbf{y})$ where $\widehat{\boldsymbol{\theta}}$ are the estimated parameters at a given iteration), or the deviance of the model $(-2[l(y|\widehat{\boldsymbol{\theta}}) - l(y|\widehat{\boldsymbol{\theta}}_{ref})]$ where $\widehat{\boldsymbol{\theta}}_{ref}$ is a saturated or other reference model estimate). (In WinBUGS the deviance is assumed to be $-2l(y|\widehat{\boldsymbol{\theta}})$). These methods look for stabilization of the probability value. This value forms a time series, and special cusum methods have been proposed (Yu and Mykland (1998)). This approach emphasizes the overall convergence of the chain rather than individual parameter convergence. Two basic statistical tools that can be used to check sequences of output, have been proposed by Geweke (1992) and Yu and Mykland (1998). For the Geweke statistic, the sequence of output is broken up into two segments following a burn-in of m length. The first and last segments of length n_b and n_a respectively are defined. Averages of the first and last segments of output are obtained:

$$\overline{\gamma}_b = \frac{1}{n_b} \sum_{j=m+1}^{m+n_b} \gamma^j$$

$$\overline{\gamma}_a = \frac{1}{n_a} \sum_{j=m+m_p-n_a+1}^{m+m_p} \gamma^j.$$

As m_p gets large then the statistic

$$G = \frac{\overline{\gamma}_a - \overline{\gamma}_b}{\sqrt{\widehat{var}(\gamma_a) + \widehat{var}(\gamma_b)}} \to N(0,1) \text{ in distribution,}$$

where $\widehat{var}(\gamma_a), \widehat{var}(\gamma_b)$ are empirical variance estimates. Usually it is assumed that $n_b = 0.1n$ and $n_a = 0.5n$. Note that we can set $\gamma^j = -2l(y|\boldsymbol{\theta}^j)$ or $\gamma^j = \log p(\boldsymbol{\theta}^j|\mathbf{y})$ and so the deviance or log posterior can be monitored as an overall measure. This test is available on R in the CODA package (geweke.diag). The second test for single sequences was proposed by Yu and Mykland (1998) and later modified by Brooks (1998). For a post-convergence sequence of length m_p an average is computed

$$\widehat{\mu} = \frac{1}{m_p} \sum_{j=m+1}^{m+m_p} \gamma^j.$$

This average is used within a cusum calculation by defining a cusum of the sequence:

$$\widehat{S}_t = \sum_{j=m+1}^{t} [\gamma^j - \widehat{\mu}] \quad \text{for } t = m+1, .., m+m_p.$$

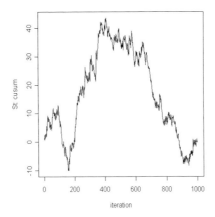

FIGURE 3.1
Cusum plot of S_t against t for 1000 length converged sample of Gamma(1,1) posterior output.

In the original proposal, a plot of $\widehat{S}_t$ against t was proposed. The interpretation of the plot relies on the identification of the hairiness or spikeyness of the cusum: a smooth cusum suggesting under-exploration of the posterior distribution, while a spikey plot represents rapid mixing. Figure 3.1 displays this plot for a 1000 length output sample from a Gamma(1,1) posterior distribution. Brooks (1998) further quantified this approach by deriving a statistic that measures the spikeyness of $\widehat{S}_t$.

Define
$$d_i = \begin{cases} 1 & \text{if } S_{i-1} > S_i \text{ and } S_i < S_{i+1} \\ & \text{or } S_{i-1} < S_i \text{ and } S_i > S_{i+1}, \\ 0 & \text{else} \end{cases}$$

for all $i = m+1, ..., m+m_p - 1$. Further define

$$D_t = \frac{1}{t-m-1} \sum_{i=m+1}^{t-1} d_i \qquad m+2 \le t \le m+m_p.$$

This statistic can be used in a number of ways. For an $i.i.d$ sequence symmetric about the mean then the expected value of d_i would be $1/2$. Further, D_t can be treated as a binomial variate with $E(D_t) = 1/2$ and $var(D_t) = \frac{1}{4(t-m-1)}$ and D_t will be approximately Gaussian with $100(1-\alpha/2)\%$ bounds

$$\frac{1}{2} \pm Z_{\alpha/2} \sqrt{\frac{1}{4(t-m-1)}}.$$

These bounds can be used as a formal tool to detect convergence. Figure 3.2 displays an example of this form of plot for the gamma output sample. This

Computational Issues

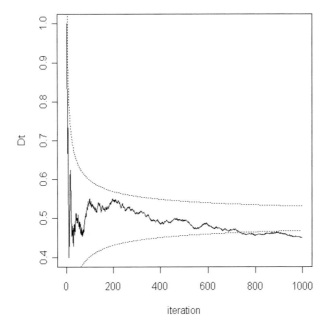

FIGURE 3.2
Plot of D_t for a single sample of 1000 from the Gamma(1, 1) posterior sample after a burn-in of 10000. The dotted lines are the asymptotic upper and lower binomial bounds.

sample deviates from the bounds somewhat and of course some assumptions about this diagnostic could be violated by output from a sampler (if it is asymmetric or approaches symmetry and independence slowly). Note that for "sticky" samplers, where values may stay for long periods (such as is possible with Metropolis–Hastings samplers), then the d_i can be modified to allow for such static behavior (see e.g., Brooks, 1998 for details and Section 3.3.7.1 below).

Second, graphical methods have been proposed which allow the comparison of the whole distribution of successive samples. Quantile-quantile plots of successive lengths of single variable output from the sampler can be used for this purpose. Figure 3.3 displays an example of such a plot. On R, with vectors out1 and out2 this can be created via commands:

```
>plot(sort(out1),sort(out2),xlab="output stream 1",ylab="output stream 2")
>lines(x,y)
> cor(sort(out1),sort(out2))
```

Further assessment of the degree of equality can be made via use of a correlation test. The Pearson correlation coefficient between the sorted sequences can be examined and compared to special tables of critical values. This adds some formality to the relatively arbitrary nature of visual inspection.

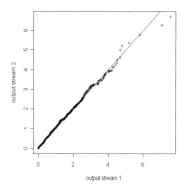

FIGURE 3.3
Quantile-quantile plot of two sequences of 1000 length of converged sample output from a gamma posterior distribution with parameters $\alpha = 1, \beta = 1$. Th equality line is marked.

3.3.6.2 Multi-chain methods

Single chain methods can, of course, be applied to each of a multiple of chains. In addition, there are methods that can only be used for multiple chains. The Gelman-Rubin statistic was proposed as a method for assessing the convergence of multiple chains via the comparison of summary measures across chains (Gelman and Rubin, 1992; Brooks and Gelman, 1998; Robert and Casella, 2005, Ch. 8).

This statistic is based on between and within chain variances. For the univariate case we have p chains and a sample of size n and a sample value of γ_i^j $j = 1, ..., n; i = 1, ..., p$. Denote the average over the sample for the i th chain as $\overline{\gamma}_i = \frac{1}{n}\sum_{j=1}^n \gamma_i^j$ and the overall average as $\overline{\gamma}_. = \frac{1}{p}\sum_{i=1}^p \overline{\gamma}_i$ and the variance of the i th chain is $\tau_i^2 = \frac{1}{n-1}\sum_{j=1}^n (\gamma_i^j - \overline{\gamma}_i)^2$. Then the between- and within-sequence variances are

$$B = \frac{n}{p-1}\sum_{i=1}^p (\overline{\gamma}_i - \overline{\gamma}_.)^2$$

$$W = \frac{1}{p}\sum_{i=1}^p \tau_i^2.$$

The marginal posterior variance of the γ is estimated as $\frac{n-1}{n}W + \frac{1}{n}B$ and this is unbiased asymptotically ($n \to \infty$). Monitoring the statistic

$$R = \sqrt{\frac{n-1}{n} + \frac{1}{n}\frac{B}{W}}$$

Computational Issues

for convergence to 1 is recommended. If the R for all parameters and functions of parameters is between 1.0 and 1.1 (Gelman et al., 2004) this is acceptable for most studies. Note that this depends on the sample size taken and closeness will be more easily achieved for large m_p. Brooks and Gelman (1998) extended this diagnostic to a multiparameter situation. On R the statistic is available in the CODA package as gelman.diag. On WinBUGS the Brooks–Gelman–Rubin (BGR) statistic is available in the Sample Monitor Tool. On WinBUGS, the width of the central 80% interval of the pooled runs and the average width of the 80% intervals within the individual runs are color-coded (green, blue), and their ratio R is red—for plotting purposes the pooled and within interval widths are normalized to have an overall maximum of one. On WinBUGS the statistics are calculated in bins of length 50. R would generally be expected to be greater than 1 if the starting values are suitably over-dispersed. Brooks and Gelman (1998) emphasize that one should be concerned both with convergence of R to 1, and with convergence of both the pooled and within interval widths to stability. One caveat should be mentioned concerning the use of between and within chain diagnostics. If the posterior distribution being approximated were to be highly multimodal, which could be the case in many mixture and spatial problems then the variability across chains could be large even when close to the posterior distribution and it could be that very large bins would need to be used for computation.

There is some debate about whether it is useful to run one long chain as opposed to multiple chains with different start points. The advantage of multiple chains is that they provide evidence for the robustness of convergence across different subspaces. However, as long as a single chain samples the parameter space adequately, then these have benefits. The reader is referred to Robert and Casella (2005), Chapter 8 for a thorough discussion of diagnostics and their use.

3.3.7 Subsampling and Thinning

McMC samplers often produce correlated samples of parameters. That is, a parameter value γ_i^j is likely to be similar to γ_i^{j-1}. This is likely to be true if successful proposals are based on proposal distributions with small variances, or where acceptances are localized to small areas of the posterior surface. In the former case, it may be that only small subsections of the posterior surface are being explored and so the sampler will not reach equilibrium for some time. Hence there may be an issue of lack of convergence when this occurs. The latter case could arise when a very spikey likelihood dominates. In themselves these correlated samples do not create problems for subsequent use of output streams, unless the sample size is very small (m_p small), or convergence has not been reached. Summary statistics could be affected by such autocorrelation. While measures of central tendency may not be much affected, the variance and other spread measures could be downward biased due to the (positive) autocorrelation in the stream. One possible remedy for this correlation is to take subsamples of the output. The simplest approach to this is to *thin* the stream by taking systematic samples at every k th iteration. By lengthening the gap between sampled units, then the more likely the correlation will be reduced or eliminated.

3.3.7.1 Monitoring Metropolis-like samplers

Samplers that don't necessarily accept a new value at each iteration cannot be monitored as easily as those that do produce new values (such as the Gibbs Sampler). With, for example, a Metropolis–Hastings algorithm the acceptance rate of new proposals is an important measure of the performance of the algorithm. The acceptance rate is defined as the number of iterations where new values are accepted out of a batch of iterations. Let's assume we have a batch size of $n_l = 100$ iterations and during that period we observe m_l accepted proposals. We assume that the number of parameters is small ($p << n_l$) so that there are potentially many transitions that could be made within n_l. The acceptance rate is just $A_r = \frac{m_l}{n_l}$. This rate could be a useful indicator of the behavior of the sampler. For example if the sampler is not mixing well then it may stick in various places failing to find acceptable proposals. This would lead to a low acceptance rate. However, on the other hand, a high acceptance rate may signify good proposals but could also mean that the sampler is "stuck" in the vicinity of a peak in the posterior surface and not searching the space in general. In both cases, the proposals may either be too small or too large to adequately search the space. Usually as a guide to a reasonable acceptance rate, for a M-H algorithm with small dimension (1 to 2 parameters) then $A_r \approx 0.5$ would be reasonable (Robert and Casella, 2005). For higher dimensions ($p > 2$) then $A_r \approx 0.25$ is reasonable. Hence for reversible jump algorithms (which are based on M-H steps with high dimension) then $A_r \approx 0.25$ might be expected. For Metropolis–Langevin or Langevin–Hastings algorithms (such as used in the R package **geoRglm**) that incorporate gradient terms then higher rates are optimal ($A_r \approx 0.6$). It should be borne in mind that in itself achievement of an optimal A_r does not necessarily imply convergence to a stationary distribution, although poor A_r could be due to lack of mixing and hence lack of convergence. It is also possible for chains to have high acceptance and very low convergence (Gamerman and Lopes, 2006). On **WinBUGS** when a Metropolis update is used then the acceptance rate can be set using the **Monitor Met** button in the **Model Menu**. This generates a plot of the acceptance rate over iteration for batches of $n_l = 100$ iterations. For user defined likelihood models using the zeroes or ones trick then A_r is always available.

The D_t statistic of Brooks (1998) can be modified for application to M–H algorithms where extended periods of "stickiness" arise:

$$d_i = \begin{cases} 1 & \text{if } S_{i-1} > S_i \text{ and } S_i < S_{i+1} \\ & \text{or } S_{i-1} < S_i \text{ and } S_i > S_{i+1} \\ & \text{or } S_{i-1} < S_i, S_{i+k} < S_i \text{ and} \\ & S_i = S_{i+1} = ... = S_{i+k}, \\ & \text{or } S_{i-1} > S_i, S_{i+k} > S_i \text{ and} \\ & S_i = S_{i+1} = ... = S_{i+k} \\ \frac{1}{2} & \text{if } S_{i-1} = S_i = S_{i+1} \\ 0 & \text{else} \end{cases}$$

for all $i = m+1, ..., m+n-1$.

In addition, for complex reversible jump samplers there may be need for forms of stratified convergence checking. For example, the dimension of the parameter set may lead to stratifying the number of parameters and this can lead to χ^2 tests and Kolmogorov-Smirnov statistics comparing a number of chains by their cumulative distribution functions (Brooks et al., 2003). Monitoring of dimension-changing algorithms is still a controversial issue.

3.4 Perfect Sampling

The idea of McMC is that simulation from a posterior distribution can be achieved over time and iterations are followed until convergence to the equilibrium is found. Propp and Wilson (1996) proposed a different approach whereby instead of iteration toward this equilibrium, a search is made to find a path from the past which will lead to coalescence at the current time. In essence a stopping time for the chain is found which corresponds to the equilibrium distribution. This is known as *coupling from the past* (CFTP). Examples of the application of such exact sampling have been made to point processes and Ising models (van Lieshout and Baddeley, 2002; Móller and Waagpetersen, 2004), case event data cluster modeling (McKeague and Loiseaux, 2002) where special McMC (reversible jump birth-death sampling) must be used, and to autologistic models for spatial and space-time data (Besag and Tantrum, 2003).

However, CFTP is not guaranteed to work for McMC transitional kernels that are not uniformly ergodic (Robert and Casella, 2005). However, perfect *slice* sampling may help toward a general algorithm that has general appeal (Mira et al., 2001).

Currently, the main problem with perfect sampling is that it is not possible to provide a general algorithm from which modeling of particular situations is immediately available. In fact, for most applications, the algorithm has to be specially designed and it is often therefore relatively difficult to adapt to changes of model form: for example, inclusion or exclusion of covariates may not be possible without significant alteration to the algorithm.

3.5 Posterior and Likelihood Approximations

From the point of view of computation it is now straightforward to examine a range of posterior distributional forms. This is certainly true for most applications of disease mapping where relative risk is estimated. However there are situations where it may be easier or more convenient to use a form of approximation to the posterior distribution or to the likelihood itself. Some approximations have been derived originally when posterior sampling was not possible and where the only way to obtain fully Bayesian estimates was to approximate (Bernardo and Smith, 1994). However other approximations arise due to the intractability of spatial integrals (for example in point process models).

3.5.1 Pseudolikelihood and Other Forms

In Section 2.1.1.3 I briefly introduced the idea of pseudolikelihood. I extend this idea here. In certain spatial problems, found in imaging and elsewhere, normalizing constants arise that are highly multidimensional. A simple example is the case of a Markov point process. Define the realization of m events within a window T as $\{s_1,......s_m\}$. Under a Markov process assumption the normalized probability density of a realization is

$$f_\theta(\mathbf{s}) = \frac{1}{c(\theta)} h_\theta(\mathbf{s})$$

$$\text{where } c(\theta) = \sum_{k=0}^{\infty} \frac{1}{k!} \int_{T^k} h_\theta(\mathbf{s}) \lambda^k(d\mathbf{s}).$$

Conditioning on the number of events (m), then the normalization of $f_m(\mathbf{s}) \propto h_m(\mathbf{s})$ then the normalization is over the m-dimensional window:

$$c(\theta) = \int_T .....\int_T h_m(\{s_1,......s_m\}) ds_1,......ds_m.$$

For a conditional Strauss process then $f_m(\mathbf{s}) \propto \gamma^{n_R(\mathbf{s})}$ and $n_R(\mathbf{s})$ is the number of R-close pairs of points to $\mathbf{s}$.

It is also true that a range of lattice models developed for image processing applications also have awkward normalization constants (auto-Poisson and autologistic models and Gaussian Markov random field models Besag and Tantrum, 2003; Rue and Held, 2005).

This has led to the use of approximate likelihood models in many cases. For example, for Markov point processes it is possible to specify a conditional intensity (Papangelou) which is independent of the normalization. This conditional intensity $\lambda^*(\xi, \mathbf{s}|\theta) = h(\xi \cup \mathbf{s})/h(\xi)$ can be used within a pseudo-likelihood function. In the case of the above Strauss process this is just $\lambda^*(\xi, \mathbf{s}|\theta) = \lambda^*(\mathbf{s}|\theta) = \gamma^{n_R(\mathbf{s})}$ and the pseudolikelihood is:

$$L_p(\{\mathbf{s}_1, \ldots \mathbf{s}_m\}|\theta) = \prod_{i=1}^{m} \lambda^*(\mathbf{s}_i|\theta) \exp(-\int_T \lambda^*(\mathbf{u}|\theta) d\mathbf{u}).$$

As this likelihood has the form of an inhomogeneous Poisson process likelihood, then this is relatively straightforward to evaluate. The only issue is the integral of the intensity over the window T. This can be handled via special numerical integration schemes (Berman and Turner, 1992; Lawson 1992a, 1992b; and Section 5.1.1). Bayesian extensions are generally straightforward. Note that once a likelihood contribution can be specified then this can be incorporated within a posterior sampling algorithm such as Metropolis–Hastings. This can be implemented on WinBUGS via a zeroes trick if the Berman-Turner weighting is used. For example the model with the i th likelihood component: $l_i = \log \lambda^*(\mathbf{s}_i|\theta) - w_i \lambda^*(\mathbf{s}_i|\theta)$ can be fitted using this method, where the weight w_i is based on the Dirichlet tile area of the i th point or a function of the Delauney triangulation around the point (see Berman and Turner, 1992, Baddeley and Turner, 2000, and Appendix C.5.3 of Lawson, 2006b).

In application to lattice models Besag and Tantrum (2003) give the example of a Markov random field of m dimension where the pseudolikelihood $L_p = \prod_{i=1}^{m} p(y_i^0|y_{-i}^0; \theta)$ is the product of the full conditional distributions. In the (auto)logistic binary case

$$p(y_i^0|y_{-i}^0; \theta) = \frac{\exp(f(\alpha_0, \{y_{-i}^0\}_{\in \partial i}, \theta))}{1 + \exp(f(\alpha_0, \{y_{-i}^0\}_{\in \partial i}, \theta))}$$

where ∂_i denotes the adjacency set of the i th site.

Other variants of these likelihoods have been proposed. Local likelihood (Tibshirani and Hastie, 1987) is a variant where a contribution to likelihood is defined within a local domain of the parameter space. In spatial problems this could be a spatial area. This has been used in a Bayesian disease mapping setting by Hossain and Lawson, 2005. Pairwise likelihood (Nott and Rydén, 1999; Heagerty and Lele, 1998) has been proposed for image restoration and for general spatial mixed models (Varin et al., 2005). All these variants of full likelihoods will lead to models that are approximately valid for real applications. It should be borne in mind however that they ignore aspects of the spatial correlation and if these are not absorbed in some part of the model hierarchy then this may affect the appropriateness of the model.

3.5.2 Asymptotic Approximations

It is possible to approximate a posterior distribution with a simpler distribution which is found asymptotically. The use of approximations lies in their often common form and also the ease with which parameters may be estimated under the approximation. Often the asymptotic approximating distribution will be a normal distribution. Here two possible approaches are examined: the asymptotic quadratic form approximation and integral approximation via Laplace's method.

3.5.2.1 Asymptotic Quadratic Form

Large sample convergence in form of the likelihood or posterior distribution is considered here. In many cases the limiting form of a likelihood or posterior distribution in large samples can be used as an approximation. The Taylor series expansion of the function $f(.)$ around vector $\mathbf{a}$ is

$$f(\mathbf{a}) + U(\mathbf{a})^T(\mathbf{x} - \mathbf{a}) + \frac{1}{2}(\mathbf{x} - \mathbf{a})^T H(\mathbf{a})(\mathbf{x} - \mathbf{a}). + R.$$

where $U(\mathbf{a})$ is the score vector evaluated at $\mathbf{a}$, R is a remainder, and $H(\mathbf{a})$ is the Hessian matrix of second derivatives of $f(.)$ evaluated at $\mathbf{a}$. For an arbitrary log likelihood with p length vector of parameters $\boldsymbol{\theta}$, then an expansion around a point is required. Usually the mode of the distribution is chosen. Define the modal vector as $\boldsymbol{\theta}^m$ and $l(\mathbf{y}|\boldsymbol{\theta}) \equiv l(\boldsymbol{\theta})$ for brevity. The expansion is defined as:

$$l(\boldsymbol{\theta}) = l(\mathbf{y}|\boldsymbol{\theta}^m) + U(\boldsymbol{\theta}^m)(\boldsymbol{\theta} - \boldsymbol{\theta}^m) - \frac{1}{2}(\boldsymbol{\theta} - \boldsymbol{\theta}^m)^T H(\boldsymbol{\theta}^m)(\boldsymbol{\theta} - \boldsymbol{\theta}^m).$$

Here $U(\boldsymbol{\theta}^m) = \mathbf{0}$ as we have expanded around the maxima and so this reduces to

$$l(\boldsymbol{\theta}) = l(\mathbf{y}|\boldsymbol{\theta}^m) - \frac{1}{2}(\boldsymbol{\theta} - \boldsymbol{\theta}^m)^T H(\boldsymbol{\theta}^m)(\boldsymbol{\theta} - \boldsymbol{\theta}^m). \quad (3.3)$$

Note that $H(\boldsymbol{\theta}^m)$ describes the local curvature of the likelihood at the maxima and is defined by

$$H(\boldsymbol{\theta}^m) = \left(-\frac{\partial^2 l(\boldsymbol{\theta})}{\partial \theta_i \partial \theta_j}\right)\bigg|_{\boldsymbol{\theta}=\boldsymbol{\theta}^m}.$$

This approximation, given $\boldsymbol{\theta}^m$, consists of a constant and a quadratic form around the maxima. In a likelihood analysis the $\boldsymbol{\theta}^m$ might be replaced by maximum likelihood estimates $\widehat{\boldsymbol{\theta}}^m$.

For a posterior distribution, it is possible to also approximate the prior distribution with a Taylor expansion. In which case a full posterior approximation would be obtained. Assume a joint prior distribution, defined by $p(\boldsymbol{\theta}|\boldsymbol{\Gamma})$ where $\boldsymbol{\Gamma}$ is a parameter vector or matrix. Assuming that $\boldsymbol{\Gamma}$ is fixed, then the approx-

Computational Issues

imation around the modal vector $\boldsymbol{\theta}^p$, again assuming the score vector is zero at the maxima, is given by

$$\log p(\boldsymbol{\theta}|\boldsymbol{\Gamma}) = \log p(\boldsymbol{\theta}^p|\boldsymbol{\Gamma}) - \frac{1}{2}(\boldsymbol{\theta} - \boldsymbol{\theta}^p)^T H_p(\boldsymbol{\theta}^p)(\boldsymbol{\theta} - \boldsymbol{\theta}^p) + R_0$$

where, R_0 is the remainder term and

$$H_p(\boldsymbol{\theta}^p) = \left(-\frac{\partial^2 \log p(\boldsymbol{\theta}|\boldsymbol{\Gamma})}{\partial \theta_i \partial \theta_j} \right)\bigg|_{\boldsymbol{\theta}=\boldsymbol{\theta}^p}.$$

Again given $\boldsymbol{\theta}^p$, this is simply a quadratic form around the maxima. There are then two posterior approximations that might be considered:

i) Likelihood approximation only:

$$p(\boldsymbol{\theta}|\mathbf{y}) \propto p(\boldsymbol{\theta}|\boldsymbol{\Gamma}) \exp\left\{ -\frac{1}{2}(\boldsymbol{\theta} - \boldsymbol{\theta}^m)^T H(\boldsymbol{\theta}^m)(\boldsymbol{\theta} - \boldsymbol{\theta}^m) \right\}$$

or

ii) full posterior approximation:

$$p(\boldsymbol{\theta}|\mathbf{y}) \propto \exp\left\{ -\frac{1}{2}(\boldsymbol{\theta} - \boldsymbol{\theta}^p)^T H_p(\boldsymbol{\theta}^p)(\boldsymbol{\theta} - \boldsymbol{\theta}^p) - \frac{1}{2}(\boldsymbol{\theta} - \boldsymbol{\theta}^m)^T H(\boldsymbol{\theta}^m)(\boldsymbol{\theta} - \boldsymbol{\theta}^m) \right\}$$

$$\propto \exp\left\{ -\frac{1}{2}(\boldsymbol{\theta} - \mathbf{m}^n)^T H_n (\boldsymbol{\theta} - \mathbf{m}^n) \right\}$$

where $H_n = H_p(\boldsymbol{\theta}^p) + H(\boldsymbol{\theta}^m)$ and $\mathbf{m}^n = H_n^{-1}(H_p(\boldsymbol{\theta}^p)\boldsymbol{\theta}^p + H(\boldsymbol{\theta}^m)\boldsymbol{\theta}^m)$. Note that $H(\boldsymbol{\theta}^m)$ is the observed information matrix. As the sample size increases this quadratic form approximation improves in its accuracy and two important results follow:

i) the posterior distribution tends toward a normal distribution, i.e.,

$$\text{as } m \to \infty \text{ then } p(\boldsymbol{\theta}|\mathbf{y}) \to N_p(\boldsymbol{\theta}|\mathbf{m}^n, H_n)$$

ii) the information matrix tends toward the Fisher (expected) information matrix in the sense that $H(\boldsymbol{\theta}^m) \to mI(\boldsymbol{\theta}^m)$ where the ijth element is

$$I(\boldsymbol{\theta})_{ij} = \int p(y|\boldsymbol{\theta}) \left(-\frac{\partial^2 l(\boldsymbol{\theta})}{\partial \theta_i \partial \theta_j} \right) dy.$$

This means that it is possible to consider further asymptotic distributional forms. For instance if the variability in the prior distribution is negligible compared to the likelihood then

$$p(\boldsymbol{\theta}|\mathbf{y}) \to N_p(\boldsymbol{\theta}|\boldsymbol{\theta}^m, H(\boldsymbol{\theta}^m))$$

or

$$p(\boldsymbol{\theta}|\mathbf{y}) \to N_p(\boldsymbol{\theta}|\boldsymbol{\theta}^m, mI(\boldsymbol{\theta}^m)).$$

Often the maximum likelihood (ML) estimates would be substituted for $\boldsymbol{\theta}^m$. If $\boldsymbol{\theta}^m$ are given or estimated via ML the posterior distribution will be multivariate normal in large samples.

Hence a normal approximation to the posterior distribution is justified at least asymptotically (as $m \to \infty$). This approximation should be reasonably good for continuous likelihood models and may be reasonable for discrete models when the rate parameter (Poisson) is large or the binomial probability is not close to 0 or 1. Of course this is likely not to hold when there is sparseness in the count data, as can arise when rare diseases are studied. Further discussion of different asymptotic results can be found in Bernardo and Smith (1994).

An example of such a likelihood approximation would be where a binomial likelihood has been assumed and $y_i|p_i \sim Bin(p_i, n_i)$ with $p_i \sim Beta(2,2)$. In this case, assume $p(\boldsymbol{\theta}|\mathbf{y}) \sim N_p(\boldsymbol{\theta}|\mathbf{m}^n, H_n)$ and $\mathbf{m}^n = \frac{\widehat{p}_i(1-\widehat{p}_i)}{n_i}\left[\frac{n_i}{\widehat{p}_i(1-\widehat{p}_i)}\widehat{p}_i\right] = \widehat{p}_i$, $H_n = 0 + \frac{n_i}{\widehat{p}_i(1-\widehat{p}_i)}$ where $\widehat{p}_i = \frac{y_i}{n_i}$ and so the distribution is $N_m(p_i|\widehat{p}_i, diag\{\frac{n_i}{\widehat{p}_i(1-\widehat{p}_i)}\})$. Hence the approximate distribution is centered around the saturated maximum likelihood estimator. In this case the prior distribution has little effect on the mean or the variance of the resulting Gaussian distribution. If, on the other hand, an asymmetric prior distribution favouring low rates of disease were assumed such as $p_i \sim Beta(1.5, 5)$, then the approximation is given by $\mathbf{m}^n = (H_p(\boldsymbol{\theta}^p) + \frac{n_i}{\widehat{p}_i(1-\widehat{p}_i)})^{-1}[H_p(\boldsymbol{\theta}^p)\boldsymbol{\theta}^p + \frac{n_i\widehat{p}_i}{\widehat{p}_i(1-\widehat{p}_i)}]$ and $H_n = H_p(\boldsymbol{\theta}^p) + \frac{n_i}{\widehat{p}_i(1-\widehat{p}_i)}$ where $H_p(\boldsymbol{\theta}^p) = 81.383$ and $\boldsymbol{\theta}^p = 0.11$. Here the mean and variance are influenced considerably.

Note that it is also possible to approximate posterior distributions with mixtures of normal distributions and this could lead to closer approximation to complex (multi-modal) distributions. Hierarchies with more than 2 levels have not been discussed here. However in principle, if a normal approximation can be made to each prior in turn (perhaps via mixtures of normals) then a quadratic form would result with a more complex form.

3.5.2.2 Laplace integral approximation

In some situations ratios of integrals must be evaluated and it is possible to employ an integral approximation method suggested by Laplace (Tierney and Kadane, 1986). For example the posterior expectation of a real valued function $g(\boldsymbol{\theta})$ is given by

$$E(g(\boldsymbol{\theta})|\mathbf{y}) = \int g(\boldsymbol{\theta}) p(\boldsymbol{\theta}|\mathbf{y}) d\boldsymbol{\theta}.$$

This can be considered as a ratio of integrals, given the normalization of the posterior distribution. The approximation is given by

$$\widehat{E}(g(\boldsymbol{\theta})|\mathbf{y}) \approx \left(\frac{\sigma^*}{\sigma}\right) \exp\{-m[h^*(\boldsymbol{\theta}^*) - h(\boldsymbol{\theta})]\}$$

Computational Issues

where $-mh(\boldsymbol{\theta}) = \log p(\boldsymbol{\theta}) + l(\mathbf{y}|\boldsymbol{\theta})$ and $-mh^*(\boldsymbol{\theta}) = \log g(\boldsymbol{\theta}) + \log p(\boldsymbol{\theta}) + l(\mathbf{y}|\boldsymbol{\theta})$ and

$$-h(\widehat{\boldsymbol{\theta}}) = \max_{\boldsymbol{\theta}}\{-h(\boldsymbol{\theta})\}, \quad -h^*(\boldsymbol{\theta}^*) = \max_{\boldsymbol{\theta}}\{-h^*(\boldsymbol{\theta})\},$$

$\widehat{\sigma} = |m\nabla^2 h(\widehat{\boldsymbol{\theta}})|^{-1/2}$ and $\widehat{\sigma} = |m\nabla^2 h^*(\boldsymbol{\theta}^*)|^{-1/2}$ where

$$[\nabla^2 h(\widehat{\boldsymbol{\theta}})]_{ij} = \left.\frac{\partial^2 h(\boldsymbol{\theta})}{\partial \theta_i \partial \theta_j}\right|_{\boldsymbol{\theta}=\widehat{\boldsymbol{\theta}}}.$$

3.6 Exercises

1) Assume a generalized linear model with Poisson likelihood: $[y_i|\theta_i] \sim Poiss(e_i\theta_i)$ with log link to linear predictor $\log \theta_i = \eta_i = \mathbf{x}'_i\boldsymbol{\beta}$ where $\mathbf{x}'_i$ is a row covariate vector of p length and $\boldsymbol{\beta}$ is a p length parameter vector. The parameter vector is assumed to have a Gaussian prior distribution with $\mathbf{0}$ mean vector and the parameters are assumed independent hence $\boldsymbol{\beta} \sim \mathbf{N}_p(\mathbf{0}, \boldsymbol{\Gamma})$ where $\boldsymbol{\Gamma} = \mathbf{diag}\{\tau_1, ..., \tau_p\}$. Show that a normal approximation to the posterior distribution of this model, given the maximum likelihood estimates $\widehat{\boldsymbol{\beta}}$, is given by $N_p(\boldsymbol{\theta}|\mathbf{m}^n, H_n)$ where $H_n = H_p(\boldsymbol{\theta}^p) + H(\boldsymbol{\theta}^m)$ and $\mathbf{m}^n = H_n^{-1}[H_p(\boldsymbol{\theta}^p)\boldsymbol{\theta}^p + H(\boldsymbol{\theta}^m)\boldsymbol{\theta}^m]$ and $\boldsymbol{\theta}^p = \mathbf{0}$, $H_p(\boldsymbol{\theta}^p) = diag\{\tau_{\beta_1}^{-1/2}....\tau_{\beta_p}^{-1/2}\}$ where σ_{β_*} is the standard deviation in the Gaussian prior distribution for β_*.

$$H(\boldsymbol{\theta}^m)_{jk} = \sum_{i=1}^{m}[A_{jk} + B_{jk}]$$

$$\text{where}$$

$$A_{jk} = \frac{y_i x_{ji} x_{ki}}{\widehat{\eta}_i (\ln \widehat{\eta}_i)^2}, \quad B_{jk} = e_i x_{ji} x_{ki} \exp\{\widehat{\eta}_i\}$$

$$\text{and } \widehat{\eta}_i = \mathbf{x}'_i \widehat{\boldsymbol{\beta}}.$$

4
Residuals and Goodness-of-Fit

Attainment of convergence of McMC algorithms does not necessarily yield good models. If a model is misspecified then it will be of limited use. There are many issues relating to model goodness-of-fit that should be of concern when evaluating models for geo-referenced disease data. In this section, I treat general issues related to the use of goodness-of-fit (GOF) measures, residual diagnostics and the use of posterior output to yield risk exceedence probabilities.

4.1 Model GOF Measures

Goodness-of-fit criteria vary depending on the properties of the criteria and the nature of the model. In conventional generalized linear modeling with fixed effects, the deviance is an important measure (McCullagh and Nelder, 1989). Usually this measure of model adequacy compares a fitted model to a saturated model. It is based on the difference between the log likelihood of the data under either model:

$$D = -2[l(y|\widehat{\theta}_{fit}) - l(y|\widehat{\theta}_{sat})].$$

The saturated model has a single parameter per observation. Often a relative measure of fit is used so that deviances are compared and the change in deviance between model 1 and model 2 is used:

$$\Delta D = -2[l(y|\widehat{\theta}_1) - l(y|\widehat{\theta}_2)].$$

Hence the saturated likelihood cancels in this relative comparison. The deviance is used in goodness-of-fit measures in Bayesian modeling, but usually without reference to a saturated model.

One disadvantage of using the deviance directly is that it does not allow for the degree of parameterization in the model: a model can be made to more closely approximate data by increasing the number of parameters. Hence attempts have been made to penalize model complexity. One example of this is the Akaike information criterion (AIC). This is defined as

$$AIC = -2l(y|\widehat{\theta}_{fit}) + 2p$$

where p is the number of parameters. The second term acts as a penalty for over parameterization of the model. The idea being that as more parameters are added then the closer the model will approximate the data. To balance this, the penalty ($2p$) is assumed. Hence the fit is penalized with a linear function of number of parameters. Model parsimony should result in the use of such a penalized form. This is widely used for fixed effect models and is the basis of the deviance information criterion discussed below.

Another variant that is commonly used as a model choice criterion is the *Bayesian information criterion (BIC)*. This is widely used in Bayesian and hierarchical models. It asymptotically approximates a Bayes factor. It is defined as
$$-2l(y|\widehat{\theta}_{fit}) + p\ln m.$$
In a model with log-likelihood $l(\theta)$ the AIC or BIC value can be estimated from the output of an McMC algorithm by
$$AIC = -2\hat{l}(\theta) + 2p$$
$$BIC = -2\hat{l}(\theta) + p\ln m,$$
where p is the number of parameters, m is the number of data points and
$$\hat{l}(\theta) = \frac{1}{G}\sum_{g=1}^{G} l(y|\theta^g),$$
the averaged log-likelihood over G posterior samples of θ. Alternatively a posterior estimate of $\widehat{\theta}$ (such as posterior expectation) must be first computed and then substituted into the AIC or BIC. Leonard and Hsu (1999) provide comparisons of these measures in a variety of examples. One disadvantage of the AIC or BIC is that in models with random effects, it is difficult to decide how many parameters are included within the model. For example, a unit level effect could be specified as $v_i \sim N(0,\tau)$. In this case there is one variance parameter, but there is also a separate value of v for each item. Hence, we have potentially super-saturation of the parameter space ($p > m$) if we count the m v_is as well as τ. Should the parameterization be 1 or $m+1$ or somewhere between these values? This quandary does not arise with random coefficient models where, for example, in a regression context we may have a p length vector of parameters, $\boldsymbol{\beta}$ say, who may have in the simplest case p variances.

4.1.1 Deviance Information Criterion

The *Deviance information criterion (DIC)* (Spiegelhalter et al., 2002) has been proposed by Spiegelhalter et al. (2002) and is widely used in Bayesian modeling. This is defined as
$$DIC = 2E_{\boldsymbol{\theta}|y}(D) - D[E_{\boldsymbol{\theta}|y}(\theta)],$$

where $D(.)$ is the deviance of the model and y is the observed data. Note that the DIC is based on a comparison of the average deviance ($\overline{D} = -2\sum_{g=1}^{G} l(y|\theta^g)/G$) and the deviance of the posterior expected parameter estimates, $\hat{\boldsymbol{\theta}}$ say: ($\widehat{D}(\hat{\boldsymbol{\theta}}) = -2l(y|\hat{\boldsymbol{\theta}})$). For any sample parameter value θ^g the deviance is just $\widehat{D}(\theta^g) = -2l(y|\theta^g)$. The effective number of parameters (pD) is estimated as $\widehat{pD} = \overline{D} - \widehat{D}(\hat{\boldsymbol{\theta}})$ and then $DIC = \overline{D} + \widehat{pD} = 2\overline{D} - \widehat{D}(\hat{\boldsymbol{\theta}})$. Unfortunately in some situations the $\widehat{pD}$ can be negative (as it can happen that $\widehat{D}(\hat{\boldsymbol{\theta}}) > \overline{D}$). Instability in pD can lead to problems in the use of this DIC. For example, mixture models, or more simply, models with multiple modes can 'trick' the pD estimate because the overdispersion in such models (when the components are not correctly estimated) leads to $\widehat{D}(\hat{\boldsymbol{\theta}}) > \overline{D}$. However it is also true that inappropriate choice of hyper-parameters for variances of parameters in hierarchical models can lead to inflation also, as can nonlinear transformations (such as changing from a Gaussian model to a log normal model). In such cases it is sometimes safer to compute the effective number of parameters from the posterior variance of the deviance. Gelman et al. (2004, 182) propose the estimator $\widetilde{pD} = \frac{1}{2}\frac{1}{G-1}\sum_{g=1}^{G}(\widehat{D}(\theta^g) - \overline{D})^2$. This value can also be computed from sample output from a chain. (It is also available directly in R2WinBUGS.) An alternative estimator of the variance is direct available from output $\widehat{var}(D) = \frac{1}{G-1}\sum_{g=1}^{G}(\widehat{D}(\theta^g) - \overline{D})^2 = 2\widetilde{pD}$. Hence a DIC based on this last variance estimate is just $DIC = \overline{D} + \widehat{var}(D)$. Note that the *expected predictive deviance* (EPD: D_{pr}) is an alternative measure of model adequacy and it is based on the out-of-sample predictive ability of the fitted model. The quantity can also be approximately estimated as $\widehat{D}_{pr} = 2\overline{D} - \widehat{D}(\hat{\boldsymbol{\theta}})$.

4.1.2 Posterior Predictive Loss

Gelfand and Ghosh (1998) proposed a loss function based approach to model adequacy which employs the predictive distribution. The approach essentially compares the observed data to predicted data from the fitted model. Define the i th predictive data item as y_i^{pr}. Note that the predictive data can easily be obtained from a converged posterior sample. Given the current parameters at iteration $j : \boldsymbol{\theta}^{(j)}$ say, then

$$p(y_i^{pr}|\mathbf{y}) = \int p(y_i^{pr}|\boldsymbol{\theta}^{(j)})p(\boldsymbol{\theta}^{(j)}|\mathbf{y})d\boldsymbol{\theta}^{(j)}.$$

Hence the j th iteration can yield y_{ij}^{pr} from $p(y_i^{pr}|\boldsymbol{\theta}^{(j)})$. The resulting predictive values has marginal distribution $p(y_i^{pr}|\mathbf{y})$. For a Poisson distribution, this simply requires generation of counts as $y_{ij}^{pr} \leftarrow Pois(e_i\theta_i^{(j)})$.

A loss function is assumed where $L_0(y, y^{pr}) = f(y, y^{pr})$. A convenient choice of loss could be the squared error loss whereby we define the loss as: $L_0(y, y^{pr}) = (y - y^{pr})^2$.

Alternative loss functions could be proposed such as absolute error loss or more complex (Quantile) forms. An overall crude measure of loss across the data is afforded by the average loss across all items: the *mean squared predictive error* (MSPE) is simply an average of the item-wise squared error loss:

$$MSPE_j = \sum_i (y_i - y_{ij}^{pr})^2 / m$$

and

$$MSPE = \sum_i \sum_j (y_i - y_{ij}^{pr})^2 / (G \times m),$$

where m is the number of observations and G is the sampler sample size. An alternative could be to specify an absolute error:

$$MAPE_j = \sum_i |y_i - y_{ij}^{pr}| / m$$

and

$$MAPE = \sum_i \sum_j |y_i - y_{ij}^{pr}| / (G \times m).$$

Gelfand and Ghosh (1998) proposed a more sophisticated form:

$$D_k = \frac{k}{k+1} A + B$$

$$= \frac{k}{k+1} \sum_{i=1}^{m} (y_i - \overline{y_i^{pr}})^2 + \frac{1}{m_p} \sum_{i=1}^{m} \sum_{j=1}^{m_p} (y_{ij}^{pr} - \overline{y_i^{pr}})^2$$

where $\overline{y_i^{pr}} = \sum_{j=1}^{m_p} y_{ij}^{pr} / m_p$

and m_p is the prediction sample size (usually $G = m_p$). Here, the k can be chosen to weight the different components. For $k = \infty$, then $D_k = A + B$. The choice of k does not usually affect the ordering of model fit. Each component measures a different feature of the fit: A represents lack of fit and B degree of smoothness. The model with lowest D_k (or $MSPE$ or $MAPE$) would be preferred.

Note that predictive data can be easily obtained from model fomulas in WinBUGS: in the Poisson case with observed data y[] and predicted data ypred[], we have for the i th item:

```
y[i]~dpois(mu[i])
ypred[i]~dpois(mu[i])
```

As the predicted values are missing they would have to be initialized.

In addition, it is possible to consider prediction-based measures as convergence diagnostics. The measures $\widehat{D}(\theta^{(j)})$ or $MSPE_j$ could be monitored using the single or multi-chain diagnostics discussed above (Section 3.3.6). Note that for any model for which a unit likelihood contribution is available, it is possible to compute a deviance-based measure such as $\widehat{D}(\theta^{(j)})$. Hence for point process (case event), as well as count-based likelihoods, deviance measures are available whereas a residual based measure (such as MSPE) is more difficult to define for a spatial event domain.

4.2 General Residuals

The analysis of residuals and summary functions of residuals forms a fundamental part of the assessment of model goodness of fit in any area of statistical application. In the case of disease mapping there is no exception, although full residual analysis is seldom presented in published work in the area. Often goodness-of-fit measures are aggregate functions of piecewise residuals, while measures relating to individual residuals are also available. A variety of methods are available when full residual analysis is to be undertaken. We define a piecewise residual as the standardized difference between the observed value and the fitted model value. Usually the standardization will be based on a measure of the variability of the difference between the two values.

It is common practice to specify a residual as

$$r_i = y_i - \widehat{y}_i \tag{4.1}$$

or

$$r_i^s = r_i / \sqrt{\widehat{var}(r_i)}$$

where $\widehat{y}_i$ is a fitted value under a given model. When complex spatial models are considered, it is often easier to examine residuals, such as $\{r_i\}$ using Monte Carlo methods. In fact it is straightforward to implement a *parametric bootstrap* (PB) approach to residual diagnostics for likelihood models (Davison and Hinkley, 1997). The simplest case, is that of tract count data, where for each tract an observed count can be compared to a fitted count. In general, when Poisson likelihood models are assumed with $y_i \sim Pois\{e_i.\theta_i\}$ then it is straightforward to employ a PB by generating a set of simulated counts $\{y_{ij}^*\}\ j = 1, ..., J$, from a Poisson distribution with mean $e_i.\widehat{\theta}_i$. In this way, a tract-wise ranking, and hence p-value, can be computed by assessing the rank of the residual within the pooled set of $J+1$ residuals:

$$\{y_i - e_i.\widehat{\theta}_i; \{y_{ij}^* - e_i.\widehat{\theta}_i\}, j = 1, ..., J\}.$$

Denote the observed standardized residual as r_i^s, and the simulated as r_{ij}^s. Note that it is now possible to compare functions of the residuals as well as making direct comparisons. For example, in a spatial context, it may be appropriate to examine the spatial autocorrelation of the observed residuals. This may provide evidence of lack of model fit. Hence, a Monte Carlo assessment of degree of residual autocorrelation could be made by comparing Moran's I statistic for the observed residuals, say, $M(\{r_i^s\})$, to that found for the simulated count residuals $M(\{r_{ij}^s\})$, where $M(\{u\}) = \frac{u^T W u}{u^T u}$ where $u_i = r_i/\sqrt{var(r_i)}$ and $r_i = (y_i - e_i.\widehat{\theta}_i)$ and W is an adjacency matrix. Note that $E[M(\{u\})]$ may not be zero and so it would be important to allow for this fact in any assessment of residual autocorrelation (see Section 5.6.1).

In the above discussion the residual definition relies on an observed dependent variable (usually a count or other discrete outcome) and a model-based fitted variable value. In the situation where case event data is modeled via point process models then the domain of interest is spatial and it is more problematic to define a residual. Note that this does not apply to conditional logistic models for case event data, as a binary outcome is modeled (see Section 5.1.2). For a spatial domain it is convenient to consider a local measure of the case density and to compare it to a model fitted density. A deviance residual was proposed by Lawson (1993a), which compared a saturated density estimate with a modeled estimate. Extension to other processes has also been made (see Baddeley et al., 2005 and discussion). The measure $\widehat{D}(\boldsymbol{\theta}) = -2l(y|\boldsymbol{\theta})$ is available for any likelihood model and so a relative comparison is possible between models with different estimated $\boldsymbol{\theta}$. Hence, $r_{d_i} = \widehat{D}_i(\boldsymbol{\theta}_s) - \overline{D}_i$ where $\overline{D}_i = -2\sum_{g=1}^{G} l(y_i|\boldsymbol{\theta}^g)/G$ and $\boldsymbol{\theta}_s$ is a saturated estimate of the parameters, or an averaged version such as $r_{d_i} = \frac{1}{G}\sum_{g=1}^{G}[\widehat{D}_i(\boldsymbol{\theta}_s) - \widehat{D}_i(\boldsymbol{\theta}^g)]$ are available. If $\boldsymbol{\theta}_s$ is fixed then these residuals are the same. A simple saturated estimate of density is $\frac{1}{A_i}$ where A_i is the area of the i th Voronoi (Dirichlet) tile. This is not a consistent estimator. Define a neighborhood set of the i th location as δ_i and the number in the set as n_{δ_i}. This set consists of all the areas that are regarded as neighbors of that point. In a tesselation the first order neighbors are usually those locations that share a common boundary with the point of interest. Hence, further averaging over tile neighbors might be useful to improve this 'local' estimate. An example could be $\widehat{\lambda}_{s_i} = \frac{n_{\delta_i}+1}{A_i+\sum_{j\in\delta_i}A_j}$. In the case event situation, for a realization of cases at locations $\{s_i\}, i = 1,...,m$, the likelihood would be a function of the intensity of the process at these locations λ_i. Hence, $\overline{D}_i = -2\sum_{g=1}^{G} l(s_i|\lambda_i^g)/G$ and $\widehat{D}_i(\boldsymbol{\theta}_s) = -2l(s_i|A_i^{-1})$ for the simple case or $\widehat{D}_i(\boldsymbol{\theta}_s) = -2l(s_i|\widehat{\lambda}_{s_i})$ where $\widehat{\lambda}_{s_i} = \frac{n_{\delta_i}+1}{A_i+\sum_{j\in\delta_i}A_j}$ for the consistent case. Hence a deviance residual can be set up for a simple non-stationary Poisson point process by computing

$$r_{d_i} = -2[l(s_i|A_i^{-1}) + \sum_{g=1}^{G} l(s_i|\lambda_i^g)/G].$$

This does not make allowance for modulation of the process however. Usually the intensity of the case event process is modulated by the 'at risk' population distribution, and the intensity of the modulated process is $\lambda_i \equiv \lambda(s_i) = \lambda_0(s_i).\lambda_1(s_i;\boldsymbol{\theta})$ where $\lambda_0(s_i)$ represents this population effect and $\lambda_1(s_i;\boldsymbol{\theta})$ is the excess risk density suitably parameterized with $\boldsymbol{\theta}$. Define $\lambda_{0i} = \lambda_0(s_i)$, and $\lambda_{1i} = \lambda_1(s_i;\boldsymbol{\theta})$. Assuming that λ_{0i} is known, then the simple saturated estimate of $\lambda(s_i)$ is $\widehat{\lambda}_i = A_i^{-1}$ (as the estimate subsumes both population and excess risk effects). Hence the residual becomes

$$r_{d_i} = -2[l(s_i|A_i^{-1}) + \sum_{g=1}^{G} l(s_i|\lambda_{0i}\lambda_{1i}^g)/G].$$

The estimation of λ_{0i} may be an issue in any application, if it is not known, and this is discussed further in Section 5.1.1.

4.3 Bayesian Residuals

In a Bayesian setting it is natural to consider the appropriate version of (4.1). Carlin and Louis (2000) describe a Bayesian residual as

$$r_i = y_i - \frac{1}{G}\sum_{g=1}^{G} E(y_i|\theta_i^{(g)}) \tag{4.2}$$

where $E(y_i|\theta_i)$ is the expected value from the posterior predictive distribution, and (in the context of McMC sampling) $\{\theta_i^{(g)}\}$ is a set of parameter values sampled from the posterior distribution.

In the tract count modeling case, with a Poisson likelihood and expectation $e_i\theta_i$, this residual can be approximated, when a constant tract rate is assumed, by:

$$r_i = y_i - \frac{1}{G}\sum_{g=1}^{G} e_i\theta_i^{(g)}. \tag{4.3}$$

This residual averages over the posterior sample. An alternative computational possibility is to average the $\{\theta_i^{(g)}\}$ sample, $\widehat{\theta}_i = \frac{1}{G}\sum_{g=1}^{G} \theta_i^{(g)}$ say, to yield a posterior expected value of y_i, say $\widehat{y}_i = e_i\widehat{\theta}_i$, and to form $r_i = y_i - \widehat{y}_i$. A further possibility is to simply form r_i at each iteration of a posterior sampler and to average these over the converged sample (Spiegelhalter et al., 1996). These residuals can provide pointwise goodness-of-fit (*gof*) measures as well as global *gof* measures, (such as mean squared error (MSE): $\frac{1}{m}\sum_{i=1}^{m} r_i^2$) and can be assessed using Monte Carlo methods. For exploratory purposes it might be useful to standardize the residuals before examination, although this is not essential for Monte Carlo assessment. To provide for a Monte Carlo assessment of individual unit residual behavior a repeated Monte Carlo simulation of independent samples from the predictive distribution would be needed. This can be achieved by taking J samples from the converged McMC stream with gaps of length p, where p is large enough to ensure independence. Ranking of the residuals in the pooled set $(J+1)$ can then be used to provide a Monte Carlo p-value for each unit.

4.4 Predictive Residuals and the Bootstrap

It is possible to disaggregate the MSPE to yield individual level residuals based on the predictive distribution. Define $y_i^{pr} \sim f_{pr}(\mathbf{y}^{pr}|\mathbf{y};\boldsymbol{\theta})$ where $f_{pr}(\mathbf{y}^{pr}|\mathbf{y};\boldsymbol{\theta}) = \int f(\mathbf{y}^{pr}|\boldsymbol{\theta})p(\boldsymbol{\theta}|\mathbf{y})d\boldsymbol{\theta}$ and $f(\mathbf{y}^{pr}|\boldsymbol{\theta})$ is the likelihood of $\mathbf{y}^{pr}$ given $\boldsymbol{\theta}$. This can be approximated within a converged sample by a draw from $f(\mathbf{y}^{pr}|\boldsymbol{\theta})$. For a Poisson likelihood, at the g th iteration, with expectation $e_i\theta_i^{(g)}$, a single value $y_i^{pr(g)}$ is generated from $Pois(e_i\theta_i^{(g)})$. Hence a predictive residual can be formed from $r_i^{pr} = y_i - y_i^{pr}$. This must be averaged over the sample. This can be done in a variety of ways. For example, we could take $r_i^{pr} = \sum_{g=1}^{G}\{y_i - y_i^{pr(g)}\}/G$. Other possibilities could be explored.

To further assess the distribution of residuals, it would be advantageous to be able to apply the equivalent of PB in the Bayesian setting. With convergence of a McMC sampler, it is possible to make subsamples of the converged output. If these samples are separated by a distance (h) which will guarantee approximate independence (Robert and Casella, 2005), then a set of J such samples could be used to generate $\{y_i^{pr}\}$ $j=1,....,J$, with, $y_i^{pr} \leftarrow Pois(e_i\widehat{\theta}_{ij})$, and the residual computed from the data r_i can be compared to the set of J residuals computed from $y_i^{pr} - \widehat{y}_i$. In turn, these residuals can be used to assess functions of the residuals and gof measures. The choice of J will usually be 99 or 999 depending on the level of accuracy required.

4.4.1 Conditional Predictive Ordinates

It is possible to consider a different approach to inference whereby individual observations are compared to the predictive distribution with observations removed. This conditional approach has a cross-validation flavour i.e., the value in the unit is predicted from the remaining data and compared to the observed data in the unit. The derived residual is defined as

$$r_i^{CPO} = y_i - y_{i,-i}^{rep}$$

where $y_{i,-i}^{rep}$ is the predicted value of y based on the data with the i th unit removed (Stern and Cressie, 2000). The value of $y_{i,-i}^{rep}$ is obtained from the cross-validated posterior predictive distribution:

$$p(y_{i,-i}^{rep}|\mathbf{Y}_{-i}) = \int p(y_{i,-i}^{rep}|\boldsymbol{\theta})p(\boldsymbol{\theta}|\mathbf{Y}_{-i})d\boldsymbol{\theta}$$

where $\boldsymbol{\theta}$ is a vector of model parameters. For a Poisson data likelihood, $p(y_{i,-i}^{rep}|\mathbf{Y}_{-i})$ is just a Poisson distribution with mean $e_i^* \theta_i$ where e_i^* is adjusted for the removal of the i th unit and θ_i is estimated under the cross-validated posterior distribution $p(\boldsymbol{\theta}|\mathbf{Y}_{-i})$.

As noted by Spiegelhalter et al. (1996), it is possible to make inference about such residuals within a conventional McMC sampler via the construction of weights. Assume draws $g = 1, ..., G$ are available and define the importance weight $w_{-i}(\boldsymbol{\theta}^g) = \frac{1}{p(y_i|e_i\theta_i^g)}$. This is just the reciprocal of the Poisson probability with mean $e_i\theta_i^g$. It is then possible to compute a Monte Carlo probability for the residual via:

$$p(y_{i,-i}^{rep} \leq y_i|\mathbf{Y}_{-i}) = p(y_{i,-i}^{rep} < y_i|\mathbf{Y}_{-i}) + \frac{1}{2}p(y_{i,-i}^{rep} = y_i|\mathbf{Y}_{-i})$$

$$\approx \sum_{l=0}^{y_i-1} \left\{ \sum_{g=1}^{G} p(y_{i,-i}^{rep} = g|\boldsymbol{\theta}^g)w_{-i}(\boldsymbol{\theta}^g) \right\} / w_i^T$$

$$+ \frac{1}{2} \left\{ \sum_{g=1}^{G} p(y_{i,-i}^{rep} = y_i|\boldsymbol{\theta}^g)w_{-i}(\boldsymbol{\theta}^g) \right\} / w_i^T$$

where $w_i^T = \sum_{g=1}^{G} \frac{1}{p(y_i|e_i\theta_i^g)}$. In general, a simple approach to computation of the CPO without recourse to refitting is to note that the conditional predictive ordinate for the i, the unit can be obtained from

$$CPO_i^{-1} = G^{-1} \sum_{g=1}^{G} p(y_i|\boldsymbol{\theta}^g)^{-1}$$

where $p(y_i|\boldsymbol{\theta}^g)$ is the data density given the current parameters.

4.5 Interpretation of Residuals in a Bayesian Setting

Diagnostics based on residuals will be indicative of a variety of model features. What should be expected from residuals from an adequate model? In general, when a model fits well, one would expect the residuals from that model to have a number of features. First they should usually be symmetric and centered around zero. Clearly variance standardization should yield a closer approximation to this, but in general this can only be approximate. Second they should not show any particular structure and should appear to be reasonably random. However, as to distribution, it is not clear that residuals should have a zero mean Gaussian form (as suggested by the use of normal quantile plots in e.g., Carlin and Louis, 2000). In the following example, I examine a dataset for congenital anomaly mortality counts for the 46 counties of South Carolina for the year 1990. The standardized mortality ratio for these data using the statewide standard population rate is shown in Section 1.1. Figure 4.1 displays Bayesian residuals from a converged Poisson–gamma model fitted to congenital anomaly mortality for South Carolina counties 1990. These residuals seem to have some structure although their QQ plot is relatively straight. Figure 4.2 displays the corresponding image and contour map of the smoothed residual surface. It is noticeable that the high positive residuals are grouped in relatively rural areas.

4.6 Exceedence Probabilities

Exceedence probabilities are important when assessing the localized spatial behavior of the model and the assessment of unusual clustering or aggregation of disease. The simplest case of an exceedence probability is $q_i^c = \Pr(\theta_i > c)$. The probability is an estimate of how frequently the relative risk exceeds the null risk value ($\theta_i = 1$) and can be regarded as an indicator of 'how unusual' the risk is in that unit. As will be discussed in Chapter 6, this leads to assessment of 'hot spot' clusters: areas of elevated risk found independently of any cluster grouping criteria. Note that under posterior sampling, a converged sample of $\{\boldsymbol{\theta}^{m+1}, ..., \boldsymbol{\theta}^{m+m_p}\}$ can yield posterior expected estimates of these probabilities as $\widehat{q}_i^c = \sum_{g=m+1}^{m+m_p} I(\theta_i^{(g)} > c)/G$ where $G = m_p$. It is straightforward on WinBUGS, for example, to compute these values. If theta[i] is set to store the current θ_i then

prexc[i]<-step(theta[i]-1)

will store the indicator of exceedence and this variable can be monitored. The posterior average of the variable will yield the posterior estimate of the

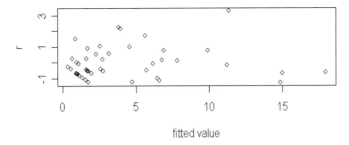

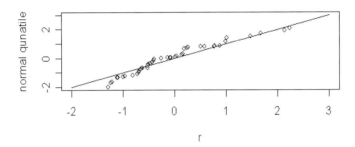

FIGURE 4.1
Residual plots of Bayesian residuals for 46 counties of South Carolina for congenital abnormality mortality data 1990. Poisson–gamma model fit with top: standardised residual versus fitted value; bottom: QQ plot of standardised residual versus standard normal quantile.

probability. Large values of $\hat{q}_i^c$ would be suggestive of unusual areas of risk. In fact this measure has been proposed as a method for detecting clusters (Richardson et al., 2004). While it certainly can be used to examine maps for individual hot spots of risk, the measure itself does not directly measure clustering of risk in terms of spatial aggregation. In fact, the measure can be applied to any underlying model when a relative risk parameter is estimated and hence may be model dependent. This is discussed more fully in the chapter on disease cluster modeling (Chapter 6).

As a comparison, the predictive residuals have also been computed for this example. Figure 4.3 displays a comparison of the density estimates of the different residuals. These appear quite similar in this example, with the predictive residual having a longer right hand tail. The associated map of the predictive residuals shows a rural-urban difference also.

There are a number of issues associated with the evaluation of $\hat{q}_i^c$. First the value of c must be specified. Second the level of probability that is regarded '*unusual*' (a say) must also be fixed. However there is a trade-off between these measures. Not only will there be different features found if you change

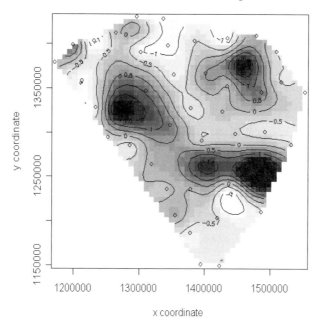

FIGURE 4.2
Bayes residual map for the South Carolina congenital anomaly mortality data. The county centroids are marked as ◇.

the threshold (c), but there will be different interpretations if threshold is changed from say $a_1 = 0.95$ to $a_2 = 0.99$ or lowered to $a_0 = 0.90$. Hence a simple rule such as, *classify as unusual any region where* $\hat{q}_i^1 > a_1$, might be equivalent to $\hat{q}_i^2 > a_0$. Whereas the null level of risk may be $c = 1$ there is not reason to assume as a threshold $\hat{q}_i^1 > 0.95$. Of course usually either a or c is fixed. It should be noted that $\hat{q}_i^c$ is a function of a model and so is not necessarily going to yield the same information as, say a residual r_i. While both depend on model elements, a residual usually also contains extra (at least) uncorrelated noise and should, if the model fits well, not contain any further structure. On the other hand, posterior estimates of relative risk will include modeled components of risk (such as trend or correlation) and should be relatively free of extra noise.

In the example of the South Carolina congenital anomaly mortality for 1990, the standardized residuals and the $\hat{q}_i^1$ seem to reflect areas of relatively unusual risk. Both reflect rural areas where incidence is marked. The $\hat{q}_i^1$ map displayed in Figure 4.4 shows these areas well. None of the areas exceed 0.95 in this case. A comparison with Figure 4.2 suggests that even though the model does reflect the elevated risk areas, in these areas there is excess risk unaccounted for.

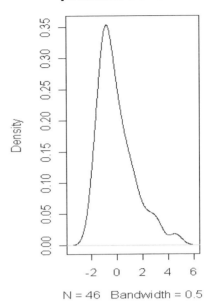

FIGURE 4.3
Residual plots using density estimation: Bayes residuals and predictive residuals for the same example: South Carolina county-level congenital anomaly mortality example.

4.7 Exercises

1) The Bernoulli model with outcome y_i is assumed for the binary label for a set of m case and n control events: $S: \{s_1, ..., s_m, s_{m+1}, ..., s_{m+n}\}$. For short, denote the outcome as $y_i \equiv y(s_i)$. The probability, conditional on a site s_i, of being a case, is $p_i \equiv p(s_i)$. Assume a logistic link to a covariate model: $logit\ p_i = \eta_i = x_i^T \boldsymbol{\beta}$ with an intercept and single covariate x_{1i}. The parameter vector is $\boldsymbol{\beta}: [\beta_0, \beta_1]'$. The parameters have zero mean Gaussian prior distributions: $\boldsymbol{\beta} \sim \mathbf{N}_2(\mathbf{0}, \Gamma)$ where $\Gamma = diag(\tau_0, \tau_1)$ and $\mathbf{N}_2$ denotes a bivariate normal distribution. Hence, each $\beta_* \sim N(0, \tau_*)$.

a) Show that the DIC for this model is given by $DIC = -2\frac{1}{G}\sum_{g=1}^{G}\sum_{i=1}^{m+n} \log[p_i^g] +$
$\frac{1}{2}\frac{1}{G-1}\sum_{g=1}^{G}\left[2(\frac{1}{G}\sum_{g=1}^{G}\sum_{i=1}^{m+n} \log[p_i^g] - \sum_{i=1}^{m+n} \log[p_i^g])\right]^2$

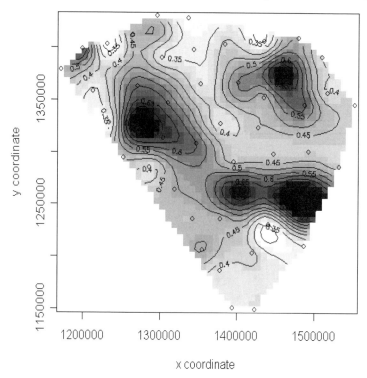

FIGURE 4.4
Exceedance probability map for C = 1. Bayes residuals and predictive residuals for the same example: South Carolina county-level congenetial anomaly mortality example.

where $p_i^g = \frac{[\exp(\beta_0^g + \beta_1^g x_{1i})]^{y_i}}{1+\exp(\beta_0^g + \beta_1^g x_{1i})}$.

b) Show that the conditional predictive ordinate (CPO) is given by $CPO_i = \left\{ \frac{1}{G} \sum_{g=1}^{G} \left[\frac{[\exp(\beta_0^g + \beta_1^g x_{1i})]^{y_i}}{1+\exp(\beta_0^g + \beta_1^g x_{1i})} \right]^{-1} \right\}^{-1}$

c) A Bayesian residual can be computed from $y_i - \frac{\exp(\widehat{\beta}_0 + \widehat{\beta}_1 x_{1i})}{1+\exp(\widehat{\beta}_0 + \widehat{\beta}_1 x_{1i})}$, where $\widehat{\beta}_0, \widehat{\beta}_1$ are posterior mean estimates. Why is this residual difficult to interpret? What remedy could be suggested to allow a more meaningful residual analysis?

2) A case event realization within a window of area T is defined as $\{s_i\}$, $i = 1, ..., m$. A modulated Poisson process model is assumed with first order intensity, conditional on a parameter vector $\boldsymbol{\theta}$, governed by $\lambda(s) = \lambda_0(s).\lambda_1(s|\boldsymbol{\theta})$. Assume that $\lambda_1(s|\boldsymbol{\theta}) = 1 + \alpha \exp\{-\beta||s - s_0||\}$ where s_0 is a fixed spatial

Residuals and Goodness-of-Fit

location. Assume also that $\lambda_0(s)$ is fixed and known. The prior distributions for the parameters are $\alpha_t = \log \alpha \sim N(0, \tau_\alpha)$ and $\beta \sim N(0, \tau_\beta)$. The log likelihood is given by

$$l(s|\alpha, \beta) = \sum_{i=1}^{m} \ln(1 + \alpha \exp\{-\beta d_i\}) - \Lambda(\alpha, \beta)$$

where $d_i = ||s_i - s_0||$ and $\Lambda(\alpha, \beta) = \int_T \lambda_0(u).[1 + \alpha \exp\{-\beta||u - s_0||\}]du$.

i) Show that, for this model under McMC iterative sampling with a converged sample of size G, the $DIC = \overline{D} + \widehat{pD}$ where $\overline{D} = -2 \sum_{g=1}^{G} [\sum_{i=1}^{m} \ln(1 + \alpha^g \exp\{-\beta^g d_i\}) - \Lambda(\alpha^g, \beta^g)]/G$

and $\widehat{pD} = \frac{1}{2} \frac{1}{G-1} \sum_{g=1}^{G} ([-2 \sum_{i=1}^{m} \ln(1 + \alpha^g \exp\{-\beta^g d_i\}) - \Lambda(\alpha^g, \beta^g)] - \overline{D})^2$.

ii) Show that a deviance residual can be computed from
$r_{d_i} = -2[l(s_i|A_i^{-1}) + \sum_{g=1}^{G} [\ln(1 + \alpha^g \exp\{-\beta^g d_i\}) - \Lambda_i^*(\alpha^g, \beta^g)]]/G]$
where $\Lambda_i^*(\alpha^g, \beta^g) = w_i \lambda_0(s_i).[1 + \alpha \exp\{-\beta||s_i - s_0||\}]$ and w_i is a Berman-Turner integration weight (see Chapter 5) for the i the unit.

iii) Suggest a total model discrepancy measure based on this residual.

3) For the model in 2) above consider an exceedence probability for the resulting estimated intensity function. The focus of interest is the relative risk $\lambda_1(s|\theta)$. Under posterior sampling the estimated risk is $\lambda_1(s|\theta) = [1 + \alpha^g \exp\{-\beta^g d_i\}]$. However usually it is assumed that $\alpha > 0$, and usually, but not necessarily, $\beta > 0$ (as the estimated risk could increase with distance from the fixed point). Hence we cannot have $\lambda_1(s|\theta) < 1$, as will be seen later in Chapter 7.3. This risk is a natural form for the function of distance. However a simpler form that is often assumed is a purely multiplicative one:

$$\lambda_1(s|\theta) = \rho. \exp\{-\beta d_i\}$$

where $\rho = \exp(\alpha)$. Show that an exceedence probability $(\widehat{g}_i)$ can be computed from:

$$\widehat{g}_i = \widehat{\Pr}(\lambda_1(s_i|\theta) > 1) = \sum_{g=1}^{G} I(\exp\{\alpha^g - \beta^g d_i\}/G).$$

Part II

Themes

5

Disease Map Reconstruction and Relative Risk Estimation

5.1 Introduction to Case Event and Count Likelihoods

5.1.1 Poisson Process Model

Define a study area as T and within that area m event of disease occur. These events are usually address locations of the cases. The case could be an incident or prevalent case or could be a death certificate address. We assume at present that the cases are geo-coded down to a point (with respect to the scale of the total study region). Hence they form a point process in space. Define $\{s_i\}, i = 1, ..., m$ as the set of all cases within T. This is called a *realization* of the disease process, in that we assume that all cases within the study area are recorded. This is a common form of data available from government agencies. Sub-samples of the spatial domain, where incomplete realizations are taken are not considered at this point.

The basic point process model assumed for such data within disease mapping is the heterogeneous Poisson process with first-order intensity $\lambda(s)$. The basic assumptions of this model are that points (case events) are independently spatially-distributed and governed by the first order intensity. Due to the independence assumption, we can derive a likelihood for a realization of a set of events within a spatial region. For the study region defined above, the unconditional likelihood of m events is just

$$L(\{\mathbf{s}\}|\psi) = \frac{1}{m!} \prod_{i=1}^{m} \lambda(s_i|\psi) \exp\{-\Lambda_T\} \qquad (5.1)$$

$$\text{where } \Lambda_T = \int_T \lambda(u|\psi) du.$$

The function Λ_T is the integral of the intensity over the study region, ψ is a parameter vector, and $\lambda(s_i|\psi)$ is the first order intensity evaluated at the case event location s_i. Denote this likelihood as $PP[\{\mathbf{s}\}|\psi]$. This likelihood can be maximized with respect to the parameters in ψ and likelihood-based inference could be pursued. The only difficulty in its evaluation is the estimation of the spatial integral. However, a variety of approaches can be used to numerical integration of this function and with suitable weighting schemes this likelihood can be evaluated even with conventional linear modeling functions within software packages (such as glm in R or S-Plus) (see e.g., Berman and Turner, 1992, Lawson, 2006b, App. C). An example of such a weighted log-likelihood approximation is:

$$l(\{\mathbf{s}\}|\psi) = \sum_{i=1}^{m} \ln \lambda(s_i|\psi) - \Lambda_T, \tag{5.2}$$

where $\Lambda_T \approx \sum_{i=1}^{m} w_i \lambda(s_i|\psi)$ and w_i is an integration weight. This scheme per se is not accurate and more weights are needed. In the more general scheme of Berman & Turner a set of additional mesh points (of size m_{aug}) are added to the data. The augmented set $(N = m + m_{aug})$ is used in the likelihood with a indicator function, I_k:

$$l(\{\mathbf{s}\}|\psi) = \sum_{k=1}^{N} w_k \{\frac{I_k}{w_k} \ln \lambda(s_k|\psi) - \lambda(s_k|\psi)\},$$

$$\text{where } \int_T \lambda(u|\psi) du = \sum_{k=1}^{N} w_k \lambda(s_k|\psi).$$

This has the form of a weighted Poisson likelihood, with $I_k = 1$ for a case and 0 otherwise. Diggle (1990) gives an example of the use of a likelihood such as (5.1) in a spatial health data problem.

In disease mapping applications, it is usual to parameterize $\lambda(s|\psi)$ as a function of two components. The first component makes allowance for the underlying population in the study region, and the second component is usually specified with the modelled components (i.e., those components describing the 'excess' risk within the study area).

A typical specification would be

$$\lambda(s|\psi) = \lambda_0(s|\psi_0).\lambda_1(s|\psi_1). \tag{5.3}$$

Here it is assumed that $\lambda_0(s|\psi_0)$ is a spatially-varying function of the population 'at risk' of the disease in question. It is parameterized by ψ_0. The second function, $\lambda_1(s|\psi_1)$, is parameterized by ψ_1 and includes any linear or non-linear predictors involving covariates or other descriptive modeling terms, thought appropriate in the application. Often we assume, for positivity, that $\lambda_1(s_i|\psi_1) = \exp\{\eta_i\}$ where η_i is a parameterized linear predictor allowing a link to covariates measured at the individual level. The covariates could include spatially-referenced functions as well as case-specific measures. Note that $\psi : \{\psi_0, \psi_1\}$. The function $\lambda_0(s|\psi_0)$ is a nuisance function which must be allowed for but which is not usually of interest from a modeling perspective.

5.1.2 Conditional Logistic Model

When a bivariate realization of cases and controls are available it is possible to make conditional inference on this joint realization. Define the case events as $s_i : i = 1, ..., m$ and the control events as $s_i : i = m+1, ..., N$ where $N = m + n$ the total number of events. Associated with each location is a binary variable (y_i) which labels the event either as a case $(y_i = 1)$ or a control $(y_i = 0)$. Assume also that the point process models governing each event type (case or control) is a heterogeneous Poisson process with intensity $\lambda(s|\psi)$ for cases and $\lambda_0(s|\psi_0)$ for controls. The superposition of the two processes is also a heterogeneous Poisson process with intensity $\lambda_0(s|\psi_0) + \lambda(s|\psi) = \lambda_0(s|\psi_0)[1+\lambda_1(s|\psi_1)]$. Conditioning on the joint realization of these processes, then it is straightforward to derive the conditional probability of a case at any location as

$$\Pr(y_i = 1) = \frac{\lambda_0(s_i|\psi_0).\lambda_1(s_i|\psi_1)}{\lambda_0(s_i|\psi_0)[1 + \lambda_1(s_i|\psi_1)]}$$
$$= \frac{\lambda_1(s_i|\psi_1)}{1 + \lambda_1(s_i|\psi_1)} = p_i \qquad (5.4)$$

and

$$\Pr(y_i = 0) = \frac{1}{1 + \lambda_1(s_i|\psi_1)} = 1 - p_i. \qquad (5.5)$$

The important implication of this result is that the background nuisance function $\lambda_0(s_i|\psi_0)$ drops out of the formulation and, further, this formulation leads to a standard logistic regression if a linear predictor is assumed within $\lambda_1(s_i|\psi_1)$. For example, a log linear formulation for $\lambda_1(s_i|\psi_1)$ leads to a logit link to p_i, i.e.,

$$p_i = \frac{\exp(\eta_i)}{1 + \exp(\eta_i)},$$

where $\eta_i = x_i'\beta$ and x_i' is the i th row of the design matrix of covariates and β is the corresponding p-length parameter vector. Note that slightly different formulations can lead to non-standard forms (see e.g., Diggle and Rowlingson, 1994). In some applications, non-linear links to certain covariates may be appropriate (see Section 7.3).

Further, if the probability model in (5.4) applies, then the likelihood of the realization of cases and controls is simply

$$L(\psi_1|\mathbf{s}) = \prod_{i \in cases} p_i \prod_{i \in controls} 1 - p_i$$
$$= \prod_{i=1}^{N} \left[\frac{\{\exp(\eta_i)\}^{y_i}}{1 + \exp(\eta_i)}\right]. \qquad (5.6)$$

Hence, in this case, the analysis reduces to that of a logistic likelihood, and this has the advantage that the 'at risk' population nuisance function does not have to be estimated. This model is ideally suited to situations where it is natural to have a control and case realization, where conditioning on the spatial pattern is reasonable.

5.1.3 Binomial Model for Count Data

In the case where we examine arbitrary small areas (such as census tracts, counties, postal zones, municipalities, health districts), usually a count of disease is observed within each spatial unit. Define this count as y_i and assume that there are m small areas. We also consider that there is a finite population within each small area out of which the count of disease has arisen. Denote this as $n_i \; \forall_i$. In this situation, we can consider a binomial model for the count data conditional on the observed population in the areas. Hence we can assume that given the probability of a case is p_i, then y_i is distributed independently as

$$y_i \sim bin(p_i, n_i)$$

and that the likelihood is given by

$$L(y_i|p_i, n_i) = \prod_{i=1}^{m} \binom{n_i}{y_i} p_i^{y_i} (1-p_i)^{(n_i - y_i)}. \qquad (5.7)$$

It is usual for a suitable link function for the probability p_i to a linear predictor to be chosen. The commonest would be a logit link so that

$$p_i = \frac{\exp(\eta_i)}{1 + \exp(\eta_i)}.$$

Here, we envisage the model specification within η_i, to include spatial and non-spatial components. Two applications which are well suited to this approach are the analysis of sex ratios of births, and the analysis of birth outcomes (e.g., birth abnormalities) compared to total births. Sex ratios are often derived from the number of female (or male) births compared to the total birth population count in an area. A ratio is often formed, though this is not necessary in our modeling context. In this case the p_i will often be close to 0.5 and spatially-localized deviations in p_i may suggest adverse environmental risk (Williams et al., 1992). The count of abnormal births (these could include any abnormality found at birth) can be related to total births in an area. Variations in abnormal birth count could relate to environmental as well as health service variability (over time and space)(Morgan et al., 2004).

5.1.4 Poisson Model for Count Data

Perhaps the most commonly encountered model for small area count data is the Poisson model. This model is appropriate when there is a relatively low count of disease and the population is relatively large in each small area.

Disease Map Reconstruction and Relative Risk Estimation

Often the disease count y_i is assumed to have a mean μ_i and is independently distributed as

$$y_i \sim Poisson(\mu_i). \tag{5.8}$$

The likelihood is given by

$$L(\mathbf{y}|\boldsymbol{\mu}) = \prod_{i=1}^{m} \mu_i^{y_i} \exp(-\mu_i)/y_i!. \tag{5.9}$$

The mean function is usually considered to consist of two components: *i)* a component representing the background population effect, and *ii)* a component representing the excess risk within an area. This second component is often termed the *relative risk*. The first component is commonly estimated or computed by comparison to rates of the disease in a standard population and a local expected rate is obtained. This is often termed *standardization* (Inskip et al. (1983)). Hence, we would usually assume that the data is independently distributed with expectation

$$E(y_i) = \mu_i = e_i \theta_i$$

where e_i is the expected rate for the i th area and θ_i is the relative risk for the i th area. As we will be developing Bayesian hierarchical models we will consider $\{y_i\}$ to be conditionally independent given knowledge of $\{\theta_i\}$. The expected rate is usually assumed to be fixed for the time period considered in the spatial example, although there is a literature on the estimation of small area rates that suggests this may be naive (Ghosh and Rao, 1994; Rao, 2003).

Usually the focus of interest will be the modeling of the relative risk. The commonest approach to this is to assume a logarithmic link to a linear predictor model:

$$\log \theta_i = \eta_i.$$

This form of model has seen widespread use in the analysis of small area count data in a range of applications (see e.g., Stevenson et al., 2005; Waller and Gotway, 2004, Chapter 9)

5.2 Specification of the Predictor in Case Event and Count Models

In all the above models a predictor function (η_i) was specified to relate to the mean of the random outcome variable, via a suitable link function. Often the predictor function is assumed to be linear and a function of fixed covariates and also possibly random effects. We define this in a general form, for p covariates, here as

$$\eta_i = x_i' \boldsymbol{\beta} + z_i' \boldsymbol{\xi},$$
$$= \beta_1 x_{1i}........\beta_p x_{pi} + \xi_1 z_{1i}......\xi_q z_{qi}$$

where x_i' is the i th row of a covariate design matrix $\mathbf{x}$ of dimension $m \times p$, $\boldsymbol{\beta}$ is a $(p \times 1)$ vector of regression parameters, $\boldsymbol{\xi}$ is a $(q \times 1)$ unit vector and z_i' is a row vector of individual level random effects, of which there are q. In this formulation the unknown parameters are $\boldsymbol{\beta}$ and $\mathbf{z}$ the $(m \times q)$ matrix of random effects for each unit. Note that in any given application it is possible to specify subsets of these covariates or random effects. Covariates for case event data could include different types of specific level measures such as an individual's age, gender, smoking status, health provider, etc., or could be environmental covariates which may have been interpolated to the address location of the individual (such as soil chemical measures or air pollution levels). For count data in small areas, it is likely that covariates will be obtained at the small area level. For example, for census tracts, there is likely to be socioeconomic variables such as poverty (percentage population below an income level), car ownership, median income level, available from the census. In addition, some variates could be included as supra-area variables such as health district in which the tract lies. Environmental covariates could also be interpolated to be used at the census tract level. For example air pollution measures could be averaged over the tract.

In some special applications non-linear link functions are used, and in others, mixtures of link functions are used. One special application area where this is found is the analysis of putative hazards (see Section 7.3), where specific distance- and/or direction-based covariates are used to assess evidence for a relation between disease risk and a fixed (putative) source of health hazard. For example, one simple example of this is the conditional logistic modeling of disease cases around a fixed source. Let distance and direction from the source to the i th location be d_i and ϕ_i, respectively, then a mixed linear and non-linear link model is commonly assumed where

$$\eta_i = \{1 + \beta_1 \exp(-\beta_2 d_i)\} \cdot \exp\{\beta_0 + \beta_3 \cos(\phi_i) + \beta_4 \sin(\phi_i)\}.$$

Here the distance effect link is nonlinear, while the overall rate (β_0) and directional components are log-linear. The explanation and justification for this formulation is deferred to Chapter 7.

Fixed covariate models can be used to make simple descriptions of disease variation. In particular it is possible to use the spatial coordinates of case events (or in the case of count data, centroids of small areas) as covariates. These can be used to model the long range variation of risk: *spatial trend*. For example, let's assume that the i th unit x-y coordinates are (x_{si}, y_{si}). We could define a polynomial trend model such as:

Disease Map Reconstruction and Relative Risk Estimation

$$\eta_i = \beta_0 + \sum_{l=1}^{L} \beta_{xl} x_{si}^l + \sum_{l=1}^{L} \beta_{yl} y_{si}^l + \sum_{k=1}^{K} \sum_{l=1}^{L} \beta_{lk} x_{si}^l y_{si}^k.$$

This form of model can describe a range of smoothly varying non-linear surface forms. However, except for very simple models, these forms are not parsimonious and also cannot capture the extra random variation that often exists in disease incidence data.

5.2.1 Bayesian Linear Model

In the Bayesian paradigm all parameters are stochastic and are therefore assumed to have prior distributions. Hence in the covariate model

$$\eta_i = x_i' \boldsymbol{\beta},$$

the $\boldsymbol{\beta}$ parameters are assumed to have prior distributions. Hence this can be formulated as

$$P(\boldsymbol{\beta}, \boldsymbol{\tau}_\beta | data) \propto L(data | \boldsymbol{\beta}, \boldsymbol{\tau}_\beta) f(\boldsymbol{\beta} | \boldsymbol{\tau}_\beta)$$

where $f(\boldsymbol{\beta}|\boldsymbol{\tau}_\beta)$ is the joint distribution of the covariate parameters conditional on the hyperparameter vector $\boldsymbol{\tau}_\beta$. Often we regard these parameters as independent and so

$$f(\boldsymbol{\beta}|\boldsymbol{\tau}_\beta) = \prod_{j=1}^{p} f_j(\beta_j | \tau_{\beta_j}).$$

More generally it is commonly assumed that the covariate parameters can be described by a Gaussian distribution and if the parameters are allowed to be correlated then we could have the multivariate Gaussian specification:

$$f(\boldsymbol{\beta}|\boldsymbol{\tau}_\beta) = \mathbf{N}_p(\mathbf{0}, \boldsymbol{\Sigma}_\beta),$$

where under this prior assumption, $E(\boldsymbol{\beta}|\boldsymbol{\tau}_\beta) = \mathbf{0}$ and $\boldsymbol{\Sigma}_\beta$ is the conditional covariance of the parameters. The commonest specification assumes prior independence and is:

$$f(\boldsymbol{\beta}|\boldsymbol{\tau}_\beta) = \prod_{j=1}^{p} N(0, \tau_{\beta_j}),$$

where $N(0, \tau_{\beta_j})$ is a zero-mean single variable Gaussian distribution with variance τ_{β_j}. At this point an assumption about variation in the hyperparameters is usually made. At the next level of the hierarchy hyperprior distributions are assumed for $\boldsymbol{\tau}_\beta$. The definition of these distributions could be important in defining the model behavior. For example if a vague hyperprior is assumed for τ_{β_j} this may lead to extra variation being present when limited learning is available from the data. This can affect computation of DIC and convergence diagnostics. While uniform hyperpriors (on a large positive range) can lead to improper posterior distributions, it has been found that a uniform distribution for the standard deviation can be useful (Gelman, 2006) i.e., $\sqrt{\tau_{\beta_j}} \sim U(0, A)$ where A has a large positive value.

Alternative suggestions are usually in the form of gamma or inverse gamma distributions with large variances. For example, Kelsall and Wakefield (2002) proposed the use of gamma (0.2, 0.0001) with expectation 2000 and variance 20,000,000, whereas Banerjee et al. (2004) examine various alternative specifications including gamma (0.001, 0.001). One common specification (Thomas et al., 2004) is gamma (0.5, 0.0005) that has expectation 1000 and variance 2,000,000.

While these prior specifications lead to relative uninformativeness, their use has been criticized by Gelman (2006) (see also Lambert et al., 2005), in favour of uniform prior distributions on the standard deviation.

To summarize the hierarchy for such covariate models, Figure 5.1 displays a directed acyclic graph for a simple Bayesian hierarchical covariate model with two covariates (x_1, x_2) and relative risk defined as $\theta_i = \exp(\beta_0 + \beta_1 x_{1i} + \beta_2 x_{2i})$ for data $\{y_i, e_i\}$ for m regions. The regression parameters are assumed to have independent zero mean Gaussian prior distributions. Figure 5.2 displays the corresponding WinBUGS code.

5.3 Simple Case and Count Data Models with Uncorrelated Random Effects

In the previous section, some simple models were developed. These consisted of functions of fixed observed covariates. In a Bayesian model formulation all parameters are stochastic and so the extension to the addition of random effects is relatively straightforward. In fact the term 'mixed' model (linear mixed model: LMM, normal linear mixed model: NLMM, or generalized linear mixed model: GLMM) is strictly inappropriate as there are no fixed effects in a Bayesian model.

The simple regression models described above often do not capture the extent of variation present in count data. Overdispersion or spatial correlation due to unobserved confounders will usually not be captured by simple covariate models and often it is appropriate to include some additional term or terms in a model which can capture such effects.

Initially, overdispersion or extra-variation can be accommodated by either a) inclusion of a prior distribution for the relative risk, (such as a Poisson -gamma model) or b) by extension of the linear or non-linear predictor term to include an extra random effect (log-normal model).

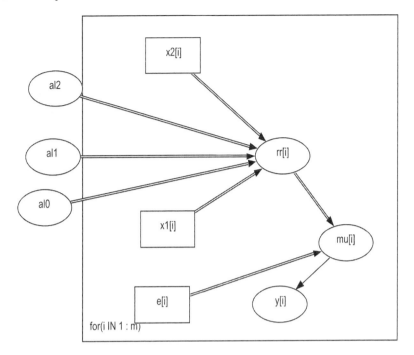

FIGURE 5.1
A directed acylic graph (in WinBUGS Doodle format) for a simple Poisson Bayesian regression with log linear relative risk and two covariates.

```
model;
{
  for( i in 1 : m ) {
    y[i] ~ dpois(mu[i])
  }
  for( i in 1 : m ) {
    mu[i] <- e[i] * rr[i]
  }
  for( i in 1 : m ) {
    rr[i] <- exp(al0 + al1 * x1[i] + al2 * x2[i])
  }
  al0 ~ dnorm( 0.0,1.0E-6)
  al1 ~ dnorm( 0.0,1.0E-6)
  al2 ~ dnorm( 0.0,1.0E-6)
}
```

FIGURE 5.2
WinBUGS odc (code) for the DAG in Figure 5.1.

5.3.1 Gamma and Beta Models

5.3.1.1 Gamma models

The simplest extension to the likelihood model that accommodates extra variation is one in which the parameter of interest in the likelihood is given a prior distribution. One case event example would be where the intensity is specified at the i th location as, $\lambda_0(s_i|\psi_0).\lambda_1(s|\psi_1) \equiv \lambda_{0i}.\lambda_{1i}$, suppressing the parameter dependence, for simplicity. Note that λ_{1i} plays the role of a relative risk parameter. This parameter can be assigned a prior distribution, such as a gamma distribution to model extra-variation. For most applications where count data are commonly found a Poisson likelihood is assumed. We will focus on these models in the remainder of this section. The Poisson parameter θ_i could be assigned a $gamma(a,b)$ prior distribution. In this case, the prior expectation and variance would be respectively a/b and a/b^2. This could allow for extra variation or overdispersion. Here we assume that parameters a, b are fixed and known. This formulation is attractive as it leads to a closed form for the posterior distribution of $\{\theta_i\}$, i.e.,

$$[\theta_i|y_i, e_i, a, b] \sim gamma(a^*, b^*) \qquad (5.10)$$

where $a^* = y_i + a, \ b^* = e_i + b$.

Hence, conjugacy leads to a gamma posterior distribution with posterior mean and variance given by $\frac{y_i+a}{e_i+b}$ and $\frac{y_i+a}{(e_i+b)^2}$ respectively. Note that a variety of θ_i estimates are found depending on the values of a and b. Lawson and Williams (2001, 78–79) demonstrate the effect of different values of a and b on relative risk maps. Samples from this posterior distribution are straightforwardly obtained (e.g., using the rgamma function on R). The prior predictive distribution of $\mathbf{y}^*$ is also relevant in this case as it leads to a distribution often used for overdispersed count data: the negative binomial. Here we have the joint distribution as:

$$[\mathbf{y}^*|\mathbf{y}, a, b] = \int f(\mathbf{y}^*|\boldsymbol{\theta})f(\boldsymbol{\theta}|a, b)d\boldsymbol{\theta} \qquad (5.11)$$

$$= \prod_{i=1}^{m} \left[\frac{b^a}{\Gamma(a)} \frac{\Gamma(y_i^* + a)}{(e_i + b)^{(y_i^*+a)}} \right]$$

5.3.1.1.1 Hyperprior distributions

One extension to the above model is to consider a set of hyperprior distributions for the parameters of the gamma prior (a and b). Often these are assumed to also have prior distributions on the positive real line such as gamma (a', b') with $a' > 0, b' = 1$ or $b' > 1$.

5.3.1.1.2 Linear parameterization

One approach to incorporating more sophisticated model components into the relative risk model is to model the parameters of the gamma prior distribution. For example, gamma linear models can be specified where

$$[\theta_i|y_i, e_i, a, b_i] \sim gamma(a, b_i)$$
where $b_i = a/\mu_i$ and
$$\mu_i = \eta_i.$$

In this formulation the prior expectation is μ_i and the prior variance is μ_i^2/a. While this formulation could be used for modeling, often the direct linkage between the variance and mean could be seen as a disadvantage. As will be seen later, a log normal parameterization is often favoured for such models.

5.3.1.2 Beta models

When Bernoulli or binomial likelihood models are assumed (such as 5.6 or 5.7) then one may need to consider prior distributions for the probability parameter p_i. Commonly a beta prior distribution is assumed for this:

$$[p_i|\alpha, \beta] \sim Beta(\alpha, \beta).$$

Here the prior expectation and variance would be $\frac{\alpha}{\alpha+\beta}$ and $\frac{\alpha\beta}{(\alpha+\beta)^2(\alpha+\beta+1)}$. This distribution can flexibly specify a range of forms of distribution from peaked ($\alpha = \beta, \beta > 1$) to uniform ($\alpha = \beta = 1$) and U-shaped ($\alpha = \beta = 0.5$) to skewed or either monotonically decreasing or increasing. In the case of the binomial distribution this prior distribution, with α, β fixed, leads to a beta posterior distribution i.e.,

$$[\mathbf{p}|\mathbf{y}, \mathbf{n}, \alpha, \beta] = B(\alpha, \beta)^{-m} \prod_{i=1}^{m} [\binom{n_i}{y_i} p_i^{y_i}(1-p_i)^{(n_i-y_i)} \cdot p_i^{\alpha-1}(1-p_i)^{\beta-1}]$$

$$= B(\alpha, \beta)^{-m} \prod_{i=1}^{m} [\binom{n_i}{y_i} p_i^{y_i+\alpha-1}(1-p_{i\cdot})^{(n_i-y_i+\beta-1)}].$$

This is the product of m independent beta distributions with parameters $y_i + \alpha$, $n_i - y_i + \beta$. Hence the beta posterior distribution for p_i has expectation $\frac{y_i+\alpha}{n_i+\beta+\alpha}$ and variance $\frac{(y_i+\alpha)(n_i-y_i+\beta)}{(n_i+\beta+\alpha)^2(n_i+\beta+\alpha+1)}$.

5.3.1.2.1 Hyperprior distributions
The parameters α and β are strictly positive and these could also have hyperprior distributions. However, unless these parameters are restricted to the unit interval, then distributions such as the gamma, exponential or inverse gamma or exponential would have to be assumed as hyperprior distributions.

5.3.1.2.2 Linear Parameterization
An alternative specification for modeling covariate effects is to specify a linear or non-linear predictor with a link to a parameter or parameters. For example, it is possible to consider

a parameterization such as $\alpha_i = \exp(\eta_i)$ and $\beta_i = \psi \alpha_i$ where ψ is a linkage parameter with prior mean given by $\frac{1}{1+\psi}$. When $\psi = 1$, then the distribution is symmetric. The disadvantage with this formulation is that a single parameter is assigned to the linear predictor and a dependence is specified between α_i and β_i. One possible alternative is to model the prior mean as $\text{logit}(\frac{\alpha_i}{\alpha_i + \beta_i}) = \eta_i$. However this also forces a dependence between α_i and β_i.

5.3.2 Log-Normal/Logistic-Normal Models

One simple device that is very popular in disease mapping applications is to assume a direct linkage between a linear or non-linear predictor (η_i) and the parameter of interest (such as θ_i or p_i). This offers a convenient method of introducing a range of covariate effects and unobserved random effects within a simple formulation. The general structure of this formulation is $\eta_i = x_i'\beta + z_i'\gamma$. The simplest form involving uncorrelated heterogeneity would be

$$\eta_i = z_{1i}$$

where z_{1i} is an uncorrelated random effect.

An example of the application of this model is given in Figure 5.3 where the WinBUGS code is presented and in Figure 5.4 where the posterior expected relative risk estimates under the Poisson-log-normal model are displayed. In this example, the counties of the State of Georgia, USA are modelled and the outcome of interest is the count of oral cancer deaths in these counties for a given year (2004). In this example the county-wise expected rate was computed from the state-wide oral cancer rate for 2004. The likelihood model assumed for this example is Poisson with $y_i \sim Poisson(e_i \theta_i)$ and $\eta_i = \alpha_0 + z_{1i}$.

Here the extra-variation is modeled as uncorrelated heterogeneity (UH) with a zero mean Gaussian prior distribution i.e., $z_{1i} \sim N(0, \tau_{z_1})$.

5.4 Correlated Heterogeneity Models

Uncorrelated heterogeneity models with gamma or beta prior distributions for the relative risk are useful but have a number of drawbacks. First, as noted above, a gamma distribution does not easily provide for extensions into covariate adjustment or modeling, and, second, there is no simple and adaptable generalization of the gamma distribution with spatially correlated parameters. Wolpert and Ickstadt (1998) provided an example of using correlated gamma field models, but these models have been shown to have poor performance under simulated evaluation (Best et al., 2005). The advantages of incorporating a Gaussian specification are many. First, a random effect which is log-Gaussian behaves in a similar way to a gamma variate, but the Gaussian model can

```
model
{
  for( i in 1 : m ) {
    y[i] ~ dpois(mu[i])
    mu[i] <- e[i] * rr[i]

    log(rr[i]) <- aderivation0 +v[i]
    v[i]~dnorm(0,1.0E-6)
    PP[i]<-step(rr[i]-1+eps)
  }

  a0~dflat()
  eps<-1.0E-6
}
```

FIGURE 5.3
Poisson-log-normal model for the Georgia county-level oral cancer mortality data. The model assumes a zero mean Gaussian prior distribution for the UH random effect. The posterior expected exceedence probability $(\widehat{\Pr}(\theta > 1))$ is computed as PP[].

include a correlation structure. Hence, for the case where it is suspected that random effects are correlated, then it is simpler to specify a log Gaussian form for *any* extra variation present. The simplest extension is to consider additive components describing different aspects of the variation thought to exist in the data.

For a spatial Gaussian process Ripley (1981, 10) any finite realization has a multivariate normal distribution with mean and covariance inherited from the process itself, i.e., $\mathbf{x} \sim \text{MVN}(\boldsymbol{\mu}, K)$, where $\boldsymbol{\mu}$ is an m length mean vector and K is an $m \times m$ positive definite covariance matrix. Note that this is not the only possible specification of a prior structure to model CH (see also Møller et al., 1998).

There are many ways of incorporating such heterogeneity in models, and some of these are reviewed here. First, it is often important to include a variety of random effects in a model. For example, both CH and UH might be included (see below: 5.5). One flexible method for the inclusion of such terms is to include a log-linear term with additive random effects. Besag et al. (1991) first suggested, for tract count effects, a rate parametrization of the form,

$$\exp\{x_i'\beta + u_i + v_i\},$$

where $x_i'\beta$ is a trend or fixed covariate component, u_i and v_i are correlated and uncorrelated heterogeneity, respectively. These components then have separate prior distributions. Often the specification of the correlated component

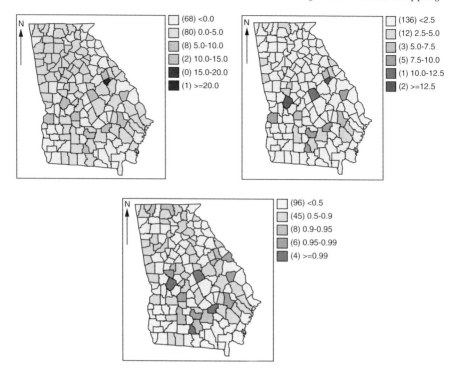

FIGURE 5.4
Georgia county level mortality counts: Oral cancer 2004. UH random effect model. County-wise expected rate computed from the state-wide oral cancer rate 2004: row-wise from top left: standardised mortality ratio; posterior expected relative risk estimates; posterior expected exceedence probability ($\widehat{Pr}(\theta_i > 1)$).

is considered to have either an intrinsic Gaussian (CAR) prior distribution or a fully specified Multivariate normal prior distribution.

5.4.1 Conditional Autoregressive (CAR) Models

5.4.1.1 Improper CAR (ICAR) models

The intrinsic autoregressions improper difference prior distribution, developed from the lattice models of Kunsch (1987), uses the definition of spatial distribution in terms of differences and allows the use of a singular normal joint distribution. This was first proposed by Besag et al. (1991). Hence, the prior for $\{u\}$ is defined as

$$p(\mathbf{u}|r) \propto \frac{1}{r^{m/2}} \exp\left\{-\frac{1}{2r}\sum_i \sum_{j\in\delta_i}(u_i - u_j)^2\right\}, \quad (5.12)$$

where δ_i is a neighborhood of the i th tract. The neighborhood δ_i was assumed to be defined for the first neighbor only. Hence, this is an example of a Markov random field model (see e.g., Rue and Held, 2005). More general weighting schemes could be used. For example neighborhoods could consist of first and second neighbors (defined by common boundary) or by a distance cut-off (for example, a region is a neighbor if the centroid is within a certain distance of the region in question). The uncorrelated heterogeneity (v_i) was defined by Besag et al. (1991) to have a conventional zero-mean Gaussian prior distribution:

$$p(v) \propto \sigma^{-m/2} \exp\left\{-\frac{1}{2\sigma}\sum_{i=1}^m v_i^2\right\}. \quad (5.13)$$

Both r and σ were assumed by Besag et al. (1991) to have improper inverse exponential hyperpriors:

$$\text{prior}(r,\sigma) \propto e^{-\epsilon/2r} e^{-\epsilon/2\sigma}, \quad \sigma, r > 0, \quad (5.14)$$

where ϵ was taken as 0.001. These prior distributions penalize the absorbing state at zero, but provide considerable indifference over a large range. Alternative hyperpriors for these parameters which are now commonly used are in the gamma and inverse gamma family, which can be defined to penalize at zero but yield considerable uniformity over a wide range. In addition, these types of hyperpriors can also provide peaked distributions if required.

The full posterior distribution for the original formulation where a Poisson likelihood is assumed for the tract counts is given by

$$P(u,v,r,\sigma|y_i) =$$
$$\prod_{i=1}^m \{\exp(-e_i\theta_i)(e_i\theta_i)^{y_i}/y_i!\}$$
$$\times \frac{1}{r^{m/2}} \exp\left\{-\frac{1}{2r}\sum_i \sum_{j\in\delta_i}(u_i - u_j)^2\right\}$$
$$\times \sigma^{-m/2} \exp\left\{-\frac{1}{2\sigma}\sum_{i=1}^m v_i^2\right\} \times \text{prior}(r,\sigma).$$

This posterior distribution can be sampled using McMC algorithms such as the Gibbs or Metropolis–Hastings samplers. A Gibbs sampler was used in the original example, as conditional distributions for the parameters were available in that formulation.

An advantage of the intrinsic Gaussian formulation is that the conditional moments are defined as simple functions of the neighboring values and number of neighbors ($n_{\delta i}$).

$$E(u_i|\ldots) = \overline{u_i} \text{ and}$$
$$var(u_i|\ldots) = r/n_{\delta i},$$

and the conditional distribution is defined as:

$$[u_i|\ldots] \sim N(\overline{u_i}, r/n_{\delta i}),$$

where $\overline{u_i} = \sum_{j \in \delta_i} u_j/n_{\delta i}$, the average over the neighborhood of the i th region.

5.4.1.2 Proper CAR (PCAR) models

While the intrinsic CAR model introduced above is useful in defining a correlated heterogeneity prior distribution, this is not the only specification of a Gaussian Markov random field (GMRF) model available. In fact, the improper CAR is a special case of a more general formulation where neighborhood dependence is admitted but which allows an additional correlation parameter (Stern and Cressie, 1999). Define the spatially-referenced vector of interest as $\{u_i\}$. One specification of the proper CAR formulation yields:

$$[u_i|\ldots] \sim N(\mu_i, r/n_{\delta i}) \tag{5.15}$$
$$\mu_i = t_i + \phi \sum_{j \in \delta_i} (u_j - t_j)/n_{\delta_i} \tag{5.16}$$

where t_i is the trend ($= x_i'\beta$), r is the variance, and ϕ is a correlation parameter. It can be shown that to ensure definiteness of the covariance matrix, ϕ must lie on a predefined range which is a function of the eigenvalues of a matrix. In detail, the range is the maximum and minimum eigenvalues ($\phi_{\min} < \phi < \phi_{\max}$) of $diag\{n_{\delta_i}^{1/2}\}.C.diag\{n_{\delta_i}^{-1/2}\}$ where $C_{ij} = c_{ij}$ and

$$c_{ij} = \begin{cases} \frac{1}{n_{\delta_i}} & \text{if } i \sim j \\ 0 & \text{otherwise} \end{cases}.$$

Of course, $\phi_{\min}$ and $\phi_{\max}$ can be precomputed before using the proper CAR as a prior distribution. It could simply be assumed that a (hyper) prior distribution for ϕ is $U(\phi_{\min}, \phi_{\max})$. As noted by Stern and Cressie (1999) this specification does lead to a weak form for the partial correlation between different sites. Note that in the simple case of no trend ($t_i = 0$) then the model reduces to

$$[u_i|\ldots] \sim N(\mu_i, r/n_{\delta i}) \tag{5.17}$$
$$\mu_i = \phi \overline{u_i}. \tag{5.18}$$

The main advantages of this model formulation are that it more closely mimics fully specified Gaussian covariance models, as it has a variance and correlation parameter specified, does not require matrix inversion within sampling algorithms, and can also be used as a data likelihood.

5.4.1.3 Case event models

For case event data, where a point process model is appropriate, it is still possible to consider a form of log Gaussian Cox process where the intensity of the process is governed by a Spatial Gaussian process and conditional on the intensity the case distribution is a Poisson process. As an approximation to the Gaussian process a CAR prior distribution can be proposed. For example, define the first order intensity of the case events as

$$\lambda(s) = \lambda_0(s) \exp\{\beta + S(s)\}$$

where $S(s)$ is the Gaussian process component. At a given case location, s_i, this will yield a likelihood contribution

$$\lambda(s_i) = \lambda_0(s_i) \exp\{\beta + S(s_i)\}.$$

By considering an intrinsic Gaussian specification for $S(s_i)$ we can proceed by assuming that the prior distribution for $\{S(s_i)\}$ is a conditional autoregressive specification i.e., for short, define $S_i \equiv S(s_i)$, and hence

$$[S_i|\ldots] \sim N(\overline{S}_{\delta_i}, r/n_{\delta i}).$$

where $\overline{S}_{\delta_i}$ is the mean of the S values in the neighborhood of S_i. Defining the neighborhood can be handled by using a Voronoi/Dirichlet tesselation to define first (or greater) order neighbors (on R the package deldir can be modified for this task). Hence a hierarchical model can be specified with the i th likelihood contribution $\lambda(s_i) = \lambda_0(s_i) \exp\{\eta_i + v_i + S_i\}$ where $\eta_i = x_i'\beta$ a linear predictor with fixed covariate vector x_i', and

$$[S_i|\ldots] \sim N(\overline{S}_{\delta_i}, r/n_{\delta i})$$
$$v_i \sim N(0, \kappa_v)$$
$$\boldsymbol{\beta} \sim \mathbf{N}(\mathbf{0}, \boldsymbol{\Gamma}_\beta).$$

Of course, in this formulation, then both $\lambda_0(s_i)$ must be estimated, and also the integral of the intensity must be computed. A Berman-Turner approximation scheme (Berman and Turner (1992)) could be used for this purpose. An example of this type of analysis is given in Hossain and Lawson (2008). The conditional logistic likelihood model (5.6) can be fitted if a control disease is available, and this obviates the necessity of estimating $\lambda_0(s_i)$ (see e.g., Lawson 2006b, ch. 8.4, App. C). However the specification of the spatial structure is different as the joint distribution of cases and controls is considered under the conditional model.

5.4.2 Fully-Specified Covariance Models

An alternative specification involves only one random effect for both CH and UH. This can be achieved by specifying a prior distribution having two parameters governing these effects. For example, the covariance matrix of a MVN prior distribution can be parametrically modeled with such terms (Diggle et al., 1998; Wikle, 2002). This approach is akin to universal Kriging (Wackernagel, 2003; Cressie, 1993), which employs covariance models including variance and covariance range parameters. It has been dubbed '*generalized linear spatial modelling*.' A software library is available in R (geoRglm). Usually, these parameters define a multiplicative relation between CH and UH. The full Bayesian analysis for this model requires the use of posterior sampling algorithms.

In the parametric approach of Diggle et al. (1998), which was originally specified for point process models, the first-order intensity of the process was specified as

$$\lambda(s) = \lambda_0(s)\exp\{\beta + S(s)\},$$

where β is a non-zero mean level of the process, and $S(s)$ is a zero mean Gaussian process with, for example, a powered exponential correlation function defined for the distance d_{ij} between the i th and j th locations as $\rho(d_{ij}) = \exp\{-(d_{ij}/\phi)^\kappa\}$ and variance σ^2. Other forms of covariance function can be specified. One popular example is the Matérn class defined for the distance (d_{ij}) as

$$\rho(d_{ij}) = (d_{ij}/\phi)^\kappa K_\kappa(d_{ij}/\phi)/[2^{\kappa-1}\Gamma(\kappa)] \qquad (5.19)$$

where $K_\kappa(.)$ is a modified Bessel function of the third kind. In this case, the parameter vector $\theta = (\beta, \sigma, \phi, \kappa)$ is updated via a Metropolis–Hastings-like (Langevin–Hastings) step, followed by pointwise updating of the S surface. Conditional simulation of S surface values at arbitrary spatial locations (non-data locations) can be achieved by inclusion of an additional step once the sampler has converged. Covariates can be included in this formulation in a variety of ways. For count data, the equivalent Poisson mean specification could be

$$\mu_i = e_i \exp\{\beta + S_i\},$$

where $\mathbf{S} \sim \mathbf{MVN}(\mathbf{0}, \mathbf{\Gamma})$ and $\mathbf{\Gamma}$ is a spatial covariance matrix (Kelsall and Wakefield, 2002). In comparisons of CAR and fully-specified covariance models there appears to be different conclusions about which are more useful in recovering relative risk in disease maps (Best et al., 2005; Henderson et al., 2002).

```
model
{
  for( i in 1 : m ) {
    y[i] ~ dpois(mu[i])
    mu[i] <- e[i] * rr[i]
  smr[i]<-(y[i]+eps2)/(e[i]+eps2)
    rr[i] <- exp(al0 +v[i]+u[i])
  v[i]~dnorm(0,1.0E-6)
  PP[i]<-step(rr[i]-1+eps)
   }

# car normal
u[1:m]~car.normal(adj[],weights[],num[],tau.u
for(k in 1:sumNumNeigh)
{
   weights[k]<-1
}
# other prior distributions
al0 ~ dflat()
eps<-1.0E-6
eps2~dnorm(0,1000)
tau.u~dgamma(0.005,0.005)
}
```

FIGURE 5.5
Code for a convolution model with both UH and CH components. CAR model used for the CH component. The posterior expected exceedence probability $(\widehat{\Pr(\theta > 1)})$ is computed in PP[]. An approximation ot the SMR is also computed.

5.5 Convolution Models

Often it is important to employ both CH and UH random effects within the specification of η_i. The rationale for this lies in the basic assumption that unobserved effects within a study area could take on a variety of forms. It is always prudent to include a UH effect to allow for uncorrelated extra variation. However, without prior knowledge of the unobserved confounding, there is no reason to exclude either effect from the analysis and it is simple to include both effects within an additive model formulation such as $\eta_i = v_i + u_i$. In general, these two random effects are not identified, but given that we are usually interested in the total effect of unobserved confounding then the sum of the effects is the important component and that is well identified. Discussion of these identifiability issues is given in Eberley and Carlin (2000). Occasionally, the computation of the relative variance contribution (intraclass correlation) can be useful. If the variance of the UH and CH component are, respectively, κ_v and κ_u, then this is just given by $\frac{\kappa_v}{\kappa_v+\kappa_u}$. Of course if these components are

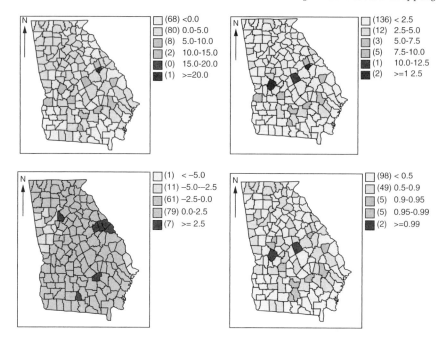

FIGURE 5.6
Georgia county level mortality counts: oral cancer 2004. Convolution model with UH and CH effects. County-wise expected rate computed from the statewide oral cancer rate 2004: row-wise from top left: standardized mortality ratio; posterior expected relative risk estimates; correlated random effect (CH) u_i ; posterior expected exceedence probability $(\widehat{Pr}(\theta_i > 1))$.

not identified then this computation will not be useful. In Figures 5.5 and 5.6, the ODC and some selected output from a convolution model for the Georgia oral cancer data are presented.

Convolution models with CAR CH random effects have been shown to be robust under simulation to a wide range of underlying true risk models (Lawson et al., 2000) and so are widely used for the analysis of relative risk in disease mapping.

5.6 Model Comparison and Goodness-of-Fit Diagnostics

Relative goodness-of-fit (gof) measures such as DIC or PPL can be applied to compare different models for the Georgia Oral Cancer dataset. Here I focus on the use of DIC and MSPE measures previously defined in Section 4.1. in relation to WinBUGS use. Within WinBUGS the DIC is available directly.

TABLE 5.1
Comparison of convolution and uncorrelataed heterogeneity models for the Georgia oral cancer dataset

Measure	Convolution model	UH model
DIC	422.68	383.30
pD	97.2	77.6
MAPE	0.9	0.818
MSPE	5.557	2.763

For posterior predictive loss it is possible to compute a measure based on values generated from the predictive distribution $\{y_i^{pred}\}$ and compare these to the observed data via a suitable loss function. For binary data an absolute value loss is useful as it measures the proportionate misclassification under the model. For positive outcomes (such as Poisson or binomial data) often a squared error loss is used, although the absolute value loss may also be useful. In general the computation involves the averaging of the loss (f) over the data and posterior sample (of size G):

$$MSPE = \sum_{j}^{m}\sum_{i} f(y_i - y_{ij}^{pred})/(m \times G). \qquad (5.20)$$

In the case of a Poisson likelihood, the following lines in an ODC, which produce squared error loss for each observation, compute a point-wise PPL:

$$ypred[i] \sim dpois(mu[i])$$
$$PPL[i] < -pow(ypred[i] - y[i], 2).$$

Both the individual values of

$$PPL_i = \sum_{j} f(y_i - y_{ij}^{pred})/G \qquad (5.21)$$

and the average over all the data (5.20) can be useful in diagnosing local and global lack-of-fit. As a comparison between the convolution and UH models we have computed the DIC, MAPE (absolute error) and MSPE (squared error) for both models. Table 5.1 below displays the results. Overall the UH model seems to yield lower DIC and is lower on both the absolute and squared error loss. Note that the DIC criterion measures how well the model fits the observed data, allowing for parameterization, while the PPL criteria compare the predictive ability of the models.

Figure 5.7 displays the point-wise PPL for squared error and absolute error loss for the Georgia Oral cancer mortality dataset under the convolution and UH models. The maps suggest a marked concentration of loss in a few regions in the northwest of the state (Fulton, Cobb, Dekalb, and Gwinett). Fulton county contains the largest urban area in the state (Atlanta), and shows the highest PPL under both squared error and absolute error. Note that under the better-fitting UH model the loss in most areas is lower than under the convolution model.

5.6.1 Residual Spatial Autocorrelation

While models for disease maps can be assessed for global fit and also at the individual unit level via residual diagnostics, there remains the question of whether any residual spatial structure has been left within the data after a model fit. One approach to this is to consider that a good model fit should leave residuals from the fit with little or no spatial correlation. Hence a test for spatial correlation in the residuals from a model fit would be a useful guide to whether the model has managed to account for the spatial variation adequately. It is possible within a posterior sampling algorithm to compute a measure of spatial autocorrelation and to average this in the final sample. This will provide an estimate of the correlation and the sample will also provide a credible interval or standard deviation of the estimate. In this way it is possible to avoid the need to consider the sampling distribution of the computed statistic, by using functionals of the posterior distribution via posterior sampling.

Various statistics could be used to measure autocorrelation but probably the commonest spatial autocorrelation statistic is Moran's I (see e.g., Cliff and Ord, 1981; Cressie, 1993, Section 6.7). This is usually defined as a ratio of quadratic forms:

$$I = \mathbf{e}'W\mathbf{e}/\mathbf{e}'\mathbf{e}$$

where $\mathbf{e} = \{e_i, ..., e_m\}$ and $e_i = (y_i - \widehat{y}_i)/\sqrt{var(\widehat{y}_i)}$ and W is the 0/1 $m \times m$ adjacency matrix for the regions with elements w_{ij}. For a Poisson data model this residual could be defined as $e_i = (y_i - \widehat{\mu}_i)/\sqrt{\widehat{\mu}_i}$. Congdon et al. (2007) following Fotheringham et al. (2002), noted that given $\{e_i\}$ and a set of 0/1 adjacencies $\{w_{ij}\}$ then a regression of e_i on e_i^* where $e_i^* = \sum_j w_{ij} e_j$ will yield a slope parameter that is an estimate of I. That is, fitting the linear model $e_i = a_0 + \rho e_i^* + \epsilon_i$ will yield the posterior average of ρ as an estimate of I. The WinBUGS code for this is given below. In this code the cumulative number of neighbors up to the ith data point is defined as cum[] with the convention that cum[1] = 0. Values of cum[] are used as an index to select neighboring residuals. In the following, y[] is the outcome and mu[] is the mean (assuming a Poisson data model), e[] is the residual and estar[] is the sum of the neighboring residuals, adj[] is the adjacency list for the regions, and sumNN is the sum of the number of neighbors:

```
for (i in 1:m){
.
y[i]~dpois(mu[i])
.
e[i]<-(y[i]-mu[i])/sqrt(mu[i])
```

Disease Map Reconstruction and Relative Risk Estimation 95

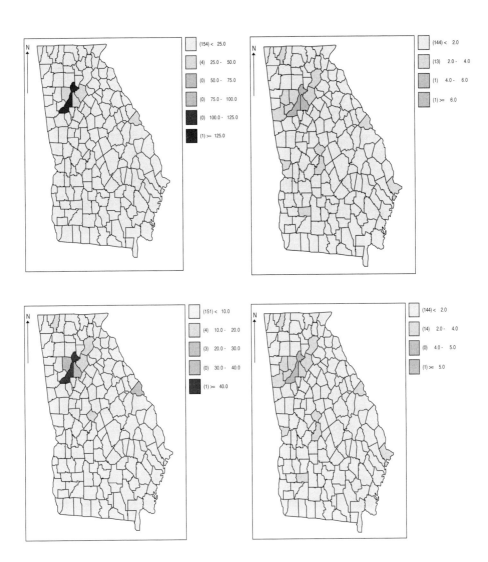

FIGURE 5.7
Georgia county-level oral cancer mortality 2004. Top row: CH model, bottom row: UH model. Point-wise posterior predictive loss for squared error loss (left panel) and absolute error loss (right panel), averaged over a converged sample of size 10000.

```
estar[i]<-sum(we[cum[i]+1:cum[i+1]])
de[i]<-e[i]-mean(e[])
d.estar[i]<-estar[i]-mean(estar[])
dt[i]<-de[i]*d.estar[i]
db[i]<-pow(d.estar[i],2)}

for (j in 1: sumNN) {we[j,i]<-e[adj[i]]}

rho<-sum(dt[])/sum(db[])
```

For the 2004 Georgia oral cancer mortality data with a fitted convolution model in Section 5.5 (see e.g., Figure 5.6), I have computed the posterior average Moran's I. In this case, the result is $\hat{\rho} = 0.0018$ (sd: 0.0334) with 95% credible interval: (-0.0692, 0.0613), which suggests that there is limited autocorrelation left after fitting the convolution model.

5.7 Alternative Risk Models

In Sections 5.4 and 5.4.2 models for risk variation were discussed which are most commonly found in applications. However there are a number of alternative modeling approaches that are less commonly applied but which might be of considerable use in certain applications. The first of these utilizes a likelihood approximation (pseudolikelihood) that was first suggested by Besag (1975). The second is a relaxation of the parametric assumptions inherent in the 2nd level prior specifications for the covariance and convolution models.

5.7.1 Autologistic Models

Besag (1975) first suggested the possibility of modeling spatially-distributed continuous variables via a conditioning on neighborhoods. Instead of considering the full conditional distributions within a likelihood function, a likelihood was proposed that is simply a product of the local conditioning within a neighborhood. Hence, for a count data example with $\{y_i\}$ observed within m arbitrary small areas we examine the $\Pr(y_i | \{y_j\}_{j \in \delta_i} | \boldsymbol{\theta})$ where $j \neq i$ and δ_i is a neighborhood of the i th small area. The pseudolikelihood is then simply assumed to be given by

$$L_p(\{y_i\}|\boldsymbol{\theta}) = \prod_{i=1}^{m} \Pr(y_i|\{y_j\}_{j \in \delta_i}|\boldsymbol{\theta}).$$

This likelihood can take a range of simple forms when specific parameterizations are assumed for $\Pr(y_i|\{y_j\}_{j\in\delta_i}|\boldsymbol{\theta})$. For example, the commonest and, possibly, most widely used example, is where the outcome in the small areas is binary, so that y_i takes the value 1/0. In this case,

$$\Pr(y_i|\{y_j\}_{j\in\delta_i}|\boldsymbol{\theta}) = \theta_i^{y_i}(1-\theta_i)^{1-y_i}$$

$$\text{and } \theta_i = \frac{\exp\{\alpha_i + \lambda S_{\delta_i}\}}{1+\exp\{\alpha_i + \lambda S_{\delta_i}\}}$$

$$\text{and } S_{\delta_i} = \sum_{j\in\delta_i} y_j.$$

Hence the likelihood is simply:

$$L_p(\{y_i\}|\boldsymbol{\theta}) = \prod_{i=1}^{m}\left\{\frac{[\exp\{\alpha_i + \lambda S_{\delta_i}\}]^{y_i}}{1+\exp\{\alpha_i + \lambda S_{\delta_i}\}}\right\}.$$

Notice that this form simply conditions on the sum of neighboring values and as this sum can be precomputed before analysis, this model could be fitted using conventional logistic regression software. One simple extension to this model is to include covariates via a linear predictor. Hence, $\theta = \frac{\exp\{\lambda S_{\delta_i}+x_i'\beta\}}{1+\exp\{\lambda S_{\delta_i}+x_i'\beta\}}$ could be specified, where $x_i'\beta$ is a linear predictor with row vector of ith covariate values x_i' and parameter vector β. Alternatively, covariate representation which is centered could be advantageous (Caragea and Kaiser, 2006). Similarly the model could be extended to include random effects. For example, we could include uncorrelated unit level random effects (ν_i) within the logit term:

$$\theta_i = \frac{\exp\{\lambda S_{\delta_i} + x_i'\beta + \nu_i\}}{1+\exp\{\lambda S_{\delta_i} + x_i'\beta + \nu_i\}}. \tag{5.22}$$

In the Bayesian context, logit $\theta_i = \lambda S_{\delta_i} + x_i'\beta + \nu_i$ can be treated as other spatial models in that both covariates and random effects can be present. Both the parameter vector β, λ and random effect ν_i (in example 5.22) will have prior distributions at the second level of the hierarchy. Further hyperprior specifications would also be available.

Note that this likelihood includes spatial correlation at the first level of the hierarchy. Hence it does not require the inclusion of any spatially-structured random effect at the second level of the hierarchy. Modifications to the model where further spatial components describing 2nd or higher order dependence could be made. For example, $S_{\delta_i} = \omega\sum_{j\in\delta_i} y_j + (1-\omega)\sum_{l\in\delta_{2i}} y_l$ where δ_{2i} is a 2nd order neighborhood (such as neighbors of first order neighbors, once counted) and ω is a weighting parameter with $0 < \omega < 1$. This would allow for some measure of distance-related dependence in the model. Note that S_{δ_i} cannot be precomputed unless ω is fixed, but the sums: $\sum_{j\in\delta_i} y_j$, and $\sum_{l\in\delta_{2i}} y_l$, can be.

TABLE 5.2
DICs for three models: autologistic with no randon effects; autlogistic with UH component; convolution model with UH and CH

Model	pD	DIC	MSPE
Autologistic	1.97	112.18	0.4268(0.0503)
Autologistic+UH	25.16	109.04	0.3142 (0.0525)
Convolution (UH+CH)	27.69	103.98	0.2844 (0.0515)

An example of the application of an autologistic model and comparison with a conventional random effect model is demonstrated in Figure 5.8 and 5.9. Here, the Ohio county level respiratory cancer mortality data set (1968) is examined. The expected rates were computed from the statewide age-gender stratified rate applied to count population strata and summed. The county level standardized mortality ratios (smr_i) of respiratory cancer for the year 1968 are available in the form of an exceedence indicator. The standardization was performed using the state-wide rate for 18 age-gender groups for that year. The data was made available as a dichotomized variate via the criteria:

$$y_i = \begin{cases} 1 \text{ if } smr_i > 2 \\ 0 \text{ otherwise} \end{cases}.$$

Interest focusses on the analysis of high risk status and hence on this binary outcome at the county level. Of course, the arbitrary dichotomization may lead to concerns about use of artificial screening levels, and this indeed should be a consideration in any analysis where thresholding has taken place. Of course, when data is only available dichotomized then it is appropriate to seek to model it directly. In Figure 5.8, the contrast between the posterior expected estimates under the simple autologistic model (top row), which are relatively uniform, and the probability of risk exceedence and UH component (v_i) (bottom row) demonstrate that there are 4 regions of elevated risk. When the convolution model with UH and CH components is fitted, as shown in Figure 5.9, it is clear that the areas of elevated risk are further highlighted and indeed the CH component seems to isolate the south eastern corner of the state as a particularly elevated region.

The difference in DIC is less marked between the different models in this case. Table 5.2 displays the DIC results for the different models. In this case, both the autologistic model with UH and the standard convolution model appear to yield better models than the simple autologistic. Interestingly, the convolution model yields a lower DIC than both autologistic models. However, the autologistic model with uncorrelated heterogeneity, is much closer to the DIC of the convolution model and is more parsimonious.

Disease Map Reconstruction and Relative Risk Estimation 99

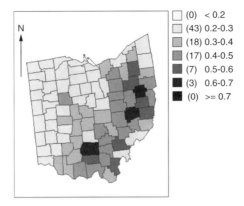

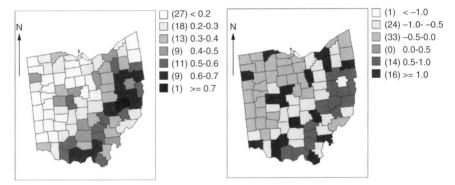

FIGURE 5.8
Autologistic model with a single binary covariate and a first order dependence on the sum of neigboring values. Top row: posterior expected probability of *smr* exceedence (>2); bottom row: same model but with a county level heterogenity random effect (v_i): UH. left: posterior expected probability of *smr* exceedence and, right: posterior expected UH component.

For this example, models with both first order and second order neighborhoods were examined, but overall the best model remained the first order neighborhood based on DIC.

Note that the difference between the autologistic model and convolution model lies mainly in the fact that the spatial correlation is at different levels of the hierarchy and that the likelihood used for the autologistic model is a pseudolikelihood. Pseudolikelihoods are known to be reasonable approximations where spatial correlation is not strong. In space-time, autologistic models can also be considered (see Chapter 11).

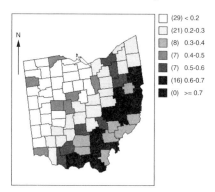
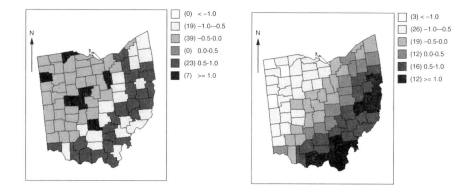

FIGURE 5.9
Standard binary convolution model with a UH and CH CAR componenent. Top row: posterior expected probability of *smr* exceedence; bottom row: left: UH component; right: CH component.

5.7.1.1 Other auto models

Besides the autologistic model, it is possible to consider other auto models based on Poisson, binomial, negative binomial, or other exponential family distributions. Usually pseudolikelihood must be used for estimation as normalizing constants are intractable. In addition, in some cases, the parameterization of the model must be constrained. In the auto-Poisson model, the most general form of the intensity is defined, for the i th region, as $\lambda_i = \exp(\alpha_i + \sum_{j=1}^{m} \eta_{ij} y_j)$ and so constraints must be put on the η_{ij}s to ensure negativity. In the case of the autobinomial, the logit of the probability of a case in the i th area is an unnormalized function of the surrounding case totals (rather than proportions):

logit $p_i = \alpha_i + \sum_{j=1}^{m} \eta_{ij} y_j$. Of the range of auto models available it is clear that the autologistic is the most popular in applications and likely to be the simplest to implement and interpret.

5.7.2 Spline-Based Models

As an alternative to strictly parametric models, it is possible to assume a semi-parametric approach to the modeling of the spatially-structured component of a disease risk model. The use of spline models is of course not limited to employment in spatial smoothing, but I will concentrate on that aspect here. The basic idea behind a semi-parametric representation of a spatial model is the assumption of a smoothing operator to represent the mean structure of the process. In the case event situation, assuming that a control disease realization is available, we could specify a conditional logistic model with

$$p_i = \frac{\exp(\eta_i)}{1 + \exp(\eta_i)}$$

with $\eta_i = x_i'\beta + S(s_i)$

where $x_i'\beta$ is a linear predictor, and $S(s_i)$ is a smoothing operator at the geo-reference s_i (location) of the i th observation. Kelsall and Diggle (1998) give an example of using this generalized additive model (GAM) methodology to a cancer mortality example. Here, focus will be given to the count data situation although many of the issues found there also apply to case vent data.

In the count data situation, make the usual assumption of observed data $\{y_i\}$, $i = 1, ..., m$ and $y_i \sim Poisson(\mu_i)$. Further define the geo-reference for i th observation as $s_i : (x_{i1}, x_{i2})$. This could be a centroid of the small area or other associated point reference. Here it is assumed that $\log \mu_i = S(s_i)$ where $S(.)$ will be defined as smoothing operator. A variety of choices are available for $S(.)$. Here, I focus on spline models which are attractive in applications and have strong links to Gaussian process models (see e.g., French and Wand, 2004). Define the mean level as

$$\log \mu_i = \alpha_0 + \sum_{j=1}^{2} \alpha_j x_{ij} + \sum_{j=1}^{n_\kappa} \psi_j C\{\|s_i - \kappa_j\|\}$$
$$= \mathbf{x}_i' \alpha + z_i' \psi,$$

where $\{\kappa_j\}$, $j = 1, ..., n_\kappa$ is a set of knots (fixed locations in space), $\{\psi_j\}$ is a Gaussian random effect, $z_i' = \{z_1, ..., z_{n_\kappa}\}$ and $\mathbf{z} = [C\{\|s_i - \kappa_j\|/\rho\}]_{1 \le i \le m,\ 1 \le j \le n_\kappa}$, the covariance function defined here as

$$C\{a\} = (1 + |a|)e^{-|a|}.$$

Define the square matrix

$$\boldsymbol{\omega} = [C\{\|\kappa_i - \kappa_j\|/\rho\}]_{1 \le i,j \le n_\kappa}$$

and the joint random effect prior distribution as

$$\psi \sim \mathbf{N}(\mathbf{0}, \tau\boldsymbol{\omega}^{-1}).$$

A reparameterization of $\mathbf{z}_* = \mathbf{z}\boldsymbol{\omega}^{-1/2}$, $\boldsymbol{\psi}_* = \boldsymbol{\omega}^{1/2}\boldsymbol{\psi}$ yields a linear mixed model with with $cov(\boldsymbol{\psi}_*) = \tau\mathbf{I}$, and then $\log \mu_i = \mathbf{x}_i'\alpha + \mathbf{z}_{*i}\boldsymbol{\psi}_*$. In French and Wand (2004), the value of ρ is fixed in advance. This allows the precomputing of the covariance matrix and reparameterization so that standard software can be used. This type of spline modeling is termed low rank Kriging. In general, it would be useful to estimate ρ as this controls the degree of smoothing. Figures 5.10 and 5.11 display the resulting posterior expected (PE) estimate maps for the two models. In Figure 5.11 the PE relative risk and CH component are shown. The model fitted to the log relative risk included a planar trend in the x, y centroids as well as additive UH and CH components. For the spline model for comparability, the log relative risk was also a function of planar trend in centroid locations but an additive spline term was also included. The covariance was assumed to be defined by $\boldsymbol{\omega} = [C\{||\kappa_i - \kappa_j||/\rho\}]_{1 \leq i,j \leq n_\kappa}$ where ρ was fixed at $\widehat{\rho} = \max(||s_i - s_j||) \; \forall_{i,j}$. This tends to produce a very smooth surface effect as can be seen in the Figure 5.11 where the top panel shows the PE relative risk with a much reduced range. The bottom panel displays the spline effect which include both spatial and non-spatial effects (beyond the trend component). Overall, the relative risk pattern is mostly similar between the two models. However, in this case the spline model did not provide a good model based on DIC. The DIC for the spline model was 761.52. Whereas for the convolution model it was 623.15. Of course, if ρ were to be estimated then it is possible that a much improved fit could be achieved. Alternative spline-based approaches have been proposed by Zhang et al. (2006) to spatiotemporal multivariate modeling, and by Macnab (2007) in a comparison of spline methods for temporal components of spatial maps.

5.7.3 Zip Regression Models

If a disease is rare, then there will be considerable sparsity in the data. The implication of this is that few cases are observed within the study area, or, for small areas, zero counts are common. In this case, the spatial distribution of cases will often form isolated clusters. A good example of this is childhood leukemia which is a rare disease but is known to cluster. The major question that is posed by this situation is whether the standard models for disease mapping hold when such sparsity of data arises. A priori the conventional log relative risk model where the log of the risk is modeled with Gaussian effects may be simply inadequate to deal with a situation where the rate is close to the boundary of its space (i.e., $\lambda_0(s) \simeq 0$, or the expected rate $e_i \simeq 0$). Singular information methods may be useful here (see e.g., Bottai et al., 2007).

Two alternatives can be immediately envisioned. First it may be possible to directly model the locations of the disease clusters via object models (Lawson

FIGURE 5.10
CAR model fit for Ohio county level respiratry cancer mortality 1979. Model includes a planar trend in centriods and both a UH and CH (CAR) component. Top row: posterior expected relative risk; bottom row: CH component.

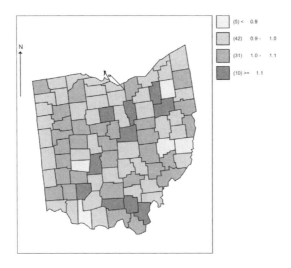

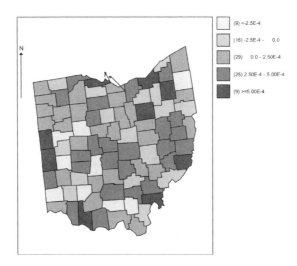

FIGURE 5.11
Spline model fit based on the low rank Kriging model as a linear mixed model with a fixed covariance parameter ρ ($\widehat{\rho} = \max(||x_i - x_j||) \ \forall ij$) where x_i is the i th centroid.

and Denison, 2002). These models do not make simple global assumptions about surface form, but rather seek to estimate locations of objects (in this case clusters). As it turns out these models can recover risk surfaces reasonably well. Examples of this are found in Lawson (2006b), Section 6.5. An alternative is to consider the marginal distribution of the concentration of cases. In the sense that any arbitrary area or mesh area will yield a local concentration of cases, it might be noted that under sparsity, many areas will have zero cases and a few will have small positive numbers. For count data in arbitrary regions this could lead to an overdispersed distribution and even multi modality in the marginal distribution. Note that this effect in count data may not be adequately modeled by an overdispersed distribution such as the negative binomial.

One solution is to consider a mixture of processes so that the low intensity is separately modeled from the peaks. For case event data we could assume

$$\lambda(s) = \lambda_0(s)[w(s) + (1 - w(s))\lambda_1(s)]$$

where $w(s)$ is spatially dependent weight which controls which process is dominant locally. This can lead to a logistic model when a control disease realization is present in that, the probability of a given location s_i being a case is:

$$\frac{w(s_i) + (1 - w(s_i))\lambda_1(s_i)}{1 + w(s_i) + (1 - w(s_i))\lambda_1(s_i)}$$

Hence it may be interesting to include covariates or other effects within both $w(s_i)$ and $\lambda_1(s_i)$. I do not pursue this approach here. Instead I will focus on the area of small area count data.

There is much literature on mixture modeling for sparse counts (Lambert (1992), Boehning et al. (1999), Agarwal et al. (2002), Ghosh et al. (2006) amongst others). When a mixture of Poisson distributions is considered the simplest case is a two component mixture where zero counts have a component $1 - p + p\exp(-\mu)$ where μ is the Poisson mean and non-zero counts have component $p\exp(-\mu).\mu^y/y!$. In general the distribution is given by

$$f(y; p|\mu) = (1-p)P_0(y,0) + pP_0(y,\mu)$$

where $P_0(y,\mu)$ is the Poisson distribution with mean μ. The inclusion of covariates can proceed as usual via link to the mean (e.g., $\mu = \exp(\mathbf{x}'\beta)$). In addition, covariates can be included in the mixture weight (p). For example we could have

$$p = \frac{\exp(\mathbf{w}'\gamma)}{1 + \exp(\mathbf{w}'\gamma)}.$$

where $\mathbf{w}'\gamma$ is a predictor with additional covariates and effects. In general, for observed data $\{y_i\}, i = 1, ..., m$ and expected counts $\{e_i\}$ the model is specified

$$[y_i|e_i, \theta_i] \sim (1-p)Pois(0) + pPois(e_i.\theta_i). \tag{5.23}$$

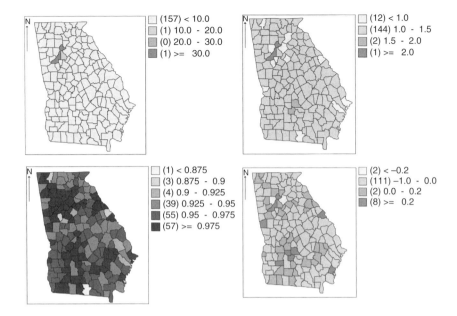

FIGURE 5.12
ZIP Bayesian model with two components applied to the Georgia asthma mortality data for 2000. Row-wise from top left: posterior expected Poisson mean, relative risk, component probability, and uncorrelated heterogeneity (UH).

A further modification clarifies the role of the components. For example, we might consider that this problem is one where an unobserved classification variable treats the zeroes as structural ($z = 1$) or usual Poisson ($z = 0$). In this case z is unobserved and must be estimated. This can be done within a data augmentation loop. In that case, the incomplete data likelihood is

$$[y_i|e_i, \theta_i, z_i = 0] \sim Pois(e_i.\theta_i) \tag{5.24}$$
$$[y_i|e_i, \theta_i, z_i = 1] \sim Pois(e_i.\theta_i^*).$$

Then the second stage would be to generate the allocation variables from $[z_i|y_i, e_i, \{\theta_i, \theta_i^*\}]$. Usually $[z_i|y_i, e_i, \{\theta_i, \theta_i^*\}] \sim Bern(p_i)$ for two components. The complete data likelihood (Marin and Robert (2007)) used to estimate the parameters would be

$$L(\{y_i\}, z) = \prod_{i=1}^{m} p_{z_i} Pois(y_i; e_i.\theta_{z_i}).$$

Figure 5.12 displays a ZIP regression analysis for the Georgia county level asthma mortality counts for the year 2000. The posterior average relative risk estimates for high and low risk counties 1 and 144 (crude SMR = 3.85, 0.0) and the posterior average relative risk estimates is, 1.322, and 0.573 under a converged convolution model, whereas under a two component ZIP model (5.23), with no spatially-correlated component, the posterior expected estimates of relative risk were 1.818, 1.094. Although these estimates seem to have similar ranges, the latter model has shifted both estimates away from zero. Interestingly, in both models the Atlanta area appears to yield a very high posterior average relative risk estimate, and also a high probability of membership in the full Poisson model. The ZIP model did not include a CH component unlike the convolution model. Some residual structure remains as evidenced by the posterior expected UH map.

Finally, it should be apparent that the idea of mixtures of components can be generalized to a wide variety of situations. The primary area of application may be the incorporation of (unobserved) multiple scales of aggregation within one analysis when it is believed that different components represent these different scales. This is discussed more extensively in Chapter 8.

5.7.4 Ordered and Unordered Multicategory Data

A special case arises when the outcome of interest is in the form of a multiple category. I previously discussed binary data as a special case of binomial data and in autologistic models (Sections 5.1.3 and 5.7.1). Extending this criteria to categorical outcomes where the levels of outcome can be > 2, there now arises the possibility of ordinal or nominal analysis. When the categories are ordered, such as disease stages, then ordinal (logistic) regression models can be assumed. If ordering is not apparent then nominal (logistic) models could be applied. Many of the concerns and issues cited in previous sections apply here with respect to use of random effects and prior distributional specification. One added issue with multi-category outcome data is whether different structures could be allowed at different levels of the category. For example, commonly available within cancer registry data is the stage at diagnosis of the cancer. This staging of the cancer is usually an ordered category. This would lead to consideration of an ordinal model. However, unstaged cancers are indeterminate in terms of stage, and so it is unclear at what level they would be best considered. This might lead one to either assume a nominal model for all the staged data or to exclude the unstaged from an ordinal analysis. Zhou et al. (2007) demonstrate the application of ordinal models: baseline category logits, proportional odds and adjacent category logits, to cancer registry data in South Carolina. They found that baseline category logits model fitted best in terms of DIC, and that a model with a spatially-correlated random effect for the regional-stage and an uncorrelated random effect for the distant stage of the cancer was best fitting in this case.

5.7.5 Latent Structure Models

An alternative to conventional modeling of the mean level of risk is to consider that the risk is composed of a combination of unobserved risk levels. These risk levels are latent and so there is no directly observed data concerning their form. These types of models really fall between areas. On the one hand, they are used to provide overall eastimates of relative risk (and so are relative risk models). On the other hand, they are also used to isolate underlying patterns of risk and so the latent risk levels may be of importance. Some of the models discussed in the Chapter 6 could be regarded as latent structure models also. For example, the hidden process/object models can be used to provide relative risk estimates, besides estimates of cluster locations (see e.g., Lawson, 2006b, Section 6.5.3). The hidden Markov models of Green and Richardson (2002) provide estimates of relative risk as do the mixture component models of Fernandez and Green (2002). Partition-based models can also be considered in this way. Here we provide a brief summary of spatial component models both latent and known. In Section 11.3.2, a brief review of space-time latent models is given.

5.7.5.1 Mixture models

A number of examples of mixture-based models have been proposed for relative risk estimation. Often these have been applied to count data and so the following discussion focusses on that data form. First of all, fixed component mixtures have been proposed. A simple example of these would in fact be ZIP regression. More generally, in a Bayesian context, one can consider fixed component models that consist of sums of random terms within the mean predictor. These models could be termed mean mixture models. The convolution model of Section 5.5 is a mean mixture with 2 fixed components. An extension of this type of model was proposed by Lawson and Clark (2002) where a mean mixture of a CAR component and an L1 norm component was used to preserve discontinuities and boundary effects (see also Congdon, 2005, Ch. 8). The mixing parameter was allowed to vary spatially and so a posterior expected mixing field was estimated. The basic model specification was

$$[y_i|e_i\theta_i] \sim Poiss(e_i\theta_i)$$
$$\log(\theta_i) = v_i + w_i u_{1,i} + (1-w_i) u_{2,i}$$
$$w_i \sim beta(\alpha, \alpha)$$

and $u_{1,i}$ and $u_{2,i}$ are CAR and $L1$ norm spatial prior distributions respectively and $v_i \sim N(0, \tau_v)$. The mixing parameter $\{w_i\}$ was allowed to vary spatially, albeit with a common exchangeable distribution, and so a posterior expected mixing field could be estimated. Figure 5.13 displays the posterior expected components of the three-component mixture fitted to North Carolina sudden infant death syndrome (SIDS) data. The L1 norm field appears quite different from the CAR component field (which seems to display a west–east trend).

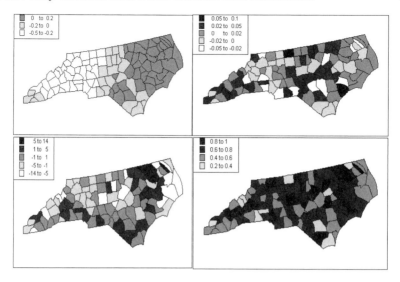

FIGURE 5.13
Three- component mixture model for SIDS in North Carolina (as reported in Lawson & Clark, 2002). Rowwise from top left: $u_{1,i}$ component, v_i component, $u_{2,i}$ component, and w_i component. Posterior averages reported.

Each field provides unique information concerning the different components of the model supported in the data.

In general an extended mixture of fixed random components could be imagined each with different prior assumptions.

An early example of hidden mixture modeling, albeit in an empirical Bayes context, was proposed by Schlattman and Böhning (1993). Their approach assumed that the distribution governing the observed data is a mixture of Poisson distributions:

$$f(y_i|\mathbf{p},e_i,\boldsymbol{\theta}) = \sum_{k=1}^{K} p_k Pois(y_i|e_i\theta_k) \tag{5.25}$$

with mixing probabilities $\{p_k\}$ and $\sum_k p_k = 1$. Both $\mathbf{p}$ and $\boldsymbol{\theta}$ are unknown. Suitable prior distributions for the components have to be specified. Besides estimation of $\mathbf{p}$ and $\boldsymbol{\theta}$, it is possible to estimate the risk in each area from posterior sampling based on

$$\widehat{\theta}_i = \frac{1}{G}\sum_{g=1}^{G}\sum_{k=1}^{K} \theta_k^g p_k^g Pois(y_i|e_i\theta_k^g) / \sum_{k=1}^{K} p_k^g Pois(y_i|e_i\theta_k^g) \tag{5.26}$$

for the case of fixed K where a posterior sample of size G is taken. When K is not fixed then a prior distribution would have to be specified for K. For

that case, (5.26) could be used with K replaced by K_g. The choice of prior specification under could be various. Clearly for the probabilities one could use a Dirichlet distribution:
$$\mathbf{p} \sim Dir(\boldsymbol{\alpha})$$
where $\{\alpha_k\}$, $k = 1, ..., K$, while for the $\{\alpha_k\}$ gamma prior distributions could be specified. In addition, prior specification for the $\{\theta_k\}$ could be based on gamma distributions. For example,
$$\theta_k \sim Ga(c_k a, a)$$
would yield a prior mean of c_k. Suitable hyperprior distributions can be assumed for the positive parameters c_k.

An ordering constraint on the components may be required if K is not fixed if the components are to be identified. Posterior sampling for the fixed K case is straightforward. For the non-fixed case, then a prior distribution must be assumed for K. This is often a Poisson with fixed rate, i.e., $K \sim Poiss(d)$, or a uniform distribution up to a fixed maximum: $K \sim U(1, K_{\max})$.

When spatial dependence is to be included, one approach has been to assume that, instead of a Dirichlet distribution for the weights, the weights have a spatial dependence structure. Fernandez and Green (2002) suggest a variety of models. One proposal, the logistic normal model, specifies that
$$f(y_i|\mathbf{p}, e_i, \boldsymbol{\theta}) = \sum_{k=1}^{K} p_{ik} Pois(y_i|e_i \theta_k)$$
$$p_{ik} = \eta_{ik}(\phi) / \sum_{l=1}^{L} \eta_{il}(\phi)$$
$$where\ \eta_{ik}(\phi) = \exp\{x_{ik}/\phi\}$$
where $\{x_{ik}\}$ is a set of spatially-correlated random field components indexed by the i th area, and ϕ is a spatial correlation parameter. The fields are given proper CAR prior distributions to ensure propriety. The relative risk estimates from posterior sampling are obtained via allocation of components.

A major alternative to these mixture type models are those that posit factorial decomposition of the risk in each area. For multiple diseases only, where y_{ij} is the observed count in the i th area and the j th disease, Wang and Wall (2003) first proposed a model where a spatial factor underlay the risk:
$$y_{ij} \sim Poiss(e_{ij}\theta_{ij})$$
$$\log(\theta_{ij}) = \log(e_{ij}) + \lambda_j f_i$$
where $\lambda_j f_i = \log(\theta_{ij}/e_{ij})$ and f_i is the spatially-referenced common risk factor. It is further assumed that
$$\mathbf{f} \sim \mathbf{N}(\mathbf{0}, \boldsymbol{\Sigma})$$

with unit variance, $\Sigma_{ij} = \exp(-d_{ij}/\phi)$, and $\sum f_i = 0$ for indentifiability. Subsequently, Liu et al. (2005) extended the proposal to structural equation models. Of course, these approaches are not univariate, and there are a wide range of dimension reduction possibilities when multivariate outcomes or multiple predictors are included within models. I do not pursue this here. In Section 11.3.2, I examine the possibility of space-time latent modeling.

Another potentially useful development is the use of Dirichlet process (DP) mixing models to provide more flexible spatial structures (Ishwaran and James, 2002, 2001; Gelfand et al., 2005; Griffin and Steel, 2006; Kim et al., 2006; Duan et al., 2007; Cai and Dunson, 2008). In addition, there is a possibility that DP mixtures could provide a flexible approach to variable dimension modeling within clustering or variable selection scenarios.

5.8 Edge Effects

The importance of the assessment of edge effects in any spatial statistical application cannot be underestimated. Edge effects play a larger role in spatial problems than in, say, time-series. Specifically, we define edge effects as "any effect upon the analysis of the observed data brought about by the proximity of the study area boundary." The effect of the edges of a study area are largely the result of the effects of *spatial censoring*. That is, the fact that observations outside the window are not observed and therefore cannot contribute to analysis within the window. This mirrors the effects of temporal censoring in say, survival analysis, where, for example, the outcome for some subjects may not be observed because the observation period has stopped prior to the outcome appearing.

Of course, all censoring depends on the idea that observations are dependent in *some* way. That is, the occurrence of observations outside the window of observation relies on observations within the window. In the spatial case, it is easily possible for individual disease response to relate to 'missing' observations outside the window. For example, it may be that an environmental health hazard is located outside or, in the case of viral etiology, an infected person or carrier is located outside. For diseases which have uncertain etiology, it could be possible that factors underlying the incidence of the disease have a spatial distribution that is spatially dependent and hence the disease incidence reflects this structure even when individual responses are independent. If, in addition, some unknown genetic etiology underpinned the disease incidence, then if this has spatial expression, the incidence of disease could relate to unobserved genetically linked subjects outwith the observation region.

In addition, such spatial censoring can affect estimation procedures, even when no explicit spatial dependence is proposed. For example, spatial smoothing methods, including geostatistical methods (Kriging), splines, or convolution random effect models, use data from different regions of the observed window in the estimation of risk at a location. Hence, if no correction is pursued for this effect at the edges, then some edge distortion will result. In other cases parametric estimation may require the computation of averages of values in neighborhoods of a chosen point. Hence, close to edges there could be considerable distortion induced by missing neighbors. This edge problem cannot only induce bias in estimation, but also tends to lead to considerable increases in estimator variance at such locations, and hence to low reliability of estimation. An example of the effect can be seen immediately when a CAR distribution is assumed. In that model, the conditional variance for the i th area is defined, in the notation of (5.18), as r/n_{δ_i}. This dependence on the number of neighbors (n_{δ_i}) implies that, for a given r, a reduction of neighbor number will increase the variance.

A number of methods have been proposed to deal with such edge effects. These methods have been in part developed within stochastic geometry, where it is often assumed that the process under study is first- and second-order stationary and isotropic (Ripley, 1988). These methods vary from (1) correction methods applied to smoothers or other estimators, for example, using weights relating to the proximity of the external boundary, (2) employing guard areas to provide external information to allow better boundary area estimation within the window, (3) simulation of missing data outside the window and iterative re-estimation or model fitting. (The use of toroidal correction is not usually appropriate in the analysis of disease incidence data, as it is not usually appropriate to make the appropriate stationarity assumptions.) This final method has significant advantages if used within iterative simulation methods such as data augmentation Gilks et al., 1996; Tanner, 1996; Robert and Casella, 2005; or general McMC algorithms, as the external data can be treated as parameters in the estimation sequence.

An example of the degree to which edge effects could affect the application of convolution models was examined under simulation by Vidal-Rodiero and Lawson (2005). In that study, counties within a large multi-state region of the US was examined and external county hulls were peeled from the observation window to examine the effect of different neighborhood 'depths' on estimation. Figure 5.14 displays the effect of stripping out a sequence of hulls of small areas around a central area. In the simulation study a large number if states within central United States were amalgamated and the counties gathered into one study area. Successive hulls of counties were then stripped and the effect of this stripping was noted. The effect of stripping on four different models (convolution (BYM), Poisson-gamma (PG), Poisson log normal (PLN), and fixed SMR (C)) was assessed. The six sets (6–11) are sets of internal regions at different depths where the relative risk was estimated. The outer sets of counties (sets 1–5) were successively stripped. Four models were fitted and the different set results are given in Figure 5.14. It is clear that sets close to the sets close to the boundary (e.g., set 6 and 7) that is stripped show bigger differences in average relative risk.

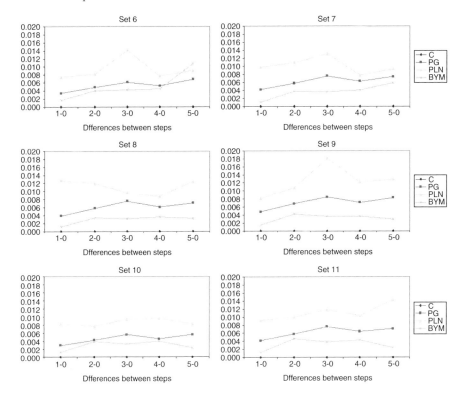

FIGURE 5.14
Model effects of hull stripping: size sets of internal regions (sets 6–11)) with four different models (convolution (BYM), Poisson-gamma (PG), Poisson log normal (PLN), and fixed SMR (C)).

5.8.1 Edge Weighting Schemes and McMC Methods

The two basic methods of dealing with edge effects are (1) the use of weighting/correction systems, which usually apply different weights to observations depending on their proximity to the study boundary, and (2) the use of guard areas, which are areas outwith the region which we analyze as our study region.

5.8.1.1 Weighting systems

Usually, it is appropriate to set up weights which relate the position of the event or tract to the external boundary. These weights, $\{w_i\}$ say, can be included in subsequent estimation and inference. Often the form will be $w_i = f(d_i)$ where d_i is the distance to the boundary, from a fixed point in a small area or in the case event situation from the case event itself. Another alternative for small area data would be to use the length of boundary of the small area in common with the study area boundary. In that case, one could propose $w_i = f(l_i)$ where l_i is the common boundary length. For example, the proportion of the total boundary length of the small area common with the study boundary might be a useful measure. The weight for an observation is usually intended to act as a surrogate for the degree of missing information at that location and so may differ depending on the nature and purpose of the analysis. Some sensitivity to the specification of these weights will inevitably occur and should be assessed in any case study. More detail on suitable weights can be found in Lawson (2006b, Ch. 5).

Defining an indicator for closeness to the boundary for each area, when in the tract count case, some external standardized rates are available, it is possible to structure an expectation-dependent weight for a particular tract, e.g., based on the ratio of the sum of all adjacent area expectations to the sum of all such expectations within the study window. Other suitable weighting schemes could be based on the proportion of the number of observed neighbors.

Guard areas An alternative approach is to employ guard areas. These areas are external to the main study window of interest. These areas could be boundary tracts of the study window itself or could be added to the window to provide a guard area, in the case of tract counts. In the case of event situation, the guard area could be some fixed distance from the external boundary, Ripley, 1988. The areas are used in the estimation process but they are excluded from the reporting stage, as they will be prone to edge effects themselves. If boundary tracts are used for this, then some loss of information must result. External guard areas have many advantages. First, they can be used *with* or *without* their related data to provide a guard area. Second, they can be used within data augmentation schemes in a Bayesian setting. These methods regard the external areas as a missing data problem (see e.g., Little and Rubin, 2002, Ch 10).

5.8.1.2 McMC and other computational methods

It is usually straightforward to adapt conventional estimation methods to accommodate edge-weighted data. In addition, if guard areas are selected and observations are available within the guard area, then it is possible to proceed with inference by using the whole data but selectively reporting those areas not within the guard area. Note that this is not the same as setting $w_i = 0$ for all guard area observations in a weighting system.

When external guard areas are available but no data are observed, resort must usually be made to missing data methods. An intermediate situation arises when in the tract count case some external standardized rates are available. In that case it is possible to structure an expectation-dependent weight for a particular tract, e.g., based on the ratio of the sum of all adjacent area expectations to the sum of all such expectations within the study window. This can be used as an edge-weight within such a weighting system. An example of a study of different edge remedies for count data can be found in Lawson et al. (1999).

5.8.2 Discussion and Extension to Space–Time

In the situation where case events are studied, then if censoring is present and could be important (i.e., when there is clustering or other correlated heterogeneity), it is advisable to use an internal guard area, or an external guard area with augmentation via McMC. In cases where only a small proportion of the study window is close to the boundaries and only general (overall) parameter estimation is concerned, then it may suffice to use edge weighting schemes. If residuals are to be weighted, then it may suffice to label the residuals only for exploratory purposes.

In the situation where counts are examined, then it is also advisable to use an internal guard area or external area with augmentation via McMC. In some cases, an external guard area of *real* data may also be available. This may often be the case when routinely collected data are being examined. In this case, analysis can proceed using the external area *only to correct internal estimates*. Edge weighting can be used also, and the simplest approach would be to use the proportion of the region *not* on the external boundary. Residuals can be labelled for exploratory purposes. The assumptions underlined in any correction method are that the model be correctly specified and that it could be extended to the areas not observed. In particular, it is questionable if an adjustment can really be obtained when ignoring the information on the outer areas. Edge-effect bias should be less prominent when an unstructured exchangeable model is chosen. Since each area relative risk would be regressed toward a grand mean, the information lacking for the unobserved external areas is very small compared to those from the observed areas. Of course, such a simple model where common expectation is found is highly unlikely to be a good model in this area.

Extending the edge-effect problem to consideration of space-time data, the situation is more complex as spatial edge effects can interact with temporal edge effects. The use of sequential weighting, based on distance from time and space boundaries, may be appropriate (Lawson and Viel, 1995). For tract counts observed in distinct time periods only, the most appropriate method is likely to be based on distance from time and space boundaries, although it may be possible to provide an external spatial and/or temporal guard area either with real data or via augmentation and McMC methods.

The use of augmentation methods can also be fruitfully employed in this context. If the external areas are known, but information concerning the disease of interest is not available in these external areas, then it is possible to regard such missing/censored data as parameters which can be estimated within an iterative sampling algorithm, such as an McMC algorithm. In addition, if partial information were known (for example the standardized rates in the external areas), then we could condition these missing data count estimates on the known information.

5.9 Exercises

5.9.1 Maximum Likelihood

To provide a back drop for the Bayesian analysis we present some basic results for likelihoods from simple mapping models.

1) A state in the United States has m counties. Within these counties, births and births with abnormalities are observed. The births with abnormalities (Ba) are a subset of all births.

The probability that a birth in the i th region is a Ba is θ_i. Each birth has an independent risk of being Ba. We observe $\{y_i\}$, $i = 1, ..., m$ Ba events in the m counties and the total births in the m counties is $\{n_i\}$.

a) A likelihood model for these data could be a binomial with probability θ_i, as in (5.7) above. Explain why this is appropriate.

b) If we assume there is a common probability across all regions, show that the maximum likelihood estimator of θ_i is given by, $\widehat{\theta} = \sum_{i=1}^{m} y_i / \sum_{i=1}^{m} n_i$.

c) A logistic linear model results if we assume a logistic link between θ_i and a linear predictor. For the model: $\theta_i = \exp(\beta_0)/\{1+\exp(\beta_0)\}$, show that the maximum likelihood estimator of β_0, either directly or by invariance, is given by

$$\widehat{\beta}_0 = \log\left\{\frac{S_y}{S_n} / \left(1 - \frac{S_y}{S_n}\right)\right\},$$

where $S_y = \sum_{i=1}^{m} y_i$, $S_n = \sum_{i=1}^{m} n_i$.

d) Show that the large sample standard error of $\widehat{\beta}_0$ is given by

$$se(\widehat{\beta}_0) = \left\{S_y - S_y^2/S_n\right\}^{-\frac{1}{2}}.$$

2) Case event data is observed within a study area W. There are m events in W and their locations are denoted by $\{s_i\}$, $i = 1, ..., m$. A realization of control events is also available in the same window: $\{s_j\}$, $j = m+1, ..., m+n$. The conditional log-likelihood for these data can be written as:

$$l = \sum_{i=1}^{m+n} y_i \eta_i - \sum_{i=1}^{m+n} \log[1 + \exp(\eta_i)],$$

Disease Map Reconstruction and Relative Risk Estimation 117

where y_i is now an indicator variable taking the value 1 for a case and 0 for a control (see 5.6 above), and $\exp(\eta_i) = \rho f(s_i; \alpha) = \exp(\alpha_0 - \alpha d_i)$ where d_i is the distance from a fixed point to the i th location and $f(s_i; \alpha) = \exp(-\alpha d_i)$. Assume we want to test for a distance effect between the cases locations and a fixed point.

a) Show that under the null hypothesis $H_0: \alpha = 0$ then the maximum likelihood estimator of ρ is just m/n.

b) If you substitute this estimator into the likelihood above, a possible test statistic to find if distance is significant is based on the first derivative of the likelihood WRT α. This is known as a score test statistic. Show that under $H_0: \alpha = 0$, the test statistic is given by:

$$\frac{m}{m+n}\left[\sum_{i=1}^{m} d_i + \sum_{j=m+1}^{m+n} d_j\right] - \sum_{i=1}^{m} d_i$$

5.9.2 Poisson–Gamma Model: Posterior and Predictive Inference

A random sample of size m from a Poisson distribution with parameter θ is denoted $x_1, \ldots, x_m$. The parameter has a prior distribution:

$$g(\theta) = \begin{cases} \lambda e^{-\lambda \theta} & \lambda > 0 \\ 0 & \text{elsewhere} \end{cases}$$

The posterior distribution of θ is given by:

$$P(\theta|\{x_i\}) = \frac{\beta^{s+1}}{\Gamma(s+1)} \theta^s e^{-\theta \beta}$$

where $\beta = m + \lambda$, and $\Gamma()$ is the gamma function and $s = \sum x_i$, assuming λ is fixed.

Derive the prior predictive distribution $\Pr(x|x_1, \ldots, x_m)$ and hence find $\Pr(x > 2|x_1, \ldots, x_m)$ when $\lambda = 2$.

5.9.3 Poisson-Gamma Model: Empirical Bayes

For the Poisson likelihood model with gamma prior distribution defined in (5.10), the unconditional distribution of y_i given a, b is negative binomial. This is also the prior predictive distribution of y_i. The marginalized log-likelihood is given by

$$L(a,b) = \sum_i \left[\log \frac{\Gamma(y_i + a)}{\Gamma(a)} + b\log(a) - (y_i + a)\log(e_i + a)\right].$$

This likelihood is free of $\{\theta_i\}$ and can be maximized to yield *empirical Bayes* estimates of a, and b (Clayton and Kaldor, 1987). Show that this leads to normal equations, which can be solved for $\widehat{a}$, and $\widehat{b}$:

$$\frac{\widehat{b}}{\widehat{a}} = \frac{1}{m}\sum_i \frac{(y_i+\widehat{a})}{(e_i+\widehat{b})}$$

$$\sum_i^m \sum_{j=0}^{y_i-1} \frac{1}{\widehat{a}+j} + m\log(\widehat{b}) - \sum_i^m \log(e_i+\widehat{b}) = 0.$$

6

Disease Cluster Detection

In the study of disease spatial distribution it is often appropriate to ask questions related to the local properties of the relative risk surface rather than models of relative risk per se. Local properties of the surface could include peaks of risk, sharp boundaries between areas of risk, or local heterogeneities in risk. These different features relate to surface properties but not directly to a value at a specific location. Relative risk estimation (AKA disease mapping; Chapter 5) concerns the 'global' smoothing of risk and estimation of true underlying risk level (height of the risk surface), whereas cluster detection is focussed on local features of the risk surface where elevations of risk or depressions of risk occur. Hence it is clear that cluster detection is fundamentally different from relative risk estimation in its focus. However the difference can become blurred, as methods that are used for risk estimation can be extended to allow certain types of cluster detection. This will be discussed more fully in later sections.

6.1 Cluster Definitions

Before discussing cluster detection/estimation methods it is important to define the nature of the clusters and/or clustering to be studied.

There are a variety of definitions of clusters and clustering. Different definitions of clusters or clustering will lead to differences in the ability of detection methods. First it should be noted that sometimes the correlated heterogeneity term in relative risk models is called a clustering term (see e.g., Clayton and Bernardinelli, 1992). This implies that the term captures aggregation in the risk and indeed this does lead to an effect where neighboring areas having similar risk levels. This is a global feature of the risk however, and also induces a smoothing of risk. This begs the question of how we define clusters or clustering: should it be a global feature or should it be local in nature?

Global clustering basically assumes that the risk surface is clustered or has areas of like elevated (reduced) risk. An uncorrelated surface, on the other hand, should display random changes in risk with changes in location and so should both be much more variable in risk level and have few contiguous areas of like risk. Figure 6.1 displays a comparison between an uncorrelated and

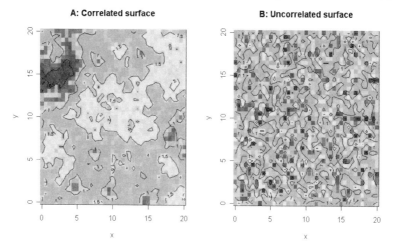

FIGURE 6.1
Simulated examples of correlated (A) and uncorrelated surfaces. Simulation using the R function GaussRF with mean 1.0.

correlated risk surface. In Figure 6.1 A: there are areas of elevated risk that may qualify as clusters (by some definition). However, modeling the overall clustering does not address their locations specifically. Hence this form of clustering does not address localized behavior or the location of clusters per se. This is often termed general clustering (Besag and Newell (1991)).

A general definition of a (spatial) cluster is: "Any spatially-bounded area of significantly elevated (reduced) risk."

This is clearly very general and requires further definition. By 'spatially-bounded' I mean that the cluster must have some spatial integrity. This could be a neighborhood criterion such as "areas must be adjoining" or "at least two adjoining areas must meet a criterion," or could be defined to have a certain type of external boundary (e.g., risk differences around the cluster must meet a criterion). A simple criterion that is often assumed is that known as *hot-spot clustering*. In hot spot clustering, any area or region can be regarded as a cluster. This is due to the assumption of a zero neighborhood criterion, i.e., no insistence on adjacency of regions within clusters. This is a convenient and simple criterion and is often assumed to be the only criterion. It is commonly used in epidemiology (see e.g., Richardson et al., 2004). Without prior knowledge of the behavior of the disease then this criterion is appealing. It could be useful for preliminary screening of data, for example.

However, this hot spot definition ignores any contiguity that may be thought to be inherent in relevant clusters. For example, it might be important that clusters of a given threshold size be investigated. This threshold size could be defined as a minimum number of contiguous areas. Hence, only groups of

Disease Cluster Detection

contiguous regions of 'unusual' risk could qualify as clusters. On the other hand, in the case of infectious diseases, it may be that a certain shape and size of cluster are important in understanding disease spread.

In this chapter I will mainly consider three different scenarios for clustering:

i. Single region hot spot relative risk detection
ii. Clusters as objects or groupings
iii. Clusters defined as residuals.

6.1.1 Hot Spot Clustering

Hot spot clustering is often the most intuitive form of clustering and may be the that which most public health professionals consider as their definition. In hot spot clustering, any area or region can be regarded as a cluster. This is due to the assumption of a zero neighborhood criterion, i.e., no insistence on adjacency of regions within clusters. Simply, any area displaying "excess" or "unusual" risk, by some criterion, is a hot spot. This is a relatively nonparametric definition.

6.1.2 Clusters as Objects or Groupings

Clustering might be considered to be apparent in a data set when a specific form of grouping is apparent. This grouping would usually be predefined. Usually the criterion would also have a neighborhood or proximity condition. That is, only neighboring or proximal areas (which meet other criteria) can be considered to be "in a cluster." Hence some parametric conditions must be met under this defintion.

6.1.3 Clusters Defined as Residuals

Often it is convenient to consider clusters as a residual feature of data. For example, lets assume that y_i is the count of disease within the i th census tract within a study area. Let's also assume that our basic model for the average count μ_i (i.e., $E(y_i) = \mu_i$) is

$$\log \mu_i = a_i + e_i.$$

Here a_i could consist of a linear or non-linear predictor as a function of covariates and could also consist of random effects of different kinds. To simplify the idea we assume that a_i is the "smooth" part of the model and e_i is the rough or residual part. The basic idea is that if we model a_i to include all relevant non-clustering confounder effects then the residual component must contain residual clustering information. Hence if we examine the estimated value of e_i then this will contain information about any clusters unaccounted for in a_i. Of course, this does not account for any pure noise that might also be found in e_i. This means, of course, that an estimate of e_i could have at least two components: clustered and unclustered (or frailty). There could, of course, be additional components depending on whether the confounding in a_i was adequately specified or estimated.

There are a number of approaches to isolating the residual clustering. First, it is possible to include a pure noise term within a_i and to consider e_i as a cluster term. For example we could assume that $a_i = f(v_i; covariates)$ where $f(.)$ is a function of a uncorrelated noise at the observation level (v_i: frailty or random effect term) and a function of covariates. Second, a smoothed version of e_i, $s(e_i)$ say, could be examined in the hope that the pure noise is smoothed out. Of course this begs the question of which component should include the clustering: should it be a model component or a residual component? If the clustering is likely to be irregular and we can be assured that no clustering confounding effects are to be found in the model component, then a residual or smoothed residual might be useful. On the other hand, if there is any prior knowledge of the form of clustering to be expected, then it may be more important to include some of that information within the model itself. The real underlying issue is the ability of models and estimation procedures to differentiate spatial scales of clustering.

6.2 Cluster Detection using Residuals

First of all, assume that we observe disease outcome data within a spatial window.

6.2.1 Case Event Data

6.2.1.1 Unconditional analysis

For the case event scenario we have $\{s_i\}, i = 1, ..., m$ events observed within the window T. Modeling here focusses on the first order intensity and its parametrization. Assume that $\lambda(s|\psi) = \lambda_0(s|\psi_0).\lambda_1(s|\psi_1)$ as defined in Chapter 5. We focus first on the specification of a residual for a point process governed by $\lambda(s|\psi)$. First, in the spirit of classical residual analysis, it is clear that we can assume that we want to compare fitted values to observed values. This is not simple as we have locations as observed data. One way to circumvent this problem is to consider a function of the observed data which can be compared with an intensity estimate at location s_i, $\lambda(s_i|\widehat{\psi})$ say. One such function could be a saturated or nonparametric intensity estimate $(\widehat{\lambda}_{loc}(s_i)$, say, where loc denotes a local estimator). Essentially this gives a slight aggregation of the data, but it allows for a direct comparison of model to data.

Disease Cluster Detection

Hence we can define a residual as

$$r_i^{loc} = \widehat{\lambda}_{loc}(s_i) - \lambda(s_i|\widehat{\psi})$$

or in the case of a saturated estimate (Lawson, 1993a)

$$r_i^{sat} = \widehat{\lambda}_{sat}(s_i) - \lambda(s_i|\widehat{\psi}).$$

Baddeley et al. (2005) discuss more general cases applied to a range of processes. An example of a local estimate of intensity could be derived from a suitably edge-weighted density estimate(Diggle, 1985). An example of the use of the saturated estimator is as follows:

First assume that an estimator is available for the background intensity $\lambda_0(s_i|\psi_0)$, $\lambda_0(s_i|\widehat{\psi}_0) \equiv \lambda_{0i}$ say. Also assume that it can be used as a plug-in estimator within $\lambda(s|\psi)$. If this is the case, then we can compute $r_i^{sat} = \lambda_{0i}[\widehat{\lambda}_{1sat}(s_i) - \lambda_1(s_i|\widehat{\psi})]$. For a simple heterogeneous Poisson process model with intensity $\lambda_0(s|\psi_0).\lambda_1(s|\psi_1)$ and using an integral weighting scheme (as described in Chapter 5), then the saturate estimate of the intensity at s_i is $1/(w_i\lambda_{0i})$. A simple weight (which provided a crude estimator of the local intensity) is $w_i = A_i$ where A_i is the Dirichlet tile area surrounding s_i, based on a tesselation of the case events. Hence a simple residual could be based on

$$r_i^{sat} = \lambda_{0i}[(w_i\lambda_{0i})^{-1} - \lambda_1(s_i|\widehat{\psi})]$$
$$= w_i^{-1} - \lambda_{0i}\lambda_1(s_i|\widehat{\psi}).$$

The use of such tile areas must be carefully considered as edge effect distortion can occur with tesselation and so boundary regions of the study window should be treated with caution. Of course both the error in estimation of the background intensity is ignored here and a crude approximation to the saturated intensity is assumed. Note that r_i^{sat} or r_i^{loc} can be computed within a posterior sampler and so a posterior expectation of the residuals can be estimated.

Figure 6.2 displays an example of the use of posterior expectation of r_i^{sat} for a model, for the well known larynx cancer data set from Lancashire, United Kingdom 1974–1983. This dataset has been analyzed many times and consists of the residential address locations of cases of larynx cancer with the residential addresses of cases of respiratory cancer as a control disease (see e.g., Diggle, 1990; Lawson, 2006b, Ch. 1), with distance decline component (variable d_i) around the fixed point (3.545, 4.140), an incinerator. The motivation for this type of analysis relates to assessment of health hazards around putative sources (putative source analysis). This is discussed more fully in Chapter 7. The model for the first order intensity is defined to depend on this distance: $\lambda_1(s_i|\theta) = \beta_0[1 + \exp(-\beta_1 d_i)]$. Appendix B contains the WinBUGS code for this example. The map displays the contours for the posterior sample estimate of $\Pr(r_i^{sat} > 0)$, the residual exceedence probability.

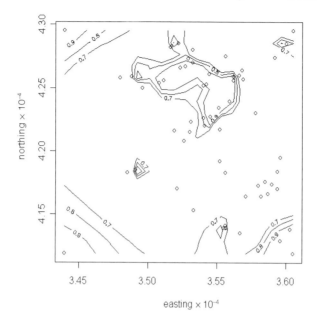

FIGURE 6.2
Map of Lancashire larynx cancer case distribution with, superimposed, a contour map of exceedence probability (0.7,0.8,0.9) for the residual (r_i^{sat}) from a Bayesian model assuming Berman–Turner Dirichlet tile integration weights and non-parametric density estimate of background risk computed from the respiratory cancer control distribution.

To allow for extra unobserved variation in this map an uncorrelated random effect term can also be included in the model. Appendix B displays the code used for this model. Figure 6.3 displays the resulting posterior average residual exceedence probability map for the model with $\lambda_1(s_i|\theta) = \beta_0[1 + \exp(-\beta_1 d_i)]\exp(v_i)$ where $v_i \sim N(0, \tau_v)$ and $\beta_* \sim N(0, \tau_{\beta_*})$. The hyperparameter specifications are given in the Appendix. Both figures suggest that there is slight evidence for an excess of aggregation in the north of the study region (where there is a large area where $\Pr(r_i^{sat} > 0) > 0.9$). There is also weaker evidence of an excess in the area to the west of the putative source (3.545,4.140), where $\Pr(r_i^{sat} > 0) > 0.8$ on average. There is also marked edge effects close to the study region corners due to the distortion of the tessellation suspension algorithm. Figure 6.3 displays a similar picture even after removal of extra noise.

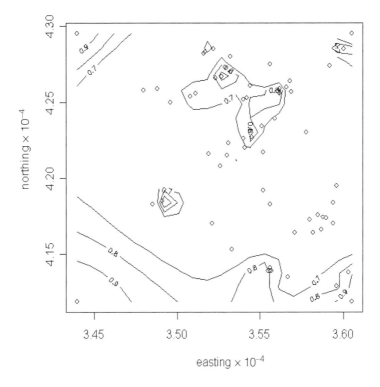

FIGURE 6.3
As per Figure 6.2 but where the model has included a random uncorrelated effect (v_i) to allow for extra variation in the risk: $\lambda_1(s_i|\theta) = \beta_0[1 + \exp(-\beta_1 d_i)]\exp(v_i)$.

6.2.1.2 Conditional logistic analysis

An alternative approach to the analysis of case event data is to consider the joint realization of cases and controls and to model the conditional probability of a case given an event has occurred at a location. This approach was discussed in Chapter 5 and has the advantage that the background effect factors out of the likelihood. Define the joint realization of m cases and n controls as $s_i : i = 1, ..., N$ with $N = m + n$. Also define a binary label variable $\{y_i\}$ which labels the event either as a case $(y_i = 1)$ or a control $(y_i = 0)$. The resulting conditional likelihood has a logistic form:

$$L(\psi_1|\mathbf{s}) = \prod_{i \in cases} p_i \prod_{i \in controls} 1 - p_i$$
$$= \prod_{i=1}^{N} \left[\frac{\{\exp(\eta_i)\}^{y_i}}{1 + \exp(\eta_i)} \right].$$

where $p_i = \frac{\exp(\eta_i)}{1+\exp(\eta_i)}$ and $\eta_i = x'_i\beta$ and x'_i is the i th row of the design matrix of covariates and β is the corresponding p-length parameter vector. Hence in this form a Bernoulli likelihood can be assumed for the data and a hierarchical model can be established for the linear predictor $\eta_i = x'_i\beta$. In general, it is straightforward to extend this formulation to the inclusion of random effects in a generalized linear mixed form. Bayesian residuals such as $r_i = y_i - \widehat{p}_i/\widehat{se}(y_i - \widehat{p}_i)$ (or directly standardized version: $r_i = (y_i - \widehat{p}_i)/\sqrt{\widehat{p}_i(1 - \widehat{p}_i)}$) are available, where $\widehat{p}_i$ is the average value of p_i from the posterior sample. Residuals from binary data models are often difficult to interpret due to the limited variation in the dependent variable $(0/1)$, and the usual recommendation for their examination is to group or aggregate the results. A wide variety of aggregation methods could be used. Spatial aggregation methods might be considered here. Figure 6.4 displays the mapped surface of the standardized Bayesian residual using $r_i = (y_i - \widehat{p}_i)/\sqrt{\widehat{p}_i(1 - \widehat{p}_i)}$ where $\widehat{p}_i$ is computed from the converged posterior sample via R2winBUGS. Appendix B displays the code for this model. The model assumed for this example also has an additive distance effect and is specified by

$$p_i = \frac{\lambda_i}{1+\lambda_i}$$
$$\lambda_i = \exp\{\alpha_0 + v_i\}.\{1 + \exp(-\alpha_1 d_i)\}.$$

Figure 6.5 displays the thresholded mapped surface of the $\Pr(r_i > 2)$ for values (0.05,0.1,0.2). This suggests some evidence of clustering or unusual aggregation in the north and also in the vicinity of the putative location in the south.

Note that all of these models assume that there is negligible clustering under the model and that any residual effects will include the clustering. These models do not explicitly model clustering, but only model long range and uncorrelated variation. Hence we make the tacit assumption that any remaining aggregation of cases will be found in the residual component. Of course other effects which were excluded from the model could be present in the residuals.

6.2.2 Count Data

For count data, it is assumed that either a Poisson data likelihood or a binomial likelihood is relevant. Note that an autologistic model could also be specified.

6.2.2.1 Poisson likelihood

In the case of a Poisson likelihood, assume that y_i, $i = 1,...,m$ are counts of cases of disease and e_i $i = 1,...,m$ are expected rates of the disease in m small areas, and so $y_i \sim Poiss(e_i\theta_i)$ given θ_i. The log relative risk is usually modelled and so $\log\theta_i$ is the modelling focus. Bayesian residuals for this likelihood

Disease Cluster Detection

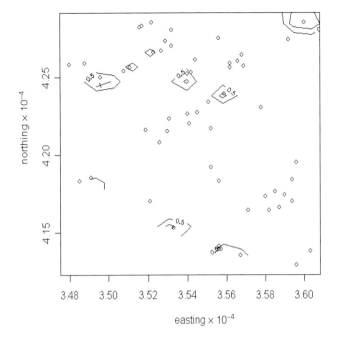

FIGURE 6.4
Contour map of the standardised Bayesian residual for the logistic case-control spatial model applied to the larynx cancer data from Lancashire UK. The display shows the posterior average residual for a sample of 5000 after burn-in.

are easily computed in standardized form as $r_i = (y_i - e_i\widehat{\theta}_i)/\sqrt{e_i\widehat{\theta}_i}$ where $\widehat{\theta}_i$ is the average value of the θ_i obtained from the converged posterior sample. In this case, the Georgia oral cancer data was examined with a Poisson data likelihood and model $\log \theta_i = \alpha_0 + v_i$ where

$$\alpha_0 \sim U(a,b)$$
$$v_i \sim N(0, \tau_v)$$

with τ_v set large and (a,b) a large negative to positive range. Appendix B has details of the WB code used. No correlated random effect is included here as it is assumed that clustering is to be found in residuals. Figure 6.6 displays the results from a converged sampler based on 10,000 burn-in and sample size of 2000. The display shows the average estimate of $\Pr(r_i > 2)$ and $\Pr(r_i > 3)$ for r_i given above. The most extreme region appears to be in the far west of Georgia. It should be noted that there is considerable noise in these residuals, particularly for $\Pr(r_i > 2)$.

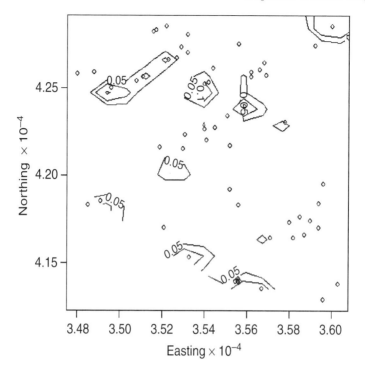

FIGURE 6.5
Map of the contoured surface of $\Pr(r_i > 2)$ estimated from the converged posterior sample for the standardised Bayesian residual in Figure 6.4.

6.2.2.2 Binomial likelihood

In the case of a binomial likelihood assume m small areas, and that in the i th area there is a finite population n_i out of which y_i disease cases occur. The probability of a case is p_i. The data model is thus $y_i \sim bin(p_i, n_i)$ given p_i, and the usual assumption is made that $\text{logit}(p_i) = f(\eta_i)$, where η_i is a linear or non-linear predictor. Of course, various ingredients can be specified for $f(\eta_i)$, including the addition of random effects to yield a binomial GLMM. A Bayesian residual for this model is given in standardized form as $r_i = (y_i - n_i\widehat{p}_i)/\sqrt{n_i\widehat{p}_i(1-\widehat{p}_i)}$ where $\widehat{p}_i$ is the average of p_i values found in the converged posterior sample.

While the above discussion has focussed on simple residual diagnostics, albeit from posterior samples, there is also the possibility of examining *predictive* residuals for any given model. A predictive residual can be computed for each observation unit as

$$r_i^{pr} = y_i - y_i^{pr}$$

Disease Cluster Detection 129

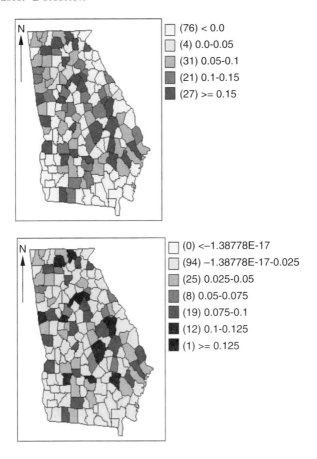

FIGURE 6.6
Georgia county maps of Bayesian residuals from a converged posterior sampler with a uncorrelated random effect term. From left to right: $Pr(r_i > 2)$ and $Pr(r_i > 3)$.

where $y_i^{pr} = \frac{1}{G}\sum_{g=1}^{G} f(y_i|\theta^g)$, and $f(y_i|\theta^g)$ is the likelihood given the current value of θ^g. Of course this will usually be small compared to the standard Bayesian residual. Note that for a given data model, y_i^{pred} can be easily generated on WinBUGS. For the binomial example above the code could be:

```
y[i]~dbin(p[i],n[i])
ypred[i]~dbin(p[i],n[i])
rpred[i]<-y[i]-ypred[i]
```

An alternative approach to residual analysis could be based in the construction of a residual envelope, based on the comparison of the Bayesian residual:

$r_i = y_i - \hat{y}_i$ with $r_i^* = y_i^{pred} - \hat{y}_i$. Unusual residuals could be assessed by assessing the ranking of r_i among the a series of B simulated $\{r_{ib}^*\}$ $b = 1, ..., B$, as discussed in Section 4.4. Further, a p-value surface can be computed from a tally of exceedences:

$$P_{v_i} = \Pr(|r_i| > |r_i^*|) = \frac{1}{B} \sum_{b=1}^{B} I(|r_i| > |r_{ib}^*|).$$

The mapped surface of P_{v_i} could be examined for areas of unusually elevated values and hence provide a tool for hot spot detection.

6.3 Cluster Detection Using Posterior Measures

Another approach to cluster detection is to consider measures of quantities monitored in the posterior that may contain clustering information. One such measure is related to estimates of first order intensity (case event data) or relative risk (Poisson count data) or case probability (binomial count data). If we have captured the clustering tendency within our estimate of any of these quantities then we could examine their posterior sample behavior. Perhaps the most commonly used example of this is the use of exceedence probability in relation to relative risk estimates for individual areas for count data (see e.g., Richardson et al., 2004). Define the exceedence probability as the probability that the relative risk θ exceeds some threshold level (c): $\Pr(\theta > c)$. This is often estimated from posterior sample values $\{\theta_i^g\}_{g=1,...,G}$ via

$$\widehat{\Pr}(\theta_i > c) = \sum_{g=1}^{G} I(\theta_i^g > c)/G$$

$$\text{where } I(a) = \begin{cases} 1 \text{ if } a \text{ true} \\ 0 \text{ otherwise} \end{cases}.$$

Of course, there are two choices that must be made when evaluating $\widehat{\Pr}(\theta_i > c)$. First, the value of c must be chosen. Second, the threshold for the probability must also be chosen, i.e., $\widehat{\Pr}(\theta_i > c) > b$ where b might be set to some conventional level such as 0.95, 0.975, 0.99, etc. In fact, there is a trade off between these two quantities and usually one must be fixed before considering the value of the other. Figure 6.7 displays the posterior expected exceedence probability maps: $\widehat{\Pr}(\theta_i > c)$ for $c = 2$, and $c = 3$ for the Georgia oral cancer data when a relative risk model with a UH component was fitted (see Section 5.3.2).

One major concern with the use of exceedence probability for single regions is that it is designed only to detect *hot spot* clusters (i.e., single regions

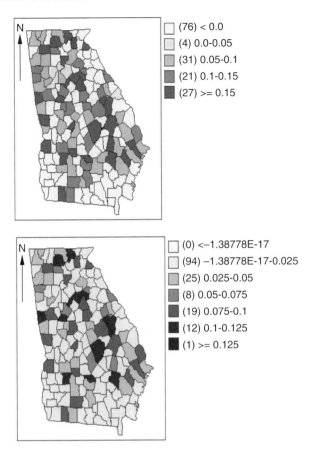

FIGURE 6.7
Georgia oral cancer: maps of $\widehat{\Pr}(\theta > c)$ for $c = 2$ and $c = 3$ for a model with a uncorrelated random effect (UH).

signalling) and does not consider any other information concerning possible forms of cluster or even neighborhood information. Some attempt has been made to enhance this post hoc measure by inclusion of neighborhoods by Hossain and Lawson (2006). For the neighborhood of the i th area defined as δ_i and the number of neighbors as n_i, then

$$\overline{q_i} = \sum_{j=0}^{n_i} q_{ij}/(n_i + 1)$$
where $q_{ij} = \Pr(\theta_j > c) \ \forall \ j \in \delta_i$
and $q_{i0} = \Pr(\theta_i > c)$.

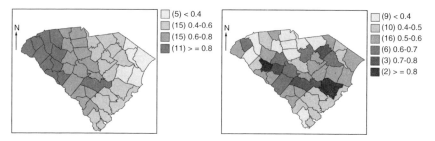

FIGURE 6.8
Display of exceedence probabilities for two models. Left panel: simple first order trend; right panel: convolution model with UH and CH only and no trend for the same data set: South Carolina county level congenital mortality 1990.

This measures $\overline{q_i}$ and q_{i0} can be used to detect different forms of clustering. Other more sophisticated measures have also been proposed (see e.g., Hossain and Lawson, 2006, for details).

A second concern with the use of exceedence probabilities is of course that the usefulness of the measure depends on the model that has been fitted to the data. It is conceivable that a poorly fitting model will not demonstrate any exceedences relate to clustering and may leave the clustering of interest in the residual noise. An extreme example of this is displayed in Figure 6.8. In that figure the same data set is examined with completely different models. The data set is South Carolina county level congenital anomaly deaths for 1990 (see also Lawson et al., 2003, Ch. 8). The expected rates were computed for an 8 year period. In the left panel a Poisson log linear trend model was assumed and in the right panel a convolution model. The trend model was $\log \theta_i = \alpha_0 + \alpha_1 x_i + \alpha_2 y_i$, where xy is the centroid location, with zero mean Gaussian prior distributions for the regression parameters whereas the right panel was $\log \theta_i = \alpha_0 + u_i + v_i$ where the u_i, v_i are correlated and uncorrelated heterogeneity terms with the usual CAR and zero mean Gaussian prior distributions. Without examination of the goodness-of-fit of these models it is clear that there could be considerable latitude for misinterpretation if exceedence probabilities are used in isolation to assess (hot spot) clustering.

As in the count data situation we can also examine exceedences for other data types and models. For example, in the case event example, intensity exceedence could be examined as: $\Pr(\widehat{\lambda}_1(s_i) > 1)$, whereas for the binary or binomial data the exceedence of the case probability could be used: $\Pr(\widehat{p}_i > 0.5)$. These can also be mapped of course. However the rider concerning the goodness-of-fit of the model as highlighted by Figure 6.8 also applies here.

6.4 Cluster Models

It is also possible to design models which explicitly describe the clustering behavior of the data. In this way parameters and functions can be defined that summarize this behavior. It should be noted that clustering behavior is often regarded as a second order feature of the data. By second order I mean, 'relating to the mutual covariation of the data.' Hence it is often assumed that covariance modeling will capture clustering in data. This is often termed *general* clustering. However, as noted in Section 6.1, while general covariance modeling can capture the overall mutual covariation (as in Figure 6.1 A) it does *not* lead to identification or detection of clusters per se. In the following I focus on the detection of clusters, rather than general clustering.

6.4.1 Case Event Data

In the analysis of point processes (PPs) there is a set of models designed to describe clustering. For an introductory overview, which focusses mainly on general cluster testing, see Diggle (2003), ch 9. Basic models often assumed for PPs, which allow clustering, are the Poisson cluster process and the Cox process. In the Poisson (Neyman-Scott) cluster process (PcP) an underlying process of parents (unobserved cluster centers) is assumed and offspring (observed points) are generated randomly in number and location. This generation is controlled by distributions. Clearly this formulation is most appropriate in examples where parent generation occurs such as seed dispersal in ecology.

An alternative to a PcP is found in the Cox process where a non-negative stochastic process ($\Lambda(s)$) governs the intensity of a heterogeneous Poisson Process (hPP). Conditional on the realization of the stochastic process the events follow a hPP. In this case

$$\lambda(s) = E[\Lambda(s)]$$

where the expectation is with respect to the process. Note that this formulation allows the inclusion of spatial correlation via a specification such as $\Lambda(s) = \exp\{S(s)\}$ where $S(s)$ is a spatial Gaussian process. This is sometimes known as a log-Gaussian Cox process (LGCP) (see e.g., Møller et al., 1998). Instead, note also that an intensity process of the form

$$\Lambda(s) = \mu \sum_{j=1}^{\infty} h(s - c_j) \tag{6.1}$$

can be assumed, where $h(s - c_j)$ is a bivariate pdf, and c_j are cluster centers. If the centers are assumed to have a homogeneous PP then this is also a PcP. Of course these models were derived mainly for ecological examples and not for disease case events. However we can take as a starting point a model for case events that includes population modulation in the first order intensity, and that also allows clustering via an unobserved process of centers.

6.4.1.1 Object models

Define the first order intensity as

$$\lambda(s|\psi) = \lambda_0(s|\psi_0).\lambda_1(s|\psi_1).$$

Assume that the case events form a hPP conditional on parameters in ψ_1. In the basic hPP likelihood, dependence on ψ_0 would also have to be considered. Often $\lambda_0(s|\psi_0)$ is estimated nonparametrically and a profile likelihood is assumed. Alternatively ψ_0 could be estimated within a posterior sampler. Here focus is made on the specification of $\lambda_1(s|\psi_1)$. Following from the basic definitions of PcPs and Cox processes it is possible to formulate a Bayesian cluster model that relies on underlying unobserved cluster center locations, but is not restricted to the restrictive assumptions of the classical PcP. Define the excess intensity at s_i as

$$\lambda_1(s_i|\psi_1) = \mu_0 \sum_{j=1}^{K} h(s_i - c_j; \tau) \qquad (6.2)$$

where a finite number of centers is considered inside (or close to) the study window. For practical purposes, K is assumed to be relatively small (usually in the range of $1-20$). The parameter τ controls the scale of the distribution. Note that in this formulation we do not insist that $\{c_j\}$ follow a homogenous PP, nor is the cluster distribution function $h(s_i - c_j; \tau)$ restricted to a pdf, although it must be non-negative. A simple extension of this allows for individual level covariates within a predictor (η_i):

$$\lambda_1(s_i|\psi_1) = \exp(\rho_0 + \eta_i).\sum_{j=1}^{K} h(s_i - c_j; \tau). \qquad (6.3)$$

where $\mu_0 = \exp(\rho_0)$.

In the following, intensity (6.2) will be examined. In general, intensity (6.2) can be regarded as a mixture intensity with unknown number of components and component values (cluster center locations). A general Bayesian model formulation can be

$$[\{s_i\}|\psi_0, \mu_0, \tau, K, \mathbf{c}] \sim \prod_{i=1}^{m} \lambda(s_i|\psi).\exp\left\{-\int_T \lambda(u|\psi) du\right\}$$

where $\psi \equiv \{\psi_0, \mu_0, K, \mathbf{c}\}$, with $\psi_1 \equiv \{\rho_0, \tau, K, \mathbf{c}\}$

$$\lambda(s_i|\psi_1) = \lambda_0(s_i|\psi_0)\lambda_1(s_i|\psi_1)$$
$$\lambda_1(s_i|\psi_1) = \exp(\rho_0). \sum_{j=1}^{K} h(s_i - c_j; \tau)$$
$$\rho_0 \sim Ga(a,b)$$
$$K \sim Pois(\gamma)$$
$$\{c_j\} \sim U(A_T)$$
$$\tau \sim Ga(c,d).$$

Here, the prior distributions reflect our beliefs concerning the nature of the parameter variation. As ρ_0 is the case event rate we assume a positive distribution (in this case a gamma distribution). The parameter γ essentially controls the parent rate (center rate) and in this case the prior for the number of centers (K) is Poisson with rate γ. Other alternatives can be assumed for this distribution. A uniform distribution on a small positive range would be possible. Another possibility is to assume that the centers are mutually inhibited and to assume a distribution that will provide this inhibition. Such a distribution could be a Markov process form such as a Strauss distribution (Móller and Waagpetersen, 2004, Ch. 6). The τ parameter is assumed to appear as a precision term in the cluster distribution function: $h(s_i - c_j; \tau)$. A typical symmetric specification for this distribution is distance based:

$$h(s_i - c_j; \tau) = \frac{\tau}{2\pi} \exp\{-\tau d_{ij}^2/2\}$$
$$\text{where } d_{ij} = \|s_i - c_j\|.$$

Other forms are of course possible including allowing the precision to vary with location and asymmetry of the directional form. Many examples exist where variants of these specifications have been applied to cluster detection problems (e.g., Lawson, 1995; Lawson and Clark, 1999b; Lawson, 2000; Cressie and Lawson, 2000; Clark and Lawson, 2002). One variant that has been assumed commonly is to change the link between the cluster term and the background risk. For example, there is some justification to assume that areas of maps could be little affected by clustering if far from a parent location. In these areas the background rate ($\lambda_0(s_i|\psi_0)$) should remain. The multiplicative link, assumed in $\lambda_1(s_i|\psi_1)$, may be improved by assuming an additive-multiplicative link as well as the introduction of linkage parameters (a, b):

$$\lambda_1(s_i|\psi_1) = \exp(\rho_0). \left\{ a + b \sum_{j=1}^{K} h(s_i - c_j; \tau) \right\}. \quad (6.4)$$

6.4.1.2 Estimation issues

The full posterior distribution for this model is proportional to

$$[\{s_i\}|\psi_0, \mu_0, a, b, \tau, K, \mathbf{c}].P_1(\psi_0, \mu_0, a, b, \tau).P_2(K, \mathbf{c})$$

where $P_*(.)$ denotes the joint prior distribution. Given the mixture form of the likelihood, it is not straightforward to develop a simple posterior sampling algorithm. Both the number of centers (K) and their locations ($\mathbf{c}$) are unknown. Hence it is not possible to use straightforward Gibbs sampling. In addition we don't require there to be assignment of data to centers and so no allocation variables are used, unlike other mixture problems (Marin and Robert, 2007, ch 6). One simple approximate approach is to evaluate a range of fixed component models with different fixed K. The model with the highest marginal posterior probability is chosen (K^*) and the sampler is rerun with fixed K^*. This two stage method is not efficient however. Another alternative would be to use the fixed dimension Metropolized Carlin–Chib algorithm (Godsill, 2001; Kuo and Mallick, 1998). Instead, for variable dimension problems such as this, resort can be made to reversible jump McMC (Green, 1995). A special form of this algorithm can be used called a spatial birth-death McMC. In this algorithm centers, at different iterations, are added, deleted or moved based on proposal and acceptance criteria. In this way the location and the number of centers can be sampled jointly. Detail of these algorithms are given in van Lieshout and Baddeley, 2002 and Lawson, 2001, Appendix C.

Figure 6.9 displays one part of the posterior output from a birth-death McMC sampler run on the Lancashire larynx cancer example. For this case the prior distributions assumed were Strauss for the joint distribution of centers and number of centers (with fixed inhibition parameter), additive-multiplicative link was used with $a = 1$, $b = 1$, a symmetric Gaussian cluster distribution was used with precision parameter κ^{-1}. The population background was estimated via a density estimation but the smoothing parameter was sampled in the posterior distribution. It was given an $InvGa(1, 100)$ prior distribution. Additional random effect terms were also included in this model. For further details of this example see Lawson, 2000). Both the number of centers and location vary over iterations in this example. Hence summarization of the posterior output is not straightforward: different distributions of parameters will be associated with different numbers of centers. One gross summary of the cluster center distribution is available whereby the density estimate surface of the centers overlain from different realizations is presented. This is simply an average over different K values. Of course this can be criticized as it ignores the possibility that markedly different spatial realizations could occur with different K values. In fact this is a general problem with mixture models. It is interesting to note that an area of elevated probability density appears close to a putative source (incinerator: location: 35450,41400).

How do these models perform and are they realistic for disease cluster detection? In general, the simplistic assumptions made by point process models

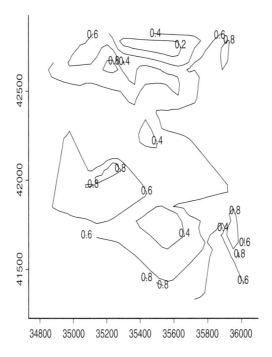

FIGURE 6.9
Lancashire larynx cancer: birth-death McMC output. The posterior expected probability density surface of the cluster center locations obtained by overlay of center realisations from different K values.

are really inadequate to describe clustering in spatial disease data. First, clustering tends to occur not as a common spatial field but often as isolated areas. Even when multiple clusters occur it is unlikely they will be of similar size or shape. In addition, clusters do not form regular shapes and any spatial time cross-section may show different stages of cluster development. For instance, there may be an infectious agent which differentially affects different areas at different times. A time-slice spatial map will then show different cluster forms in different areas. Another factor is that scales of clustering can appear on spatial maps. This is not considered in simple cluster PP models.

Given the possibility that unobserved confounders are present then the resulting clustering will be: a) unlikely to be summarized by a common model with global clustering components; b) cluster distribution functions with regular forms may not fit the irregular variation found.

The use of birth-death McMC with cluster models is not as limited as it may at first seem however. The disadvantages of this form of modeling are a) tuning of reversible jump McMC is often needed and so the method is not readily available, b) interpretation of output is more difficult due to the sampling over a joint distribution of centers and number of centers, and c) possible rigidity of the model specification.

However there are a number of advantages. First, it can easily be modified to include variants such as spatially dependent cluster variances (thereby allowing different sizes of clusters in different areas) and even a semiparametric definition of $h(s_i - c_j; \tau)$ which would allow some adaptation to local conditions. Second, it is also important to realize that by posterior sampling and averaging over posterior samples it is possible to gain flexibility: even with a rigid symmetric form such as $\frac{\tau}{2\pi} \exp\{-\tau d_{ij}^2/2\}$ it is easy to see that the resulting cluster density map does not reflect a common global form (Figure 6.9) and indeed highlights the irregularity in the data. This of course is quite unlike the rigidity found in commonly-used cluster testing methods like SatScan (http://www.satscan.org/). In addition there is a wealth of information provided from a posterior sampler that can even include additional clustering information. For instance, the posterior marginal distribution of number of centers can yield information about multiple scales of clustering (even when these are not included in the model specification). Figure 6.10 displays a histogram of the posterior marginal center rate parameter for a different data example. In that example there appears to be a major peak at 6–7 centers whereas subsidiary peaks appear at 10–11 and also at 13. This may suggest different scales of processes operating in the study window.

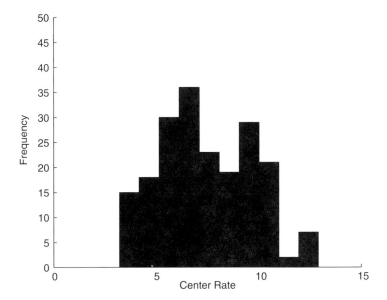

FIGURE 6.10
Posterior expected distribution of number of centers from a converged sampler.

Finally, it is also possible to increase the flexibility of the model by introduction of extra noise in the cluster sum. For example, the introduction of a random effect parameter for each of the centers: $\sum_{j=1}^{K} \exp(\psi_j).h(s_i - c_j; \tau)$ with $\psi_j \sim N(0, \tau_\psi)$, can lead to improved estimation of the overall intensity of the process. Another option that could be exploited which allows the sampling of mixtures more nonparametrically is the use of Dirichlet process prior distributions for mixtures (Ishwaran and James, 2002; Ishwaran and James, 2001; Kim et al., 2006). This has so far not been explored.

6.4.1.3 Data dependent models

Another possible approach to modeling is to consider models that do not assume a hidden process of centers but model the data interdependence directly. Such data-dependent models have various forms depending on assumptions.

6.4.1.3.1 Partition models and regression trees Partition models attempt to divide up the space of the point process into segments or partitions. Each partition has a parameter or parameters associated with it. The partitions are usually disjoint and provide complete coverage of the study domain (T). An example of disjoint partition (or tiling) is the Dirichlet tesselation which is constructed around each point of the process. Each tile consists of allocations closer to the associated point than to any other. Figure 6.11 displays such a tesselation of the Lancashire larynx cancer data set. It is clear from the display that small tiles (small tile area) are associated with aggregations of cases. The area in the south of the study region is particularly marked. The formal statistical properties of such a tesselation are known (Barndorff-Nielsen et al., 1999) for most processes (such as the marginal distribution of tile areas).

However in partition modeling, the tesselation is used in a different manner. Byers and Raftery (2002) describe an approach where a Dirichlet Tesselation is used to group events together. Hence a tiling consisting of K tiles with areas a_k is superimposed on the points and the number of events within a tile (n_k) are recorded. The first order intensity of the process is discretized to be constant within tiles (λ_k). The tile centers are defined to be $\{c_k\}$. Based on this definition a posterior distribution can be defined where

$$L(\mathbf{n}|\boldsymbol{\lambda}) \propto \prod_{k=1}^{K} \lambda_k^{n_k} \exp\{-\lambda_k a_k\}$$
$$K \sim Poiss(\nu)$$
$$\{\lambda_k\}, k = 1, ..., K | K \text{ iid } Ga(a,b)$$
$$\{c_k\}, k = 1, ..., K | K \text{ iid } U(T).$$

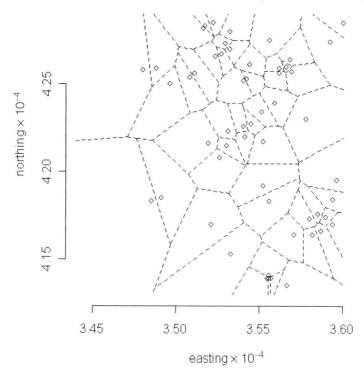

FIGURE 6.11
Lancashire larynx cancer data: Dirichlet tesselation produced with 4 external dummy points using the DELDIR package on R.

In this definition, the centers and areas are not given any stochastic dependency, whereas the areas are really dependent on the center locations. In addition, the number of centers is not fixed in general. This lead to a posterior distribution, within the general case, which does not have fixed dimension, but assuming ν, a, b fixed, is proportional to

$$\frac{\nu^K}{K!} \prod_{k=1}^{K} \lambda_k^{(n_k+a-1)} \exp\{-\lambda_k(a_k + b)\}.$$

In general, a reversible jump McMC algorithm or Metropolized Carlin–Chib algorithm must be used to sample from this posterior distribution unless K is fixed. The focus of this work was the estimation of λ_k. Of course, in general, λ_k will vary over iterations of a converged posterior sample and won't be allocated to the same areas. Hence, any summarization of the output would have to overlay the realizations of λ_k for a predefined grid mesh of sites (possibly the data points), at which the average intensity would be estimated. Hence a smoothly varying estimate of the intensity would result. In addition to simple intensity estimation, the authors also include a binary inclusion variable (d_k), which has a Bernoulli prior, and catergorizes the tile as being in a high intensity area ($d_k = 1$) or not. This allows a form of crude intensity segmentation (between areas of high and low intensity). In that sense the method provides a clustering algorithm, albeit where only two states of intensity delineate the "clusters." Mixing over the posterior allows for gradation of risk in the converged posterior sample (see, for example, Figure 6.3 b) of Byers and Raftery, 2002).

Note that in their application, Byers and Raftery (2002) have no background (population) effect which would be needed in an epidemiological example. In application to disease cases it may be possible to estimate a background effect using a control disease and to use this as a plug-in estimate (i.e., replace λ_k by $\widehat{\lambda}_{0k}\lambda_k$ in the likelihood, where $\widehat{\lambda}_{0k}$ is a background rate estimate in the k th tile). Alternatively, if counts of the control disease are available within tiles then it would be possible to construct a joint model for both counts. Hegarty and Barry (2008) have also introduced a variant where product partitions are used to model risk.

6.4.1.3.2 Local likelihood An alternative view, considers the use of a grouping variable which relates to a sampling window. The sampling window is a subset of the study window. For example, a sampling window (lasso) is defined to be controlled by a parameter (δ). This parameter controls the size of the window. Usually, (but not necessarily) the window is circular so that δ is a radius. First consider cases of disease collected within a window of size δ, and denote these as n_δ. Second, denote cases of a control disease as e_δ within the lasso. Now assume that the case disease and control disease are observed at a set of locations and denote these as $\{x_i\}$, $i = 1, ..., n$ and $\{x_i\}$, $i = n+1, ..., n+m$. The joint set of $\{x_i\}$ can be described jointly by a Bernoulli distribution with case probability $p(x_i) = \lambda(x_i)/(\lambda_0(x_i) + \lambda(x_i))$, conditional on $\lambda_0(x_i), \lambda(x_i)$ and their parameters. Further, assume that within the lasso there is a risk parameter θ_{δ_i} and that $\lambda(x_i) = \rho\theta_{\delta_i} \ \forall x_i \in \delta_i$. Now assume that within the lasso the probability of a case or control is constant. In that case we can write down a local likelihood of the form

$$\prod_{i=1}^{n+m} \left[\frac{\rho\theta_{\delta_i}}{1+\rho\theta_{\delta_i}}\right]^{n_{\delta_i}} \left[\frac{1}{1+\rho\theta_{\delta_i}}\right]^{e_{\delta_i}}.$$

Note that the lasso depends on δ_i, defined at the i th location, and different assumptions about these can be made. Attention focusses on the estimation of θ_{δ_i}, rather than δ_i, which can yield information about clustering behavior. Based on this local likelihood (Kauermann and Opsomer, 2003), it is possible to consider a posterior distribution with suitable prior distribution for parameters. For example, the $\boldsymbol{\delta}$ can have a correlated prior distribution (either a fully specified Gaussian covariance model or a CAR model). Alternatively it has been found that assuming an exchangeable gamma prior appears to work reasonably well. In addition the dependence of θ_{δ_i} on δ_i across a range of $\delta_i s$ should be weak a priori and so we assume a uniform distribution. Assuming a CAR specification, the prior distributions are then:

$$[\theta_{\delta_i}|\delta_i] \sim U(a,b) \; \forall i \tag{6.5}$$
$$\rho \sim IGa(3, 0.01)$$
$$[\delta_i|\boldsymbol{\delta}_{-i};\tau] \sim N(\overline{\delta}_{\Delta_i}, \tau/n_{\Delta_i})$$
$$\tau \sim IGa(3, 0.01)$$

where Δ_i is a neighborhood of the i th point, $\overline{\delta}_{\Delta_i}$ is the mean of the neighborhood δs and n_{Δ_i} is the number of neighbors, $N(,)$ is an (improper) Gaussian distribution, τ is a variance parameter, and ρ is a rate both with reasonably vague inverse gamma (IGa) distributions. An advantage of this approach is that a fixed dimension posterior distribution can be specified, albeit with a local likelihood. Figure 6.12 displays a smoothed version of the posterior aver-

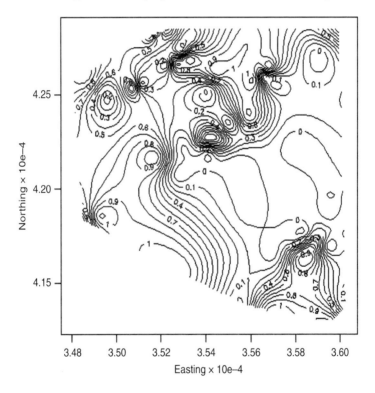

FIGURE 6.12
Lancashire Larynx cancer data: exceedence probability map. Shown is the smoothed posterior average value, from a convreged sampler, of $1 - \widehat{\Pr}(\theta_{\delta_i} > 1)$ for the case data only. The smoothing was done using the MBA (R) package.

age value of the exceedence value of θ_{δ_i}: the function shown is $1 - \widehat{\Pr}(\theta_{\delta_i} > 1)$ for the converged sampler with the prior specification shown (6.5). The case only map is shown with only the convex hull of the case distribution contoured. Further details of this model are given in Lawson (2006a). Note that it is clear that the southern area in the vicinity of (3.55×10e-4, 4.15×10e-4) demonstrates a very high exceedence probability (<0.01).

The main advantage of this approach is in the simplicity of programming (compared to birth-death McMC methods) and the ease with which interpretation of output can be made. The disadvantage is that in one sense the methods do not directly model clustering, and may be regarded as smoothing methods, although their cluster detection performance in simulations is impressive (see e.g., Hossain and Lawson, 2006)

6.4.2 Count Data

6.4.2.1 Hidden process/object models

As in the situation where case events are analyzed, it is possible to specify object models for count data. Essentially these assume a hidden process exists and this must be estimated. The hidden process could take a variety of forms. In the earliest examples, these forms followed those for case events where a hidden process of cluster centers was posited. The aggregation effects of accumulating case events into small areas (census tracts, zip codes, etc.) leads to integrals of the first order case intensity in the expectation of the count. Hence, denoting the count of disease in the i th area (within area a_i) as n_i, then

$$E(n_i) = \int_{a_i} \lambda(u) du.$$

and an 'exact' hidden process model (HPM) would be defined as

$$n_i \sim Pois(\int_{a_i} \lambda(u) du) \qquad (6.6)$$

$$\lambda(u) = \lambda_0(u|\psi_0)\lambda_1(u|\psi_1)$$

$$\lambda_1(u|\psi_1) = \sum_{j=1}^{K} \exp(\phi_j).h(u - c_j; \tau_h)$$

$$\phi_j \sim N(0, \tau_\psi).$$

Here, $h(u - c_j; \tau_h)$ is a cluster distribution function as before and $\{c_j\}$ is a set of hidden cluster centers and the background population function $\lambda_0(u|\psi_0)$ must be estimated as before (Lawson and Clark, 1999a). Often, a simplified (approximate) version of (6.6) is assumed where the expectation is simply a function of an expected rate (e_i) and relative risk parameter (θ_i):

$$n_i \sim Pois(e_i\theta_i),$$
$$\log \theta_i = \alpha_0 + \alpha_1 \sum_{j=1}^{K} \phi_j h(C_i - c_j; \tau_h),$$

where $\{C_i\}$ is a set of centroids of the small areas, α_0 is an intercept and α_1 is a linking parameter. The $\{\phi_j\}$ play the same role as before as random effects and the $\{c_j\}$ is hidden centers to be estimated. The expected rates are usually calculated based on an external standard population. An example of a variant of this model is given in Lawson (2006b), ch 6, where $\theta_i = \exp(\phi_j).[1 + \sum_{j=1}^{K} h(C_i - c_j; \tau_h)]$ with $\phi_j \sim N(0, \tau_\psi)$. Another variant of this model was specified by Gangnon (2006) where the log relative risk in a small area is assumed to be specified by $\log \theta_i = \alpha_0 + \Gamma_i + e_i$, where

$$\Gamma_i = \sum_{j=1}^{K} \phi_j h(C_i - c_j; \tau_h) \quad (6.7)$$

and ϕ_j is now a relative risk component associated with the j th cluster and $h(.)$ is a cluster membership function assumed to be uniform on a disc. The intercept term is α_0, while the e_i term is a random effect with zero mean Gaussian prior distribution. The function $h(.)$ associates small areas to centers. The cluster centers are assumed to be centroids of the small areas (to avoid empty clusters which could arise due to the membership function definition). The main difference between these two variants is the inclusion of relative risk components in the latter model (6.7).

6.4.2.2 Data dependent models

As in case event applications, it is possible for count data to be modelled without recourse to HP models. Two alternatives to such modeling are the use of partition models and splitting methods, and the local likelihood. A common feature of these methods is that spatial membership must be computed within an area. In either case the methods require computation of distances within the spatial configuration of the data. Within McMC algorithms these distances have to be recomputed as parameters are sampled.

6.4.2.2.1 Partition models and regression trees One of the first examples of applying partition modeling to small area count data was proposed by Knorr-Held and Rasser (2000). In their modeling approach, at the first level of the hierarchy, the small area count y_i is assumed to be conditionally independent Poisson with expectation $e_i h_j$ where e_i is the usual standardized expected rate. The relative risk for a given area is chosen from a discrete set of risk levels (which are called 'clusters' by the authors). These 'clusters' are a set of contiguous regions $\{C_j\}$ $j = 1, ..., n$, which have associated constant risk $\{h_j\}$. The relative risk assigned to the i th area is just h_j if $i \in C_j$. Hence, $y_i \sim Pois(e_i h_j)$. The number of clusters (k) is treated as unknown: $k \in \{1, ..., n\}$. Hence this method seeks a non-overlapping partition of the map into areas of like risk. In this definition of clustering, clusters are areas of *constant* risk and of course this is a different definition of clustering from the more usual definitions using "elevated risk." The essential difference is that the clusters are discrete partitions in this case. However, as in other cases, posterior averaging can lead to a more continuous relative risk estimate. Computational issues related to sampling partitions led the authors to consider reversible jump McMC. In the examples cited the number of partitions were quite large (40-45) in the posterior realizations for male oral cavity cancer in 544 districts in Germany.

Partition models have since been extended (Denison and Holmes, 2001; Denison et al., 2002; Ferreira et al., 2002). In application to count data in small areas, the usual Poisson assumption is made at first level: $y_i \sim Pois(e_i \varepsilon(C_i))$. The partition is defined to be $\varepsilon(C_i) = \mu_{r(i)}$ the relative risk level associated with a region of a tesselation that the centroid C_i of the small area lies in. In this approach a discrete tesselation is used as a partition. The prior distribution of the levels is assumed to be independent gamma: $[\mu_j|\gamma_1, \gamma_2] \sim Ga(\gamma_1, \gamma_2)$, $j = 1, ..., k$. A prior model is also assumed for the location of tile centers:

$$p(\mathbf{c}) = \frac{1}{K} \frac{1}{\{Area(T)\}^k},$$

where K is the maximum number of centers/tiles and k is the current number. The marginal likelihood of the data given the centers is

$$\text{constant} + \sum_{j=1}^{k} \{\log \Gamma(\gamma_1 + n_j \bar{y}_j) - (\gamma_1 + n_j \bar{y}_j) \log(\gamma_2 + n_j \overline{N}_j)\}$$

where n_j is the number of points in the j th region and $\bar{y}_j$ is the mean of the observations in that tile and $\overline{N}_j$ is the mean expected rate in the j th tile. An inclusion rule must be specified for the small areas being included within any given tile. The usual rule is to include if the centroid falls within the tile. Unlike the Knorr–Held–Rasser model this allows there to be different partition shapes. Computation for the tesselation model is based on a birth-death McMC algorithm where the centers are added, deleted, or moved sequentially.

Of course, there are considerable edge effect issues with partition models and these apply to all partition models that use tesselations (rather than groupings): at the external study region boundary tesselations will inevitably be distorted due to censoring adjacent regions *outside* the study window. Grouping algorithms (such as proposed by Knorr-Held and Rasser, 2000) could also be affected as no information outside the boundaries have been reported and this could distort the edge region allocations. These seem to be largely ignored in the literature on partition modeling.

Extensions to tree models could be imagined (Denison et al., 2002). For example, all the "clustering" partition models proposed assume a flat level of partitioning. However we could conceive of a form of splitting where a tree-based hierarchical cluster formation is conceived. It could be possible to construct a multi-level partition by allocation of higher levels of partition based on residuals from lower levels. This could be conceived as a multivariate partition function with L levels and K_l partitions at the l th level. There could be ordering constraints on the different levels. This could provide an approach to multiscale feature identification in the data.

Connections between partition models and mixture models for relative risk are also evident. For example, the model of Green and Richardson (2002) assumes that there are a small number of risk components $\{\theta_j, j = 1, 2, ..., k\}$ and the small areas are allocated one of these levels via the allocation variables $\{z_i, i = 1, ..., n\}$. Hence $y_i \sim Pois(e_i \theta_{z_i})$ where θ_{z_i} is the allocated risk level for the i th area. Unlike other approaches, the allocation variables are given a spatially-structured Potts prior distribution. This allows grouping at a higher level in the hierarchy. In contrast, Fernandez and Green (2002) describe a mixture model where the count is assumed to be governed by a weighted sum of Poisson distributions and the weights have spatial structure.

One general question concerning these methods is whether they really can be considered 'clustering' methods or not. While all these methods could yield a variety of posterior information concerning risk, and some part of that information could be employed to detect clusters, they do not directly address the detection of clusters; rather they seek to find underlying discrete risk levels that characterize the map. Often, the authors regard their methods as competitors with disease map smoothing models such as the convolution model of Besag et al. (1991) (see for example discussions in Knorr-Held and Rasser, 2000, Ferreira et al., 2002, or Fernandez and Green, 2002) and they have been compared in simulations as such (Best et al., 2005). Essentially, mixture models are closely identified with latent structure models and these are discussed more fully in Sections 5.7.5 and 11.3.2.

6.4.2.2.2 Local likelihood As in the case event situation (6.4.1.3.2), local likelihood models could also be applied. For small area counts, then within a lasso, centered at the centroid of the area, of dimension δ_i, with a Poisson assumption, then the probability of y_{δ_i} counts with expected rate e_{δ_i}, is just:

$$f(y_{\delta_i}|e_{\delta_i}, \theta_{\delta_i}; \delta_i) = [e_{\delta_i}\theta_{\delta_i}]^{y_{\delta_i}} \exp(-e_{\delta_i}\theta_{\delta_i})/y_{\delta_i}!.$$

This can be employed within a local likelihood as:

$$L(\boldsymbol{\theta}|\mathbf{y}) = \prod_{i=1}^{m} f(y_{\delta_i}|e_{\delta_i}, \theta_{\delta_i}, \delta_i).$$

Note that, in this case, the factorial term ($y_{\delta_i}!$) must be included in the likelihood as it varies with δ_i. In this case the data and expected rate are accumulated within the lasso. A rule must be assumed for the inclusion of small areas. Centroids falling within the lasso is a common assumption. Then $y_{\delta_i} = \sum_{i \in \delta_i} y_i$, and $e_{\delta_i} = \sum_{i \in \delta_i} e_i$. In this definition the counts and expected rates within neighboring lassos can overlap and hence they can be correlated. Usually, the focus is on the estimation of θ_{δ_i} rather than δ_i. The linkage between the risk and the lasso is specified via:

$$\log \theta_{\delta_i} = \delta_i + \varepsilon_i$$

where $\varepsilon_i \sim N(0, \tau_\epsilon)$ an unstructured component and δ_i is assumed to have a spatially-structured prior distribution. In examples, a CAR prior distribution has been assumed for this purpose i.e.,

$$[\delta_i | \delta_{-i}; \beta_\delta] \sim N(\overline{\delta}_i, \beta_\delta / d_i)$$

where $d_i = \sum_{j}^{m} I(j \in \delta_i)$, and $\overline{\delta}_i = \sum_{j \in \delta_i} \delta_j / d_i$. This allows the correlation inherent in the local likelihood to be modeled at a higher level of the hierarchy. Covariates can also be added to this formulation if required (see Hossain and Lawson, 2005 for discussion). Figure 6.13 displays an example of the

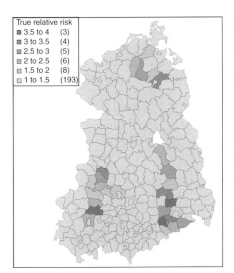
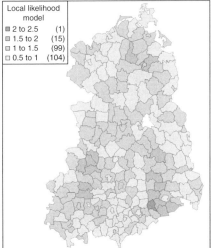

FIGURE 6.13
East German lip cancer mortality (1980-1989) data. Simulated "true" risk map (left panel) and local likelihood posterior expected relative risk estimates (right panel).

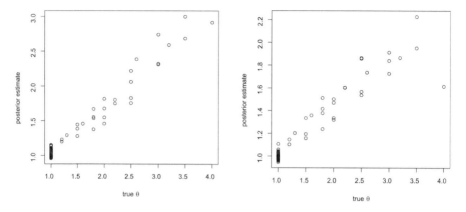

FIGURE 6.14
Comparison of true and posterior estimated relative risks for the simulated data for the East Germany district geographies. Left panel: BYM model; right panel: local likelihood model.

application of the local likelihood model to a simulated realization where clusters of risk are introduced into the district geographies of the former East Germany. The counts are generated from a Poisson model with mean e_i. θ_i^{true} where the θ_i^{true} are the true risks. The left-hand panel depicts the true risks and the right-hand panel depicts the estimated risks under the local likelihood model with CAR prior distribution for the lasso parameters and inverse gamma distributions for the precision parameters (τ). A comparison of the estimation ability of a random effect convolution model compared to the local likelihood model is also made in Figure 6.14. It is noticeable that the local likelihood model produces a greater shrinkage than the convolution model. In full simulation comparisons based on operating characteristic curves the LL model performs well in recovering true risk as well as some features of clusters (see e.g., Hossain and Lawson, 2006).

6.4.3 Markov Connected Component Field (MCCF) Models

A number of alternative approaches to count data clustering have been proposed and should be mentioned as they relate to methods previously discussed. First a cluster modelling approach based on Markov connected component fields has been proposed (Gangnon and Clayton, 2000). This approach also seeks to find a grouping of small areas into "clusters." The cluster set k clusters has associated a risk level λ_j, $j = 1, .., k$. The membership of the clusters is defined for each of the N small areas ($i = 1, ..., N$), via a membership (allocation) variable: $\mathbf{z} = (z_1, ..., z_N)$. Hence the counts in the N areas have distribution, conditional on the cluster assignment (**z**):

$$y_i \sim Pois(e_i \lambda_{z_i})$$

where λ_{z_i} is the disease risk for the cluster associated with the i th area. Clusters are assumed to be aggregations of the basic small areas, and so the allocation assigns different disease risks to each area depending on prior distributional constraints. This setup was later used by Green and Richardson (2002) but they assumed a spatially structured prior distribution for the **z**.

Here, the disease risks are assumed to have a $Ga(\alpha, \beta)$ prior distribution given the **z**, where $Ga(\alpha, \beta)$ has mean α/β. However the **z** are assumed to have a Markov connected component field prior governing their form (Mól1er and Waagepetersen, 1998). This construction allows there to be various cluster form specifications included in the prior structure. This construction allows a potential function to describe the clustering:

$$p(\mathbf{z}) \propto \exp\{-\sum_{j=1}^{k} S_j\}$$

where the S_j is a score function for the j th cluster. The authors use properties of the clusters such as circularity (shape) and size to yield a composite score: $S_j = \alpha + S_{j1}(size) + S_{j2}(shape)$. Once specified the authors use a randomized model search criteria to evaluate different models. In a similar development, but different application, Móller and Skare (2001) further apply these priors to imaging problems.

Note that this MCCF approach has many advantages over purely spatially-structured distributions for component assignment. It also allows the models to address specific features of the clusters that are not considered by other partitioning approaches.

6.5 Edge Detection and Wombling

A closely related area of concern to disease clustering is the idea that discontinuities in maps are the focus. These could be the edge of some uniform risk area or the edge of a variable risk cluster of some kind. In cluster modeling usually there is some criteria defining the cluster and its boundary. In edge detection the boundary is usually defined by a jump or discontinuity in risk, and so this in a sense takes the approach whereby the focus is the differences in risk rather than finding the area of elevated risk.

Wombling, is closely related to the edge effect problem of Section 5.8, in that the focus is on boundaries of regions. There is a growing interest in the ability to locate boundaries within a spatial domain. This has been developing within

Geography for some time (see e.g., Oden et al., 1993 amongst others). These boundaries often have some natural context, such as catchments of health providers (Ma et al., 2006) or areas which vote predominantly for one party in elections (O'Loughlin, 2002). Basically these methods are edge detection methods, and many such methods are commonly found in the image processing literature (see e.g., the Canny operator, Mars-Hilldreth, Gaussian derivative kernels; Bankman, 2007). Often they are intended for object recognition and use derivative-based methods for locating discontinuities.

The discovery of discontinuities in disease maps may be of interest in detection of natural boundaries for health provision (recent Bayesian examples are Ma et al., 2006, Lu and Carlin, 2005). One simple approach to this problem can be examined within a Bayesian hierarchical model. Assume for small areas with observed tract counts y_i and $y_i \sim Poiss(e_i \theta_i)$.

Within a posterior sampling algorithm, θ_i^g is the estimate of θ_i at the g th iteration. It is possible to estimate the posterior expected value of the absolute difference between relative risks, $\Delta \theta_{ij}$, by simply computing:

$$\Delta \theta_{ij} = \sum_{g \in d} |\theta_i^g - \theta_j^g|/n(d),$$

where d denotes the converged sample set and $n(d)$ is the number in that set. Hence this estimator is available for all region boundaries. The parameters $\{\Delta \theta_{ij}\}$ really detect discontinuities between regions and there is no information in this approach that informs the edge detection beyond the average difference between *modelled* estimates of adjoining area risks. Hence cluster or object information that could 'tie' the areas together is not used. Pre-smoothing the estimates of risk (e.g., via a convolution model) may lead to reduction in discontinuity, of course, and so the choice of model for the θ could be crucially important. More sophisticated approaches have also been proposed (see e.g., Ma et al., 2006).

It is not clear how discontinuities are to be modeled for applications in disease incidence studies. Underlying most models of risk is the assumption of continuity of risk over space. The convolution smoothing models assume this as do most other risk models. The fixed mixture model of Lawson and Clark (2002) and also the mixture models of Green and Richardson (2002) do attempt to honor discontinuities within a general framework of smooth risk and allows the smoothness to be selected differentially over space. It is a matter of debate whether pure discontinuity models have advantages in incidence studies.

7
Ecological Analysis

The term, ecological analysis, is used loosely here to denote a situation where an aggregated disease outcome variable is geo-referenced and is to be related to predictors/covariates. These predictors can be observed at different levels of spatial aggregation. In what follows, I do not discuss misaligned (MIDP) or modifiable areal unit problems (MAUP) directly as these are discussed in Chapter 8.

7.1 General Case of Regression

The setting to be discussed here, in the simplest case, assumes a dependent variable (outcome) for a small area as y_i and this is to be related to predictors. The predictors can be geo-referenced, or not, depending on context. For example, we might measure a disease rate within a census tract and want to relate it to the level of poverty in the tract. Hence both the dependent variable is an aggregate count of disease (y_i) and the covariate is an aggregate measure (percentage) of numbers of family units or people who fall below the defined poverty level in the small area (x_{1i}). This is an example of ecological regression: the disease count is thought to relate to poverty but we have no direct measure at a lower level of aggregation (e.g., the individual level) whether poverty is associated with the disease outcome.

A typical mean model would be of the form

$$y_i \sim f(\mu_i)$$
$$\text{where} \quad E(y_i) = \mu_i$$

with

$$g(\mu_i) = S(x_{1i})$$

where $g(\mu_i)$ is a link function and $S(x_{1i})$ is a linear or non-linear function of the covariate. In a simple case, with a Poisson data model and $\mu_i = e_i \theta_i$, with

151

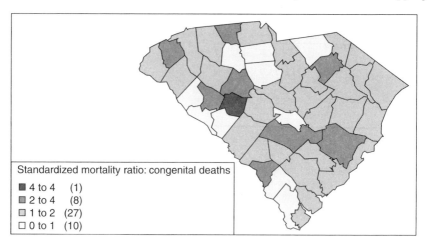

FIGURE 7.1
South Carolina congenital death standardized mortality ratio 1990. Standardisation based on statewide rate for same year.

a log-linear setup, with expected rate as offset, we would have

$$y_i \sim Pois(\mu_i)$$
$$\mu_i = e_i \theta_i$$
$$= e_i \exp\{\alpha_0 + \alpha_1 x_{1i}\}.$$

In this formulation, $exp\{\alpha_0\}$ acts as an overall scaling parameter, and $\exp\{\alpha_1 x_{1i}\}$ adds the modulation of the covariate. Inference would focus on the parameters α_0 and α_1. This is just a log-linear Poisson regression model and in a Bayesian context we would assign prior distributions for α_0 and α_1 to complete the specification. As the focus is only on regression parameter inference rather than on estimation of the relative risk per se, then it may be tempting to fit this log-linear model as defined, in the belief that, for example, unobserved confounding would have little effect on such parameter estimates. County level counts of congenital deaths for the US state of South Carolina are available for a number of years. Here I focus on one year: 1990. Figure 7.1 displays the standardized mortality ratio (smr) for the congenital anomaly counts for the counties of South Carolina in 1990.

The expected rate was computed using the age × sex standardized rate for South Carolina. Also available for this example, from the 1990 US census, is the percentage of family units living in poverty. This is determined by comparison to a threshold income level. It is reasonable to suppose that some relationship might hold between congenital deaths and poverty or deprivation in a population. Of course this is an ecological relation as individual poverty and disease outcome is not measured directly. Figure 7.2 displays the rela-

Ecological Analysis

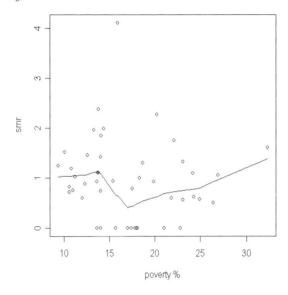

FIGURE 7.2
Scatterplot of the standardized mortality ratio for congenital abnormality outcome for the counties of South Carolina versus percentage poverty from the US census 1990. A loess line fit has been added.

tionship found. While there is some evidence for a positive relation at higher poverty levels, there is considerable noise in the relation and at lower levels there may appear to be an inverse relation. This form of relation may be apparent for a number of reasons. First, many areas in South Carolina are largely rural and these may display a positive relation between the smr and poverty given they may be predominantly areas of low income. However some urban fringe areas have high concentrations of high income units mixed with rural poor populations. In these fringe areas the *smr* may be relatively high but the income will average out to be relatively high also. Second, in areas with a small percentage of poverty units it may be that health "bootstrapping" can take place in the sense that in predominantly high income areas the health experience of low income units may be improved so that there could be a reduction in the positive relation. A negative relation might even arise. This is termed "ecological inversion." There is some evidence for ecological inversion in this display.

An appropriate model for these data might include a spline fit to mimic the relation. However as a first model, I will assume a basic log-linear Poisson regression model with diffuse Gaussian prior distributions for the regression parameters and uniform prior distributions for the associated standard deviations (Gelman, 2006). Model results are found in Table 7.1. The models fitted were as follows. Model 1 is a simple Poisson log-linear regression, model 2 is

TABLE 7.1
Model fitting results for a variety of models for the South Carolina congenital mortality data, (See text for details).

	Model 1	Model 2	Model 3	Model 4
α_0	0.0859(0.248)	0.0236(0.191)	0.068(0.243)	0.192(0.209)
α_1	-0.0034(0.016)	0.0002(0.0119)	-0.0032(0.0155)	0.044(0.1198)
α_2	-	-	-	-
τ_0	43.87(393.2)	325.2(4159)	107.6(1338)	3425.(63910)
τ_1	35100.0(4.6E+4)	15210.0(3.5E+5)	48510.0(8.0E+5)	2611(49710)
τ_v	-	242.8(321.5)	68.43(143.9)	4.845(9.03)
τ_u	-	-	525.3(678.1)	-
DIC	171.17	171.01	173.16	167.95

the same but with an added uncorrelated zero mean Gaussian random effect (with precision τ_v) and model 3 is the same as model 2 but with an added correlated random effect with a CAR prior distribution and precision τ_u. Note that for the simple regression none of the parameters are well estimated and even when conventional random effects are added the model is not improved significantly. In fact, the addition of the correlated effect increases the DIC. Overall the random effects do not appear to improve the model fit to any significant degree. Instead, noting that a non-linear model seems to be more appropriate as a description of the relation, I consider a low rank spline model fitted to the predictor:

$$\log \theta_i = \alpha_0 + \alpha_1 x_{1i} + ... + \alpha_p x_{1i}^p + S(x_{1i}) \qquad (7.1)$$

$$S(x_{1i}) = \sum_{k=1}^{K} \psi_k (x_{1i} - c_k)_+^p$$

where $\{c_k\}$, $k = 1, ..., K$ is a set of knots and $\{\psi_k\} \sim N(0, \tau_\psi)$

Model 4 is a simplified version of (7.1) which is simply $\log \theta_i = \alpha_0 + \alpha_1 \sum_{k=1}^{K} \psi_k (x_{1i} - c_k)_+$. Further model fits using spline only and a full second degree polynomial with added second degree spline ($p = 2$) gave DICs of 169.8 and 173.8 respectively. This supports the contention that Model 4 provides the best fit in this case, for the models considered.

It is also interesting to note that when a convolution model was fitted to the data with a linear predictor (Model 3), a large outlier appeared in the estimated $\widehat{\theta}_i$. This was for Berkeley county on the urban-rural fringe of Charleston (see Figure 7.3), where the exceedence probability $(\widehat{\Pr}(\theta_i > 1))$ is 0.903. This is considerably higher than any other county. For the best fitting spline model none of the counties have excessive $\widehat{\theta}_i$s nor $\widehat{\Pr}(\theta_i > 1)$ exceeding 0.7.

To further highlight this effect, a simulation has been carried out, where, at the individual level, there is a strong positive relation between a binary

Ecological Analysis

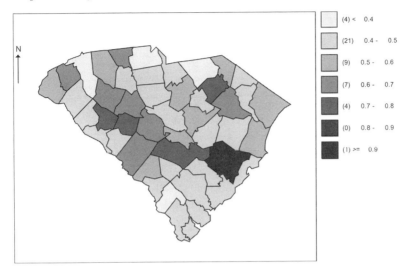

FIGURE 7.3
South Carolina congenital deaths and percentage of poverty by county 1990: probability of exceedence of the posterior expected relative risk.

disease outcome and socioeconomic status (income below or above poverty threshold). It is assumed that there were 100 individuals within 100 areas. An income distribution was assumed for each region and based on that individuals were categorized as below or above poverty level ($30,000). Conditional on this given binary variable (x_{ij}: poor or not) the probability of disease was simulated via a logistic transform with added binomial noise. This transform allows the specification of the individual relation between poverty and outcome: $y_{ij}^s < -bin(1, p_{ij})$ where logit $p_{ij} = \alpha_0 + \alpha_1 x_{ij}$. For the simulation shown the relation was logit $p_{ij} = 0.2 + 1.5 x_{ij}$, a reasonably strong positive relation between outcome and poverty state. Counts of disease were then aggregated across the individuals within areas ($j = 1, ..., 100$), as was the number of poor (to yield percentage of poverty). To demonstrate the ability of such aggregation to yield a relatively complex relationship between outcome and poverty, Figure 7.4 displays the simulated aggregate relation between average income and disease count and poverty proportion and disease count. It is noticeable that while a general increase in poverty seems to relate to an increase in disease count, this relation does not hold strongly. For some levels of poverty the relation is not strong and also apparently reversed. The noise in the relation is quite high of course.

In further simulations, uncorrelated heterogeneity (zero mean Gaussian noise) was introduced to the linear predictor, and further variation in income distribution. These changes lead to greater noise in the relation and changes in the overall gradient or linear form of the relation. Certainly from this output it

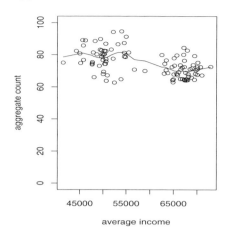 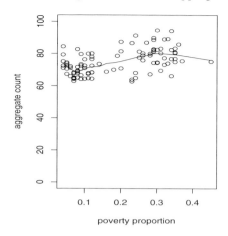

FIGURE 7.4
Simulation-based aggregate relation between total disease count and percentage of poverty for 100 areas with 100 individuals in each area. The loess fit is shown at the aggregate level. Left panel: count versus average income, right panel: count versus poverty proportion.

would appear that there is no strong indication of a non-linear positive relation at the individual level. However, if the focus is the estimation of the aggregate level relation then it is clear that the overall relation at the aggregate level is weakly positive (or weakly negative with average income). It is also clear that a simple log linear model may not represent the variation well. The addition of extra variation in a model in the form of random effects may help to reduce the noise but this does not necessarily improve the estimation of the covariate relation. If the covariate relation is mis-specified then addition of modeled heterogeneity may not lead to a better model. This was demonstrated in the data example above where a spline model, without random effects, yielded a better empirical model fit, based on DIC, than a convolution model with log linear predictor. Some authors (e.g., Clayton et al., 1993) have advocated the inclusion of spatially-structured(CH) random effects to make allowance for biases induced by the ecological nature of the analysis. In some cases this may be important, especially when making inferences at a different aggregation level. However, the above example demonstrates that the use of convolution models (which include CH and UH effects), can yield poor empirical fits when the aggregate relation is mis-specified.

In addition to this warning, there is now some evidence that convolution models, particularly those which have an (improper) CAR model specification for the CH effect, can lead to very poor estimation of certain covariate effects (Ma. et al., 2007). In fact, the use of CAR random effects in linear combination with linear predictors with spatially-referenced covariates

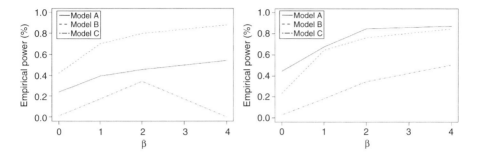

FIGURE 7.5
Empirical power curves estimated from credible intervals for the β parameter from a distance covariate model which includes additive random effect terms: Model A covariate model only; Model B as model A with UH term added; Model C as model B with CH term added. Left panel: no background heterogeneity; right panel: log Gaussian Cox process with a spatial Gaussian generating process.

(trend surface components), can yield very poor estimates of the linear parameters even under a strong linear relation. Figure 7.5 displays two examples of the empirical power within a simulation of the estimation of a distance covariate parameter. Case event data were simulated under a variety of models and then a fine grid mesh was used to bin the events to form counts. The count in the i th bin (y_i) was assumed to have a Poisson distribution with expectation $e_i \theta_i$. Various models were assumed for θ_i. In the display below, Model A is $\log \theta_i = \alpha + \log(1 + \exp(-\beta d_i))$ where d_i is the distance from a fixed location to the centroid of the i th bin; model B is $\log \theta_i = \alpha + \log(1 + \exp(-\beta d_i)) + v_i$ where $v_i \sim N(0, \tau_v)$ a UH component; and model C is $\log \theta_i = \alpha + \log(1 + \exp(-\beta d_i)) + v_i + u_i$ where $u_i | u_{-i} \sim N(\overline{u}_{\delta_i}, \tau/n_{\delta_i})$, a CAR prior distribution. This latter model is a convolution model with added covariate term. The simulations were carried out under a variety of scenarios. Two of these involved variant forms of background heterogeneity in risk. The left panel in Figure 7.5 has no additional heterogeneity while the right panel is under a binned log Gaussian Cox model with a generating process that is a spatial Gaussian process with exponential covariance: $\sigma \exp\{-d_{ij}/\omega\}$ with $\sigma = 0.1$ and $\omega = 0.5$ where d_{ij} is the distance between i, j th points. It is noticeable the convolution model appears to have poor performance under either scenarios in the estimation of the distance effect compared to either a simple log linear model or an UH component model. (This performance is also found when a Poisson simulation is made directly into the bins, and also when a multiplicative log link is defined.) Hence, it is also important to consider carefully the use of CAR-based convolution models when covariates are to be estimated. While convolution models are robust against misspecification and are useful for general relative risk estimation (see e.g., Lawson et al. 2000; Best et al. 2005), there can be considerable aliasing of long-range spatial effects. Of course, covariates that are not directly spatial in form but are aliased with such spatial effects (which can be mimicked by CAR models) may also be affected.

7.2 Biases and Misclassification Error

Besides the considerations discussed above, it remains important to consider aggregate ecological analysis simply to provide a description of aggregate level relations. Often the biases apparent when aggregate inference is to be pursued are much reduced (see e.g., Greenland 1992; Greenland and Robins 1994). When inference at different aggregation levels is to be considered, however, additional problems arise. The "gold standard" for inference in medical studies is often the individual level, in that it is often the aim to be able to infer an outcome from individual level data. This is true for clinical or intervention trials where individual responses are used as the basis of group (aggregate/population) summarization or inference. A distinction should be drawn here between inference to be applied to an individual and inference made from individual data. The former can be attempted from various levels of data aggregation (with varying levels of success), whereas inference from individual level data can be used to make inferences about individuals and also aggregated levels in the population.

On the other hand, with aggregated data (such as county-level disease count data) is it possible to make individual inference? This is a much more difficult undertaking.

7.2.1 Ecological Biases

When inference is made from an aggregated study to a lower level of aggregation then bias can occur. This bias can have a number of component biases. A good example of the extent of such bias is given (for a non-medical example). In the simplest example, assume a linear regression relation with $j = 1, ..., n_i$ individuals in $i = 1, ..., m$ units (areas). Assume first that at the individual level for the j th individual in the i th unit the response model is

$$y_{ij} = a + bx_{ij} + e_{ij}$$
$$= f(a, b, x_{ij}) + e_{ij}$$
$$\text{where } E(e_{ij}) = 0$$

Ecological Analysis

If we aggregate over the n_i individuals then with $y_i = \sum_j y_{ij}$ and $x_i = \sum_j x_{ij}$ and $e_i = \sum_j e_{ij}$. In this case, a linear model might be

$$y_i = a^* + b^* x_i + e_i^*.$$
$$= f^*(a^*, b^*, x_i) + e_i^*.$$

Now the question essentially is, can we make inferences from $a^*, b^*, \{e_i^*\}$ at the disaggregated individual level. In one view, this can be interpreted as an example of the modifiable areal unit problem (see Section 8.1). However, here inference at a lower level of aggregation is the sole focus.

In general, it is important to consider the model

$$E(y_i) = E(f^*(a^*, b^*, x_i)) \qquad (7.2)$$

where expectation is with respect to y. Here it is often assumed naively that $E(y_i) = f(x_i) = a + bx_i$. However if the within-area distribution of x_{ij} is heterogeneous then this will not hold. This also assumes that the individual relationship has the same *form* (as well as parameter values) as the aggregate relationship. This is the naive ecological model of Salway and Wakefield (2005). We are interested in how (a^*, b^*) relates to (a, b), in particular how the slope parameter b^* relates to b. Under a logistic model when the response is binary then b would be the *odds ratio* for the exposure. In general the ecological bias is the difference between the estimated b^*, say $\hat{b}^*$, and the true individual parameter b. If the individuals did not vary with their exposures i.e., $x_{ij} = x_i$ then there is no bias.

Biases can arise from a variety of sources:

1) Bias due to confounding: either variables missing on individuals or group/area level

2) Bias due to effect modification: exposure effect varying between groups/areas

3) Contextual effects: areal/group level variables which are unmeasured (Greenland and Robins, 1994)

4) Measurement error: there may be error in the classification of discrete exposures or measured confounders, as well as error in continuous covariates

Within area variation in exposure/confounders is a major contributor to ecological biases. Other sources of bias are of course not unique to ecological studies (measurement error and unobserved confounding). These will be discussed later.

7.2.1.1 Within-area exposure distribution

If its possible to specify the within-area (group) exposure distribution then it is possible to try to assess the bias of this source. The aggregated model that corresponds with the basic individual level model integrates over the distribution of x (assuming the exposures are independent) i.e.,

$$f^*(a^*, b^*, x_i) = E_x\{E_y(y_{ij})\} = \int f(a, b, x_{ij})p(x)dx \qquad (7.3)$$

where $p(x)$ is the distribution of the exposure. Salway and Wakefield (2005) cite a range of approximations to within-area distributions when there is independence or no spatial correlation in the data. Of course, in spatial applications exposures could easily be correlated and we would instead be interested in aggregation of events from (say) a point process to a count process with small areas. It has been shown that for a log Gaussian Cox process with stationary covariance and log linear model for covariates (in this case assume that the intensity is $exp(x(s)\beta)$ where $x(s)$ is a spatially-referenced covariate value at s), then the spatial moments of the within-area distribution can be computed if the within-area distribution of the covariate is known, i.e.,

$$E^A = \frac{1}{|A|} \int_A exp(x(\mathbf{s})\beta)d\mathbf{s} = \int_{-\infty}^{\infty} exp(x\beta)dF^A(x)$$

$$\text{where } F^A(x) = \frac{1}{|A|} \int_A I(x(\mathbf{s}) \leq x)d\mathbf{s} \qquad (-\infty < x < \infty)$$

the spatial cdf of $x(\mathbf{s})$ (Cressie et al., 2004). This implies that partial knowledge of F^A (e.g., bounds, mean, variance) could be used to characterize the within-area distribution. For a single binary covariate (x_{ij}), and the area average $\bar{x}_i$ ($\bar{x}_i \approx x_i/n_i$) then the expected count for the i th area is

$$E^A = 1 + \bar{x}_i(e^\beta - 1).$$

More complex situations could arise (with, for example, continuous spatial fields). Assume the case of a bivariate continuous covariate with $exp(x_1(s)\beta_1 + x_2(s)\beta_2)$, and separate spatial means and sample variances are available from surveys $(\bar{x}_1, \bar{x}_2, S_1^2, S_2^2)$ with the spatial covariance given by:

$$C_{12}(A) = \frac{1}{|A|} \int_A (x_1(\mathbf{s}) - \bar{x}_1)(x_2(\mathbf{s}) - \bar{x}_2)d\mathbf{s}.$$

The approximation to E^A is then

$$E_0^A = \exp(\bar{\mathbf{x}}\boldsymbol{\beta} + \frac{1}{2}\boldsymbol{\beta}^T\boldsymbol{\Gamma}\boldsymbol{\beta})$$

where $\bar{\mathbf{x}} = \{\bar{x}_1, \bar{x}_2\}, \boldsymbol{\beta} = \{\beta_1 + \beta_2\}$, and

$$\boldsymbol{\Gamma} = \begin{bmatrix} S_1^2 & C_{12}(A) \\ C_{12}(A) & S_2^2 \end{bmatrix}.$$

Ecological Analysis

Hence as long as the means, variances, and covariances are known, then it is possible to improve on $\exp(\overline{\mathbf{x}}\boldsymbol{\beta})$ by addition of covariation information.

Can adjustments be made in these cases? If the within-area distribution is known or can be approximated to a reasonable level then the ecological model can be used directly with these ingredients. For highly skewed distributions then numerical integration may be required (see e. g. Salway and Wakefield, 2005). For spatial dependence then the maximum entropy approach appears to work well.

Another approach to dealing with a range of ecological problems is to try to include individual level data within the model so that the linkage between the aggregated and disaggregated data is modeled.

An example of the use of the spatial approximation was applied to the percentage poverty variable (which is an average of the binary covariate at the individual level) and the total count of abnormalities by county. For this situation, we assumed a Poisson data likelihood for the county anomaly count with expectation $e_i \theta_i$ with $\log \theta_i = \beta_0 + \log(1 + \overline{x}_i(e^{\beta_1} - 1))$ where $\overline{x}_i$ is the percentage of poverty and β_1 is the slope parameter. This model was fitted with prior distributions as follows:

$$\beta_0 \sim N(0, \tau_0)$$
$$\beta_1 \sim Ga(1, 1)$$
$$\tau_0 = 1/\sigma_0^2$$
$$\sigma_0 \sim U(0, 100)$$

with the restriction placed on the distribution of β_1 due to the possible sampling singularity when $\overline{x}_i(e^{\beta_1} - 1) < -1$. This model yielded a DIC of 165.68 which is lower than the spline model already cited. Of course, this does not necessarily imply that this is the best model for these data. The addition of a term for uncorrelated heterogeneity (UH), however, yields a higher DIC: 172.54, and in this case does not improve the model fit.

7.2.1.2 Measurement error (ME)

Clearly another source of considerable error in regression models is the possibility that predictors/covariates/exposures are measured with error. In the case of a discrete covariate this is called misclassification error (Gustafson, 2004). For example, if the data on disease outcome is related to individual income (as in the example above) with income dichotomized into below or above poverty level then we have a binary covariate. If someone was wrongly categorized as poor: (1) when they should be "not poor" (0) then this would be a misclassification. Of course if the outcome is also binary (disease or no disease) then misclassification could occur if diagnosis was prone to false positives or false negatives. A number of methods are available for such discrete error problems within a Bayesian paradigm and they are discussed in detail in Gustafson (2004). For continuous variables it is usual to assume different types of error depending on the form of error appropriate in context. In a simple binomial formulation assume that and individual has distribution given by $y_{ij} \sim bin(1, p_{ij})$ and $logit(p_{ij}) = \alpha_0 + \alpha_1 x_{ij}$. Assume that exposure variable x_{ij} is observed with error. In our example, assume that in this case x_{ij} is the self-reported income for the individual. In a self-report context error may creep in due to various psycho-social (contextual) effects. Under-reporting of income may happen when someone does not want to appear "too well-off," on the other hand someone else may want to brag and exaggerate their income. We might assume this error is additive as a first assumption. Hence, a model for the observed income x_{ij} could be

$$x_{ij} = x_{ij}^T + e_j$$

where x_{ij}^T is the true income and also note that the error (e_j) has a person-specific component. This is regarded as the classical ME specification. Now the relationship with the outcome y_{ij} is via a logit link to the covariate. However we would usually assume that the outcome is related to the true covariate (and not the error corrupted version. Hence we would want a model such as

$$y_{ij} \sim bin(1, p_{ij}) \qquad (7.4)$$
$$logit(p_{ij}) = \alpha_0 + \alpha_1 x_{ij}^T. \qquad (7.5)$$

Now ME could be included in this model in a number of ways. First we could assume a reverse model for error where $x_{ij}^T = x_{ij} + e_j$, which assumes that by adding noise to the observed variable the true value will be obtained. Substituting this into (7.5) we have

$$logit(p_{ij}) = \alpha_0 + \alpha_1(x_{ij} + e_j). \qquad (7.6a)$$

This is known as *Berkson* error (see e.g., Carroll et al., 2006). A suitable distributional assumption for the random effect e_j would be $e_j \sim N(0, \tau_e)$. Two other alternatives can be considered for this error. One is a general random effect model which decouples the random effect from the covariate to yield a simple frailty model:

$$logit(p_{ij}) = \alpha_0 + \alpha_1 x_{ij} + e_j.$$

While this model has less justification than the Berkson model with respect to ME, it does appear to often demonstrate better goodness-of-fit, presumably because there is much noise within the model fit in general between y_{ij} and x_{ij}, compared to the noise in x_{ij} itself. The final option is to jointly model the covariate and the outcome in the sense that both the disease outcome and the observed covariate depend on the unobserved true value of the covariate. For example, if the disease outcome were binary and a binomial likelihood

Ecological Analysis

model was assumed with logit link then, if the observed data are regarded as having classic ME, a reasonable model would be

$$y_{ij} \sim bin(1, p_{ij})$$
$$logit(p_{ij}) = \alpha_0 + \alpha_1 x_{ij}^T,$$
$$x_{ij} \sim N(x_{ij}^T, \tau_x),$$

where τ_x is a variance term. In this case we now have a latent variable x_{ij}^T underlying both likelihoods and this can be regarded as an example of a latent variable or structural equation model (SEM). Hence, a Bayesian SE model (Stern and Jeon, 2004) could be defined once prior distributions for the parameters $\{\alpha_0, \alpha_1, \tau_x\}$ were defined. Further hyperprior distributions could be assumed for parameters in the prior distributions defined.

7.2.1.3 Unobserved confounding and contextual effects

Clearly, one major source of error in any regression study, let alone ecological study is the possibility of unobserved confounding. Confounding variables could be those that create different responses in the outcome and so, if not accounted for, may influence the result. For example, environmental insults (such as air pollution) could affect asthma outcomes. In addition, smoking could affect this outcome. Hence a study which did not look at smoking or other respiratory-challenging lifestyle variables but simply looked at the relation between air pollution and asthma might draw erroneous conclusions. Such confounders could act to elevate the risk of the disease outcome, possibly in tandem with the exposure of interest (air pollution). There are two situations that should be considered.

First of all, direct correlation with the exposure variable may serve to alter the relation observed. For example, the combination of an observed enhanced exposure (e.g., air pollution) and (say) low socioeconomic status (via unobserved average income or an unobserved smoking indicator) could lead to spurious disease elevation due to the combination of effects (one of which is unmeasured). This often happens when, for example, industrial sites are studied and in the vicinity of these sites elevated disease risk is found. However the vicinity is often also a low socioeconomic status area. This of course supports the use of deprivation indices (Diggle and Elliott, 1995) or other indices of risk to make allowance for such effects in environmental epidemiology studies.

Second, it is quite common for unobserved confounders to leave a degree of variation in risk unexplained in the resulting model fit. It may be that a confounder present in an area leads to higher disease risk, but the exposure is low in that area. For instance, areas with high numbers of smokers could yield high asthma mortality but could be far from air pollution sources. If smoking status was not measured these areas would appear as large residuals or outliers. To combat these unobserved confounder problems it has been suggested that random effects should be introduced into the analysis to "soak up" this extra variation (see e.g., Lawson, 1996). In general, this supports the use of generalized linear mixed models (GLMMs) in these analyses, and these are quite commonly applied now.

Third, unobserved confounders could induce spatially-correlated effects in the risk variation (Clayton et al., 1993) and so the extension to spatially correlated (CH) random effects has been recommended. In general, the recommendation would be that both UH and CH effects should be added in any study, to allow for different possible forms of extra variation. It has also been emphasized that forms of ecological bias can be, to a degree, accommodated by the inclusion of CH effects (Clayton et al., 1993). Hence for a Poisson data likelihood model for a small area disease count we would have

$$y_i \sim Pois(e_i \theta_i)$$
$$\log(\theta_i) = \beta_0 + x_i' \beta + u_i + v_i$$

where u_i, v_i are the CH and UH random effects, respectively. While this is now a general panacea, the caution must be given that a) CH and UH terms may not improve overall model fit; b) can lead to highly biased estimates of covariate terms (depending on the prior model assumptions), especially if aliased with the long range spatial variation; and c) ecological within area distributional considerations can lead to better aggregate models (which can fit better than random effect models).

Finally, *contextual effects* (Goldstein and Leyland, 2001; Voss, 2004; Chaix et al., 2006) are considered to be variables that specify the socio-environmental context of an individual are a special case of confounding. Contextual effects are defined as "aspects of the social and economic milieux of an area which engender an area outcome effect." Often these are found at an aggregate level. For instance, an individuals's outcome on a clinical trial might be related to the area they live in. Hence, for example, the county of residence might be a contextual variable for that individual. Another important example would be the ecological inversion example. If a person of poor socioeconomic (*se*) status lives in a high *se* area, that can lead to reduced health risk to that individual. Hence the person's outcome may be pulled toward the area level expected outcome. Hence the *se* status of the area of residence could be an important variable in explaining health outcome at the individual level and may explain ecological inversion. A typical model for an individual binary outcome for the i th individual, y_i, might be modeled via a logit link to a probability such as

$$logit(p_i) = \beta_0 + \beta_1 x_{1i} + \beta_2 x_{cj} \atop i \in j$$

where x_{1i} is the individual *se* status and x_{cj}, $i \in j$ is the *se* status of the jth small area to which the i th person belongs. Of course a range of such effects could be envisaged where hierarchies of regional or other clustering effects could be added to an individual level model.

7.3 Putative Hazard Models

In this section, I focus on the analysis of a specific application area: the modeling of disease risk around a known location or locations. This focus is a particular example of a regression application which can have ecological elements. Some of the discussion will focus on case event level modeling, which is not at an aggregated level. However, many of the issues discussed above are relevant to aspects of this modeling and so for completeness it is included here.

In putative source analysis, the location(s) of potential (putative) source(s) of health hazard (pollution or other insult) are known, and it is the task of the analysis to attempt to find out if the source or sources affect health risk in their vicinity. Hence the term *putative* is used to mean 'suspected' in this case. Many examples come from environmental epidemiology where a location is the focus of the risk assessment (Lawson, 2002). Putative sources are related to exposure pathways. For example, if the air pollution is thought to be important then locations of sources of air pollutants would be the focus (e.g., incinerators, chimneys, road networks). If the exposure pathway were water ingestion then the focus might be water sources or supply networks (e.g., groundwater wells, rivers). Usually the risk is assumed to be related to location of residence of the population. This is termed residential exposure or risk, and measures of the relation between the source location and residence is used in the analysis.

Analyses will be formulated depending on whether a disease is of interest or whether a source is of potential interest. For instance, if we are interested in acute asthma risk, then we might monitor emergency room admissions for asthma in the vicinity of an air pollution putative source (Anto and Sunyer, 1990). On the other hand, if a public report of a general (non-specific) fear of an elevation of disease risk in the vicinity of a putative source is made, then focus may be on the source characteristics and diseases that may be affected. Hence multiple diseases may finally be analyzed in this case. For example, the Sellarfield nuclear reprocessing plant in NW England, United Kingdom, was the focus of studies in late 1980s. This lead to a variety of radiation-related disease studies (mainly for radiation-related outcomes e.g., childhood leukemia) (see e.g., Gardner, 1989). However, risk from such a site may be from a variety of sources (water pollution, air pollution, occupational radiation risk etc.) besides simply residential air pollution exposure. Hence it is not always clear what the main effects are that should be modeled. In a study in NW England of larynx cancer around a putative source (incinerator), evidence for residential exposure was assessed via a distance covariate measured from residential address of death certificate and putative source (Diggle, 1990; Diggle and Rowlingson, 1994). In Figure 7.6, the left panel displays the residential locations of incident cases of larynx cancer for the period 1974–1983.

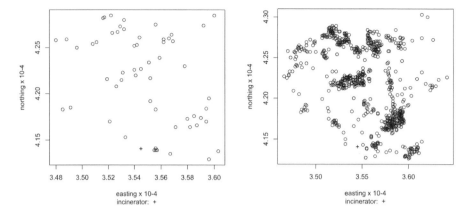

FIGURE 7.6
Larynx cancer and lung cancer incident cases in the vicinity of an incinerator in NW England for the period 1974–1983. The incinerator is marked with "+."

The right panel displays the distribution of lung cancer cases for the same period. At location (35450, 41400) is an incinerator which is the putative focus in this case. It could be considered that an incinerator could elevate disease risk around it and respiratory disease could be a target. The evidence for the effect of the incinerator could be manifold. The primary effect might be elevated incidence near the site of the putative source. Hence one might be tempted to consider a distance decline effect around the source. This would be a primary form of evidence for a linkage. Of course, confounding due to correlation between deprivation and distance would need to be considered if such information were available. For a variety of source type and exposure pathways distance decline is a fundamental piece of evidence. A secondary form of evidence is the directional in nature. With air pollution as the primary putative exposure the effect of wind direction and strength should be considered. There are many examples where directional effects can be important. Often within putative health studies a retrospective analysis of incident cases or mortality events is carried out. This is often needed as the existence of a putative source is often only noted after some exposure period has taken place.

7.3.1 Case Event Data

Diggle (1990) used larynx cancer case residential addresses as the outcome of interest in a post hoc study of that disease around a putative source (incinerator). The study is "post hoc" as elevated incidence of larynx cancer was registered as a concern by the local residents in the vicinity of the incinerator.

Ecological Analysis

This concern motivated the study. The impact of the "post hoc" nature of the study is largely a design issue and is discussed more fully in Lawson (2006b, Ch. 7). The original data used in that study is shown in Figure 7.6. The cases of larynx cancer (58) within a rectangular study window for the period of 1974–1983 are shown in the left panel. As part of the study, case residential addresses of respiratory cancer for the same study period were collected. These were to be used as a type of *control* disease that could allow for the spatial distribution of the background "at risk" population. This essentially acts as a geographical control at a fine resolution level. Any areas where there are lots of *at risk* people are more likely to yield cases and so we must adjust for this effect. The right hand panel of the Figure 7.6 displays the map of these 978 control cases. Some discussion has focused on whether respiratory cancer is a valid control disease for larynx cancer in a putative air pollution study. Here we assume that the control is valid, but in general the issue of choice of control disease is important in any particular application.

Assume we observe within a study region (W), a set of m cases, with residential addresses given as $\{s_i\}$, $i=1,...,m$. Here the random variable is the *spatial location*, and so we must employ models that can describe the distribution of locations. Often the natural likelihood model for such data is a heterogeneous Poisson Process (PP). In this model, the distribution of the cases (points) is governed by a first order intensity function. This function, $\lambda(s)$ say, describes the variation across space of the intensity (density) of cases. This function is the basis for modeling the spatial distribution of cases. Denote this model as

$$\mathbf{s} \sim \mathbf{PP}(\boldsymbol{\lambda}(\mathbf{s})).$$

The unconditional likelihood associated with this model is given, bar a constant, by:

$$L = \prod_{i=1}^{m} \lambda(s_i) \exp\{-\int_W \boldsymbol{\lambda}(\mathbf{u}) d\mathbf{u}\}$$

where $\lambda(s_i)$ is the first order intensity evaluated at the sample locations $\{s_i\}$. This likelihood involves an integral of $\boldsymbol{\lambda}(\mathbf{u})$ over the study region.

The definition of the intensity of cases must make allowance for the effect of the background at risk population. Often the intensity is specified with a multiplicative link between these components:

$$\lambda(s) = \lambda_0(s)\lambda_1(s|\boldsymbol{\theta})$$

Here the *at risk* background is represented by $\lambda_0(s)$ while the modeled excess risk of the disease is defined to be $\lambda_1(s|\boldsymbol{\theta})$, where $\boldsymbol{\theta}$ is a vector of parameters. In putative source modeling we usually specify a parametric form for $\lambda_1(s|\boldsymbol{\theta})$ and treat $\lambda_0(s)$ as a nuisance effect that must be included. Usually some external data is used to estimate $\lambda_0(s)$ nonparametrically (leading to profile likelihood). In the larynx cancer example, the respiratory cancer distribution would be used to estimate $\lambda_0(s)$.

It is possible to reformulate this problem by viewing the joint realization of cases and controls and, conditional on that realization, examining the probability that the binary label on a point is either (1: case) or (0: control). If this approach is taken the background nuisance function disappears from the problem (Diggle and Rowlingson, 1994). This depends implicitly on a control disease being available and relevant (i.e., matched well) to the problem.

Assume the problem can be reformulated as a binary logistic regression where $\lambda_0(s)$ drops out of the likelihood. Denote the control disease locations as $\{s_j\}, j = m+1, ..., m+n$, and with $N = n+m$, a binary indicator function can be defined:

$$y_i = \begin{cases} 1 \text{ if } i \in 1, .., m \\ 0 \text{ otherwise} \end{cases}$$

$$\forall i, \ i = 1, ..., N$$

and the resulting likelihood is just given by

$$L(\mathbf{s}|\boldsymbol{\theta}) = \prod_{i=1}^{N} \frac{[\lambda_1(s_i)]^{y_i}}{1 + \lambda_1(s_i)}.$$

By conditioning on the joint set of cases and controls the resulting logistic likelihood does not require the evaluation a spatial integral nor the estimation of a background population function. The definition of the form of $\lambda_1(s_i)$ will be important in inference concerning putative sources of hazard.

Parametric Forms

Often we can define a suitable model for excess risk within $\lambda_1(s)$. In the case where we want to relate the excess risk to a known location (e.g., a putative source of pollution) then a distance-based definition might be considered, first of all. For example,

$$\lambda_1(s) = \rho \exp\{\mathbf{F}(s)\boldsymbol{\alpha} + \gamma d_s\} \tag{7.7}$$

where ρ is an overall rate parameter, d_s is a distance measured from s to a fixed location (source) and γ is a regression parameter, $\mathbf{F}(s)$ is a design vector with columns representing spatially-varying covariates, and $\boldsymbol{\alpha}$ is a parameter vector. The variables in $\mathbf{F}(s)$ could be site-specific or could be measures on the individual (age, gender, etc.). In addition this definition could be extended to include other effects. For example, we could have

$$\lambda_1(s) = \rho \exp\{\mathbf{F}(s)\boldsymbol{\alpha} + \eta v(s) + \gamma d_s\} \tag{7.8}$$

where $v(s)$ is a spatial process, and η is a parameter. This process can be regarded as a random component and can include within its specification spatial correlation between sites. One common assumption concerning $v(s)$ is that it is a random field defined to be a spatial Gaussian process.

An example of the kind of specification typical in a putative source example would involve a range of variables or functions of variables thought to be indicative of risk association with the source. The variables included depend on the context. In retrospective studies where no information or direct measures of emission patterns are available then resource must be made to exposure surrogates (i.e., variables that may show a retrospective linkage with the source). Distance from source is a prime example of a variable that might yield such information. Direction from source to residence may also be indicative of wind-related effects (particularly in air pollution studies). For prospective studies, direct measures of pollutant outfall (such as soil-sampled or air-sampled chemical or particulate concentrations) could be monitored over time. Without these direct measurements, surrogates would be required and often these would have to represent historical time-averaged effects in retrospective studies.

What form would a relevant exposure model take? The definition for $\lambda_1(s|\boldsymbol{\theta})$ often assumed is as follows (see Diggle, 1990; Diggle and Rowlingson, 1994; Lawson, 1995; Diggle et al., 2000; Wakefield and Morris, 2001, Lawson, 2006b for variants):

$$\lambda_1(s_i|\boldsymbol{\theta}) = \exp\{A_{1i}\}.\exp\{A_{2i}\}.A_{3i} \qquad (7.9)$$
$$A_{1i} = x_i'\boldsymbol{\beta} + z_i'\boldsymbol{\gamma}$$
$$A_{2i} = \rho_1\cos(\phi_i) + \rho_2\sin(\phi_i)$$
$$A_{3i} = [1 + \alpha_0 e^{-\alpha_1 d_i}]$$

$\boldsymbol{\theta} = \{\boldsymbol{\beta},\boldsymbol{\gamma},\boldsymbol{\alpha},\boldsymbol{\rho}\}, \boldsymbol{\alpha} = \{\alpha_0,\alpha_1\}, \boldsymbol{\rho} = \{\rho_1,\rho_2\}$. Here the distance variable is defined as $d_i = ||s_i,c||$ where c is the putative source location and the angle to the source is defined as ϕ_i. A generalization allows there to be multiple sources and we can include these in one model by adding further distance or direction variables with parameters. This is not pursued here. The rationale for each of the terms (A_1,A_2,A_3) is as follows. All terms are exponentiated to ensure positivity, although term A_{3i} has a link parameter (α_0) which requires a constraint so that $\alpha_0 e^{-\alpha_1 d_i} < 1$. The term A_{1i} consists of covariates and random effects. The row vector of covariates (x_i') can consist of personal covariates (although within a logistic likelihood model these would have to be available for the control as well as case disease. The corresponding regression parameters are the vector $\boldsymbol{\beta}$. The covariates can include functions of cartesian coordinates for trend estimation and these are available for all locations. The individual level random effects can be included via the row vector z_i' with the corresponding unit vector $\boldsymbol{\gamma}$. These effects could include individual frailty terms (with, for example, zero mean Gaussian prior distributions) or correlated effects where the prior distribution includes some form of spatial correlation. The general specification above in (7.7) and (7.8) demonstrates a variant of this specification. The term A_{2i} specifies the directional dependence in the outcome. By including functions of the trigonometric functions (cos, sin) it is possible to recover the mean angle of the exposure. In this case only linear functions are assumed.

More complex variants are possible (see e.g., Lawson, 1993b) that allow for angular-distance correlation or peaked distance effects. An alternative specification for the distance effect could be $A_{2i} = \rho_1 \cos(\phi_i - \mu_0)$. Here the ρ_1 plays the role of an angular concentration parameter and the angle is measured relative to an overall mean (μ_0). If a predominant time-averaged wind direction is found to affect a source then the estimation of μ_0 might be important in determining a link to a source. Finally, the term A_{3i} defines the distance effect. The rationale for the hybrid-additive form, $[1+\alpha_0 e^{-\alpha_1 d_i}]$, is the idea that risk at distance from the source should not affect the background disease risk. If a multiplicative model were assumed (such as $A_{3i} = e^{-\alpha_1 d_i}$) this would lead to a reduction in risk at great distances which is not appropriate. It should be mentioned however that often it is much more difficult to estimate α_0, α_1 under the hybrid-additive model as the parameters are not well identified and constraints must be placed on $\alpha_0 e^{-\alpha_1 d_i}$ (see also Ma. et al., 2007). More details of possible model variants are given in Lawson (2006b, Ch. 7). Step function forms have been proposed by Diggle et al. (1997), but the underlying rationale for these, that there could be a zone of constant risk around a source, is not bourne out by dispersal models or empirical studies of source dispersion. On the other hand, peak-decline models are supported by time-averaged dispersal models (see e.g., Arya, 1998). A simple example of this general approach is given in Wakefield and Morris (2001), albeit for an aggregated small area application. In that work the $A_{2i} = 0$ with no assumed directional effects, and $A_{1i} = \beta_0 + \beta_1 x_{1i} + u_i + v_i$ a single deprivation index covariate (x_{1i}) and two random effects (one correlated u_i and one uncorrelated v_i). The third term is defined as $A_{3i} = [1 + \alpha_0 e^{-(d_i/\alpha_1)^2}]$ which gives a Gaussian distance effect rather than exponential. However the comments above also apply to this model form.

A general specification for the logistic example applied to the larynx cancer data has been specified in 6.2.1.2. In that section the focus was on cluster detection. However the underlying model used there is also relevant here. The model assumed for the case probability was

$$p_i = \frac{\lambda(\mathbf{s}_i|\boldsymbol{\theta})}{1+\lambda(\mathbf{s}_i|\boldsymbol{\theta})}$$
$$\lambda(\mathbf{s}_i|\boldsymbol{\theta}) = \exp\{\beta_0 + v_i\}.\{1 + \exp(-\alpha_1 d_i)\}$$

where d_i is distance from the incinerator, β_0 was an intercept term, and $v_i \sim N(0, \tau_v)$ an uncorrelated random effect, and zero mean Gaussian prior distributions for the β_0 and α_1 parameters. There is no directional term. In this case this is justified given the choice of study area: a rectangle with the putative source close to one region boundary. This means that much of the directional data are censored (outside the boundary of the region). Hence there is limited use in including a directional model here. A correlated random effect could also be included within this model, though this is not reported here. For case-control data this is possible either by assuming a full multivariate Gaussian process prior distribution for the correlation (with covariance specified as a function of inter-point distances). It is also possible to specify neighborhoods via the construction of a Dirichlet tesselation of the complete realization and the derivation of tile neighbors. Care must be taken in this latter case to avoid edge effects, although these should not be great for the definition of neighborhoods (rather than distances). In our example the posterior expected estimates of β_0 and α_1 (with *sds* in brackets) were -6.35 (0.831) and 0.695 (3.054). Hence, in this example the overall rate was well estimated whereas the distance effect is not. The posterior expected estimate of the precision of the uncorrelated random effect was 0.1229 (0.07413) in the model where $v_i \sim N(0, \tau_v)$ and $\tau_v = 1/a^2$ where $a \sim U(0, 100)$ following the suggestion of Gelman (2006).

Some comments concerning analyses of putative source data should be made in light of the general discussion above concerning ecological bias, confounding, contextual effects and measurement error. While some of these comments are most appropriate to the aggregated data situation, we discuss many issues here which are common to both.

First of all, it is important to critique the model components included above. Should correlated random effects be included? Would they absorb the effects of unobserved confounders? In general it may be important to include both uncorrelated and correlated effects, from the standpoint that confounders could induce noise effects of both kinds. However it should be borne in mind that confounders correlated with the distance or directional effects are not likely to be removed by random effect inclusion. The inclusion of variables in the analysis that inform about context could also be important. For example deprivation indices available at a level aggregated above residence (such as at census tract or zip code) could help to inform about regional excess risk. Of course deprivation might be correlated with distance or direction. In Wakefield and Morris (2001) this was certainly true. In the same work, less smoothers of the distance effect suggests that an irregular decline occurs and it may be more appropriate to consider spline models for the distance and or direction. In fact a 2-D spline model for the distance and directional effect could be a useful inferential tool. Of course splines may not yield unequivocal evidence for a risk gradient. The possibility that ecological bias exists in aggregate data will be discussed in the next section. The possibility that measurement error (ME) exists in outcome or covariates can also be important. For example misdiagnosis could occur where a control could in fact be a case or vice-versa. This would be more likely if the two diseases were linked by progression. For example, early stage breast cancer could be used as a control for late stage breast cancer. Clearly the staging could be subject to misclassification. ME could exist in any covariates whether its the location of an address or the socioeconomic status of an individual or the deprivation status of a region. One solution for covariates is to either assume Berkson error and a model such as

$$\beta_1(x_{1i} + \varepsilon_i)$$
$$\varepsilon_i \sim N(0, \tau_\varepsilon)$$

where x_{1i} is a covariate or to utilize the SEM approach and to specify a joint model for the covariate and the outcome.

7.3.2 Aggregated Count Data

It is often relevant or feasible to consider the analysis of count data within aggregated spatial units (small areas). These units will usually be arbitrary political administrative units (e.g., census tracts, zip codes, counties, municipalities, postal zones etc.). The definition of these units should have little or no impact on the health outcome observed.

Assume we observe counts $\{y_i\}$, $i = 1, ..., m$ in m small areas and, we also observe expected rates $\{e_i\}$, $i = 1, ..., m$. While we usually assume the expected rates to be fixed for our purposes, it could be useful to consider them to be random quantities also (see e.g., Best and Wakefield, 1999). Here we mainly focus on fixed expected rates.

A typical model at the data level is often

$$y_i \sim Pois(\mu_i)$$
$$\mu_i = e_i \theta_i$$

and the focus is on the modeling of the relative risks $\{\theta_i\}$. Usually the log relative risk is the focus and we often formulate a model akin to that in (7.9) where the i th small area is 'located' at its centroid. Of course this assumes an average effect over the small area rather than direct modeling of the risk aggregated from the point process model. Direct aggregation from a Poisson process would give $y_i \sim Pois(\int_{a_i} \lambda(\mathbf{u}|\boldsymbol{\theta})d\mathbf{u})$, where a_i is the physical extent of the i th small area. Now if both $\lambda_0(s)$ and $\lambda_1(s|\boldsymbol{\theta})$ were constant over the area (a strong assumption) then this would result in $\lambda_{0i}.\lambda_{1i}.|a_i|$ (where here $|.|$ denotes 'area of') which is almost the same as $e_i \theta_i$ bar the area effect. The expected rate is usually standardized over the population rather than area. However if you make the (strong) assumption that the population is uniform of course then if the e_i is specified for the local population then the assumption is that $\lambda_{0i}|a_i| \approx e_i^* n_i = e_i$ where e_i^* is the externally standardized unit population rate. This *decoupling approximation* as it's called is often made as the starting point of an analysis. It is not usually unreasonable when non-spatial region-specific covariates are included but it can be important when spatially-dependent covariates (such as interpolated pollution measures) are involved.

Making the simple assumption of $\mu_i = e_i \theta_i$, then we can specify the general model as

$$\theta_i = \exp\{A_{1i} + A_{2i} + \log(A_{3i})\}$$
$$A_{1i} = x_i'\beta + z_i'\gamma$$
$$A_{2i} = \rho_1 \cos(\phi_i) + \rho_2 \sin(\phi_i)$$
$$A_{3i} = [1 + \alpha_0 e^{-\alpha_1 d_i}]$$

where the i th small area is located at the centroid or other suitable associated point, d_i is the distance from the centroid to the source location, ϕ_i is the angle from the centroid to the source location. Often it is assumed that $A_{1i} = \beta_0 + u_i + v_i$ where the typical convolution model with a CAR prior distribution of Chapter 5 is assumed:

$$u_i | u_{-i} \sim N(\bar{u}_{\delta_i}, \tau_u/n_{\delta_i})$$
$$v_i \sim N(0, \tau_v)$$

With small area data and neighborhoods defined then a CAR is a convenient and reasonable assumption. An alternative specification could be of the form of a full multivariate Gaussian with a covariance matrix, thus

$$\mathbf{u} \sim \mathbf{N}(\mathbf{0}, \sigma^2 \mathbf{\Gamma})$$

where the i, j th element of the covariance matrix is $\gamma_{i,j} = \exp(-d_{ij}/\phi)$. This has the advantage of directly modeling distance effects, has a distance-dependent covariance and also has a zero mean vector, and so models a stationary process.

The CAR specification is not stationary and can suffer from aliasing with long range spatial effects (see e.g., Ma. et al., 2007). One option is to use a proper CAR with trend specification. Unfortunately, the full MVN specification requires inversion of an $m \times m$ covariance matrix whenever new parameters are evaluated, e.g., within a posterior sampling algorithm. This could be a major computational disadvantage.

In the example below, respiratory cancer incidence for the year 1988 in the counties of Ohio was examined. The U.S. Department of Energy Fernald Materials Processing Center is located in southwest Ohio (Hamilton County). The Fernald facility recycles depleted uranium fuel from U.S. Department of Energy and Department of Defense nuclear facilities. The facility is located 25 miles northwest of Cincinatti. The recycling process can create a large amount of uranium dust which is radioactive. The period of greatest emission activity was between 1951 and the early 1960s and during that period some dust may have been accidentally released into air. Respiratory cancer is of interest in relation to a potential environmental health hazard. Exposure to radioactive contaminated air in the vicinity of the facility could, over a period of years, lead to increased risk for a variety of diseases. Exposure risk can be considered to be increased if residence were proximal to the facility during the highest activity years or in subsequent decades.

One disease of concern to evaluate would be respiratory cancer as it is the most prevalent form of cancer potentially associated with this exposure. An exposure pathway via inhalation would be considered. Available is data for Ohio counties for the period 1988 for counts of respiratory cancer. This period is sufficiently lagged from the peak emission time that the cancer lag time (20–25 years) should have passed. The expected rates used for standardization here are the Ohio state standardized for age, gender breakdowns of each county. Two covariates at the county level are also available. The first covariate is the percentage of poverty for each county from the 1990 census. The 1990 census is used as it is the nearest to the year in question and the level should remain reasonably stable over two years. This covariate would be useful in allowing for deprivation effects that could confound the respiratory cancer outcome. This may include general health outcomes but also behavioral effects such as smoking or use of alcohol in lifestyle. The second covariate is the simplest exposure surrogate variable: distance from the site. This distance was computed to the centroids of the counties.

A sequence of models was fitted to these data with different assumptions. First of all a basic model with a convolution prior distribution for spatial effects, and measurement error for both covariates was considered with the form:

$$y_i \sim Pois(e_i.\theta_i)$$
$$\theta_i = \exp\{\alpha_0 + \alpha_1(x_{1i} + \epsilon_{1i}) + \alpha_2 \log(f_i) + u_i + v_i\}$$
$$f_i = 1 + \exp\{-\alpha_3(d_i + \epsilon_{2i})\}.$$

The random effects $\epsilon_{1i}, \epsilon_{2i}$ have zero mean Gaussian prior distributions with standard deviations with uniform distributions on the range (0,10) (Gelman, 2006). These represent Berkson error in the covariates. We also considered different prior distributional assumptions for the random effects in the convolution component. The first option (Model 1 in Table 7.2) was with fixed but very small precisions (0.0001) for the uncorrelated random effects $(v_i, \epsilon_{1i}, \epsilon_{2i})$ and second with variance hyperprior distributions $(\tau_* = 1/\sigma_*^2; \sigma \sim U(0,10))$.

Also considered was a variant of the CH effect: a proper CAR model (Section 5) with $c_{ij} = \frac{1}{n_{\delta_i}}$ if $i \sim j$ and $c_{ij} = 0$ if $i \nsim j$. This model with no measurement error yielded $\widehat{\gamma} = 0.729(0.142)$, $\widehat{\tau}_{pc} = 8902.0(26010.0)$ with a DIC of 557.03 for model with fixed precision on the UH effect (0.0001). Overall, different precision specifications seem to affect the model fits considerably in that Model 2 is much superior to Model 1. The inclusion of ME appears also to be important as is the distance effect, even when it is not well estimated (Model 4). Interestingly, and as a caution, model 1 yields a significant distance effect $(\widehat{\alpha_1}, \widehat{\alpha_2})$ and supports the idea of a possible source effect. However out of the models fitted, the lowest DIC is for the model with a CAR component

TABLE 7.2
Results for a variety of models fitted to 1988 respiratory cancer incident counts for counties of Ohio

Model	DIC	α_0	α_1	α_2	α_3
1 (fixed precisions)	656.23	-0.506(0.275)	0.111(0.002)	-2.043(0.065)	2.296(0.063)
2 (variance H-priors)	520.49	-0.362(0.108)	0.030(0.009)	-0.242(0.279)	4.993(5.252)
3 (no ME)	616.75	-0.935(0.350)	0.034(0.006)	0.526 (0.400)	-0.156(0.180)
4 (no distance)	619.35	-0.490(0.074)	0.034(0.006)	-	-
5 Proper CAR	557.03	-0.817(0.059)	0.061(0.006)	0.024(0.004)	-6.204(1.072)

and precision hyper-prior distributions, with measurement error, where the distance effect is not significant.

Other analyses of these data, especially in the more general space–time context, are found in Zia et al. (1997), Waller et al. (1997), Carlin and Louis (2000), Knorr-Held and Besag (1998), Knorr-Held (2000). Measurement error was considered by Zia et al. (1997) in a space–time context.

7.3.3 Spatiotemporal Effects

When data are observed with a time label then it is possible to extend modeling by considering spatiotemporal effects. Perhaps the most convenient way to do this is to consider a breakdown of effects between *main* effects of space and time separately and the interaction between space and time.

In Section 11 a more general review of disease mapping models is made. Here we briefly consider how space–time data can be modeled with putative sources of hazard as the main focus. The extension of methods for spatial applications to where we have data observed in space and time is immediate.

7.3.3.1 Case event data

Assume we observe within a study region (W) and a time period (T), a set of m cases, with residential addresses given as $\{s_i\}$, $i = 1,...,m$, and also time labels $\{t_i\}$, $i = 1,...,m$. Here the random variables are the *spatial location and the time of occurrence*, and so we must employ models that can describe the distribution of locations and times. A recent review of a wide range of approaches to space–time point process data appears in Diggle (2007). Time here could be a diagnosis date, date of death or cure. The heterogeneous Poisson process (hPP) model assumed for spatial data can be extended to space–time readily. In this model, the distribution of the cases (points and times) is governed by a first order intensity function. This function, $\lambda(s,t)$, describes the variation across space and time of the intensity of cases. This function is the basis for modeling the spatiotemporal distribution of cases. Denote this model as

$$\mathbf{PP}(\lambda(s,t)).$$

As in spatial applications, the unconditional likelihood associated with this model is given, bar a constant, by:

$$L = \prod_{i=1}^{m} \lambda(s_i, t_i) \exp\{-\int_W \int_T \lambda(\mathbf{u},\mathbf{v}) d\mathbf{u} d\mathbf{v}\}$$

where $\lambda(s_i, t_i)$ is the first order intensity evaluated at the sample locations $\{s_i, t_i\}$. This likelihood involves an integral of $\lambda(\mathbf{u}, \mathbf{v})$ over the study region and time period.

The definition of the intensity of cases must make allowance for the effect of the background at risk population, which in this case will be time-varying.

Often the intensity is specified with a multiplicative link between these components:
$$\lambda(s,t) = \lambda_0(s,t)\lambda_1(s,t|\boldsymbol{\theta}).$$

Here the *at risk* background is represented by $\lambda_0(s,t)$ while the modeled excess risk of the disease is defined to be $\lambda_1(s,t|\boldsymbol{\theta})$, where $\boldsymbol{\theta}$ is a vector of parameters. As before a likelihood model can be derived from this likelihood and Bayesian methods could be based on this form. An example of using this form in cluster detection was given by Clark and Lawson (2002). The disadvantage of using this form is the need to integrate the intensity over space and time. This can be avoided if a control disease were available within the study region over the same time period. Once again the conditional logistic model could be derived. Assume that a control disease is governed by intensity $\lambda_0(s,t)$. The joint realization of case and control diseases for a Poisson process with intensity $\lambda_0(s,t)[1 + \lambda_1(s,t|\boldsymbol{\theta})]$. Then by conditioning on the joint realization, the binary labeling of the points will be governed by the case probability
$$p_i = \frac{\lambda_1(s_i,t_i|\boldsymbol{\theta})}{[1 + \lambda_1(s_i,t_i|\boldsymbol{\theta})]}$$

Then given the set of locations, the point labels (y_i) can be considered at the data level to be independently distributed with a binomial distribution:
$$y_i \sim bin(1, p_i).$$

This is, again, just a logistic model for the binary outcome, where the probability is a function of space and time. Interest will focus on the definition of the excess or relative risk function $\lambda_1(s_i,t_i|\boldsymbol{\theta})$. The specification of $\lambda_1(s_i,t_i|\boldsymbol{\theta})$ will depend on the context and will be important to include covariates, ideally time-varying, as well as random effects. Variates that pertain to evidence for a link to a putative source of hazard could be various. First a general formulation could be as follows:
$$\lambda_1(s_i,t_i|\boldsymbol{\theta}) = \exp\{A_{1i} + A_{2i} + A_{3i}\}$$
$$A_{1i} = \mathbf{x}_i'\boldsymbol{\beta} + \mathbf{z}_i'(t)\boldsymbol{\gamma}$$
$$A_{2i} = f(d_i, \phi_i, t_i)$$
$$A_{3i} = \Sigma_i + \xi_i + \psi_i.$$

Within A_{1i} are terms depending on fixed constant covariates $(\mathbf{x}_i')$ and their parameters $\boldsymbol{\beta}$, and also terms depending on time-varying covariates $\mathbf{z}_i'(t)$ and their parameters $\boldsymbol{\gamma}$. Time varying covariates could be very important in these studies. For example in a prospective study, if pollutant concentration were available at different times then these would be time-varying. For term A_{2i} functions of distance to source (d_i) and angle to source could be important (as in the spatial case). However, because a source may vary its output over time, the resulting spatial risk field would vary over time, and so time varying effects should be included in A_{2i}. The final term includes random effects that can allow for spatial (χ_i), temporal (ξ_i), and spatiotemporal interaction (ψ_i). Note that in all analyses, covariates would have to be available for all case and control locations. Special methods would have to be developed when this was not the case.

An example of a possible model for a time-vary emission source (air pollutant) could be, for data given with polar coordinates (ϕ_i, d_i):

$$\lambda_1(s_i, t_i | \boldsymbol{\theta}) = \exp\{\boldsymbol{\beta}_0 + \beta_1 x_{1i} + \beta_2 x_{2i} + f_i(t) + \chi_i\}$$

where x_{1i} is the age of the person, x_{2i} is the socioeconomic status of the person, and

$$f_i(t_i) = \log(1 + \alpha_0(t_i) \exp\{-\alpha_1(t_i) d_i\}) + \kappa(t_i) \cos(\phi_i - \mu(t_i))$$

$$\alpha_0(t_i) \sim Gamma(c_0 \mu(t_i), c_0)$$

$$\alpha_1(t_i) \sim N(\alpha_1(t_{i-1})/\Delta(t_i, t_{i-1}), \tau_{\alpha_1})$$

where $\mu(t_i)$ could be defined as a time-varying risk function for example, and $\alpha_1(t_i)$ is a form of Gaussian process and $\Delta(t_i, t_{i-1})$ is the time difference between the i th and the previous case/control. Note also that the directional component with precision $\kappa(t_i)$, and mean angle $\mu(t_i)$ will also in general vary with time. Essentially the time-averaging that is assumed for a static spatial model must be dropped here in favour of a parsimonious dynamic model. While of course a convolution of Gaussian distributions could be employed for a directional component around a source (see e.g., Esman and Marsh, 1996; Arya, 1998), this is not parsimonious compared to a Von Mises-type formulation such as $exp\{\kappa(t_i) \cos(\phi_i - \mu(t_i))\}$. The final component of the risk function could consist of spatial and temporal random effects and interaction effects. Care should be taken in the choice of such effects as the ability to detect exposure effects may depend on the specification of the random components. First we could consider a separate spatial component, such as χ_i where spatial dependence (fixed in time) could be specified either via a Gaussian process specification with a distance-based spatial covariance (i.e., $\boldsymbol{\chi} \sim \mathbf{MVN}(\mathbf{0}, \boldsymbol{\Gamma})$) where $\Gamma_{ij} = \sigma^2 \exp\{-\alpha d_{ij}\}$. Further CAR alternatives could be considered if a suitable neighborhood structure were assumed. Second temporal effects could be assumed whereby a conditional autoregressive Gaussian dependence is defined on the time lag between events:

$$\xi_i \sim N(f(\xi_{i-1}), \tau_\xi)$$
$$f(\xi_{i-1}) = \alpha \xi_{i-1} / \Delta(t_i, t_{i-1}).$$

Finally, a space-time interaction could be assumed. Various specifications could be imagined for this ranging from nonseparable dependence structures (see e.g., Knorr-Held, 2000; Gneiting et al., 2007) to independent effects. The simplest and most parsimonious form might be

$$\psi_i \sim N(0, \tau_\psi).$$

Ecological Analysis

This would at least reassure that aliasing between covariate effects varying over time would be minimized.

Finally it should be noted that often the binary outcome in space–time is an 'end-point' event, say, for example, in an infectious disease situation where infection spreads within a finite population. In that case, special survival-based methods can be used to examine the progression of the disease (Lawson and Leimich, 2000; Lawson and Zhou, 2005). Examples of space–time analysis around sources of pollution are few and this area is one that could be much further developed.

7.3.3.2 Count data

In the situation where small area counts are recorded within fixed time periods in a sequence, the modeling approach is a relatively straightforward extension of the spatial case. Define the counts of disease within $i = 1, ..., m$ spatial small areas and $j = 1, ..., J$ disjoint and adjacent time periods as $\{y_{ij}\}$. The corresponding expected rates with these space–time units are $\{e_{ij}\}$. Also assume relative risk parameters for each unit: $\{\theta_{ij}\}$. The basic data model is often again Poisson with

$$y_{ij} \sim Pois(e_{ij}.\theta_{ij}).$$

Inference focuses on terms within the specification of θ_{ij}. Also assume that the distance and direction (angle) from a source is known and can be computed as (d_i, ϕ_i). Assume a log linear form:

$$\theta_{ij} = \exp\{A_{1i} + A_{2j} + A_{3ij}\}.$$

Here the terms have an explicit spatial (i), temporal (j), and interaction (ij) label. An example of a typical specification could be:

$$A_{1i} = f(\mathbf{x}'_i \boldsymbol{\beta}) + u_i + v_i \tag{7.10}$$
$$A_{2j} = \xi_j + g(\alpha_j d_i) + \kappa_j \cos(\phi_i - \mu_{\phi j}) \tag{7.11}$$
$$A_{3ij} = \psi_{ij}.$$

Here, the first term includes fixed covariates within small areas (including distance and direction), and so $f(\mathbf{x}'_i \boldsymbol{\beta})$ could include functions of fixed areal covariates (poverty, SEs, distance, direction) whereas u_i, v_i could be the usual CH and UH random effects (see e.g., Heisterkamp et al., 2000 for an early example). The second term has the temporally dependent components. The random effect (ξ_j) often has an autoregressive dependence. The term $g(\alpha_j d_i)$ would be a function of the time dependent parameter α_j which relates to distance. Again an autoregressive dependence could be assumed for this. The directional parameters κ_j and $\mu_{\phi j}$ also can have dependence on previous times. Time variation of output from sources can possibly be modeled in this way.

Finally the interaction can have prior independence or can have prior non-separable structure (Knorr-Held, 2000).

In the example that follows I have applied a quite general model to the variation over 10 years (1979–1988) of respiratory cancer in Ohio. I have assumed a general model of the form:

$$y_{ij} \sim Pois(e_{ij}.\theta_{ij})$$
$$\theta_{ij} = \exp\{A_{1i} + A_{2j} + A_{3ij}\}.$$

Here, we assume no directional effect as the spatial scale of the county level data is quite large and it is unlikely that a directional effect could be manifest at this scale. We also assume that a distance effect could still remain even via occupational exposure and so we model this here:

$$A_{1i} = \alpha_0 + \alpha_1 \log[1 + \alpha_2 \exp\{-\alpha_3 d_i\}] + u_i + v_i$$
$$A_{2j} = \xi_j$$
$$A_{3ij} = 0$$

with

$$u_i|u_{-i} \sim N(\overline{u}_{\delta_i}, \tau_u)$$
$$v_i \sim N(0, \tau_v)$$
$$\xi_j \sim N(\xi_{j-1}, \tau_\xi).$$

The variance parameter (τ_*) distributions are assumed to be defined with $\sqrt{\tau_*} \sim U(0, 10)$. The regression parameters are all assumed to have zero mean Gaussian distributions with large variances (1/0.00001).

Alternative models have been considered. First the addition of $A_{3ij} = \psi_{ij}$ with $\psi_{ij} \sim N(0, \tau_\psi)$ was examined (Model 2). Finally, temporal dependence in the regression parameters was considered. Specifically, an auto-regressive prior distribution was assumed for α_3. of the form $\alpha_{3j} \sim N(\alpha_{3j}, \tau_3)$ which leads to the

$$A_{1ij} = \alpha_0 + \alpha_1 \log[1 + \alpha_2 \exp\{-\alpha_{3j} d_i\}] + u_i + v_i.$$

Table 7.3 displays the results of the fitting process. The model displaying the best fit overall is Model 3 with no space–time interaction with fixed α_1, α_2 but with temporally dependent α_{3j}.

Figure 7.7 displays the estimated temporal random effect (ξ_j) and 95% credible interval for model 1. Figure 7.8 displayed the results for the same effect but under Model 2 with zero-mean Gaussian space–time interaction. Figure 7.9 displays the posterior averaged temporally dependent distance regression effect for Model 3. Figure 7.10 displays the corresponding posterior averaged time dependent random effect for all years for Model 3. Note that in Model 3 it was necessary to fix $\alpha_1 = \alpha_2 = 1$, due to the identifiability issues when

TABLE 7.3
Ohio respiratory cancer (1979–1988): putative source model fits

Model	DIC	α_0	α_1	α_2	α_3
1	5762.6	-0.393(0.082)	129.8(22.8)	0.003(3.69E-4)	0.047(0.072)
2	5759.8	27.16(0.086)	-362.4(3.14)	0.101(4.36E-4)	0.089(9.35E-4)
3	5739.9	-0.625(0.063)	1	1	-

time dependence is allowed for these parameters. It is clear out of the limited number of models that have been fitted here, that a time-varying regression on a simple model of distance is considerably better (in terms of DIC) than constant parameters. The time dependent distance effect, α_{3j}, remains well estimated under this model whereas the temporal random effect is negligible. Model 3 did not include an interaction term and it would also be interesting to examine the effect of inclusion of such a term, though this is not pursued here. Note also that for many applications it would also be important to include a directional effect (possible time-varying) in the model (such as in 7.11). Finally, we have not presented the mapped output for the posterior averaged spatially-expressed random effects in this model (CH and UH). These may be of interest for the examination of unusual aggregations of risk as they appear or disappear over time. Of course, Bayesian residuals, predictive residuals, or even exceedence probabilities can be computed for space-time models in the form: $q_{ij} = \widehat{\Pr}(\theta_{ij} > 1) = \sum_{k=1}^{K} I(\theta_{ij}^k > 1)/K$ where $\{\theta_{ij}^k\}$, $k = 1, ..., K$ is the posterior sampled values of the relative risk for each region and time period. Residual maps or maps of q_{ij} could also be very informative. Of course, as noted earlier, the reliability of q_{ij} heavily depends on the correctness of the model.

The main emphasis in this section has been in demonstrating the modeling of spatiotemporal effects when time is included. Other issues that are not addressed here, but which could be important are measurement error in covariates or outcomes, ecological bias when making inference at lower aggregation levels from space-time data, and contextual or confounder effects.

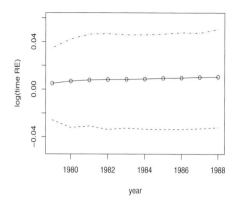

FIGURE 7.7
Ohio respiratory cancer: 1979–1988; estimated temporal random effect with 95% credible interval for Model 1.

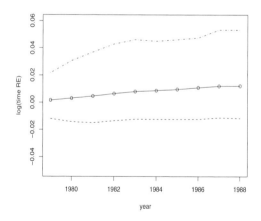

FIGURE 7.8
Ohio respiratory cancer 1979–1988: posterior average temporal random effect with 95% credible interval under Model 2 with zero-mean Gaussian interaction.

Ecological Analysis

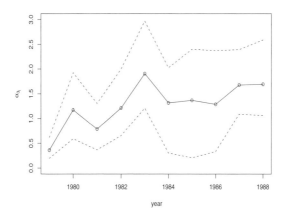

FIGURE 7.9
Ohio respiratory cancer 1979–1988: space–time model with time-dependent diance effects. Plot of posterior average distance effect over years with 95% credible interval

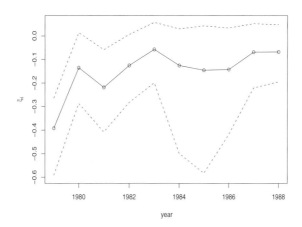

FIGURE 7.10
Ohio respiratory cancer 1979–1988: posterior average time random effect (ξ_j) with 95% credible interval.

8
Multiple Scale Analysis

The spatial analysis of single diseases is often sufficient. However, in some applications there is a need to consider different scales of aggregation within an analysis.

One such situation arises when it is of interest to consider a relationship at different aggregation levels. For example, if the relation of an outcome at county level to a covariate is examined, will the relationship hold true at lower aggregation levels (e.g., census tract) or at higher levels (e.g., state or country)? In general, it is unlikely that this would be the case as, if it were, there would be little need to consider different levels of analysis. In fact ecological bias would not occur. Scale change issues are often known as the modifiable areal unit problem (MAUP), whereby modification of the areal units could lead to different inferences. In geostatistics, this is known as the change of support problem (Cressie, 1996; Banerjee et al., 2004).

8.1 Modifiable Areal Unit Problem (MAUP)

The MAUP can be considered to have a variety of special cases. One of these is that of ecological bias (seen in Chapter 7) where the issue is whether inference can be made at a lower level of aggregation (individual level usually) from aggregate data. For example, can we make inferences from county or region level analysis to the individual level? We saw in that case there are various aspects of this problem. These include measurement error, knowledge of the within area distribution of exposures, contextual effects, and unobserved confounding.

8.1.1 Scaling Up

By scaling up, I mean trying to make inferences at a higher aggregation level than that used in the analysis. In general, aggregation leads to smoothing or averaging of data. For example, a spatial process is present at location s, $z(s)$ say, and when observed over a larger area A, the process will be a smoothed version i.e., $z(A) = \int_A z(u)du$. The integration is with respect to the extent of A. Note that the mean of the process in A can be defined as $\mu(A) = \int_A z(u)du/|A|$, where $|A| = \int_A du$. This can be regarded as an average over the area. Note that this integration leads to a reduction in variability, and so at the aggregate level we would expect there to be less variability. Cressie (1993, Section 5.2), noted this aspect in a geostatistical context. In an analogy with GIS operations, this is equivalent to zooming out in a map operation. One problem that this leads to is that processes operating at different aggregation levels may appear, or become important, at different scales. The possibility that the process observed at different scales will behave differently is clear. This suggest that 'scale labeling' is useful when dealing with changes in support or aggregation. By scale labeling, I mean the allocation of a scale of operation of a process. Methods for incorporation of scale effects within a Bayesian analysis of small area health data are various.

First, it is clear that it is possible to consider aggregated variables as confounders within an analysis. Multilevel modeling (Leyland and Goldstein, 2001) often addresses the issue of multiple levels within an analysis and these can include spatially-aggregated covariates. This of course includes contextual effects as a primary example (see Section 7.2.1.3; also Goldstein and Leyland, 2001). In general, the scaling up of health outcomes has been described for individual (point process) to small area (count) in previous chapters. Denote a scale level (integer) variable: $l_k, k = 1, ..., K$ where K is the number of levels. It is assumed here that aggregation levels can be discretized into such levels. Hence, an aggregation involves an outcome model indexed by the level: $f_k(y_{ik}; \mu_{ik}, l_k)$. Here, y_{ik} $i = 1, ..., m$ is the outcome variable for m units. (Assume here that this could be binary or a count or less often continuous.) Note that at different levels there could be different models and so subscripted f_k is appropriate. We need to establish the relation between levels of the scaling. Given a set of levels it is tempting to consider a general model formulation which links levels in the analysis. Assume that all units are aligned and that k is ranked from lowest to highest aggregation level. Define the set alignment as follows: there are m_k regions at the k th level and for $k = 2, .., K$ there are m_{k-1} regions at the lower aggregation. The allocation of the regions at m_{k-1} to the m_k regions is defined by $S_{i,k}$ which is the set of regions at the $k-1$ level uniquely within the i th region at the k th level. For count data this would mean that $y_{ik} = \sum_{l \in S_{i,k}} y_l$, $y_{ik-1} = \sum_{l \in S_{i,k-1}} y_l$, ...for $k = 2, ..., K$. For example, we could specify a vector model of the form

$$\mathbf{y} = \begin{cases} \{y_{i1}\}, i = 1, ..., m_1 \\ \{y_{i2}\}, i = 1, ..., m_2 \\ \{y_{i3}\}, i = 1, ..., m_3 \\ \cdot \\ \cdot \\ \{y_{iK}\}, i = 1, ..., m_K \end{cases} \sim \begin{cases} f_1(\boldsymbol{\mu}_1, l_1) \\ f_2(\boldsymbol{\mu}_2, l_2) \\ \cdot \\ \cdot \\ f_K(\boldsymbol{\mu}_K, l_K) \end{cases}$$

Often the distribution at each level is the same and so $\mathbf{y}_k \sim f(\boldsymbol{\mu}_k, l_k)$. An example of this approach is given in Section 8.1.3. Often the nesting of the data leads not only to a single distribution but also to a single likelihood. For example with nested Poisson counts then

$$L(\boldsymbol{\mu}_k|\mathbf{y}_k) = \prod_1 f(\mathbf{y}_1; \boldsymbol{\mu}_1) \cdots \prod_K f(\mathbf{y}_K; \boldsymbol{\mu}_K).$$

Linkage between the scales can be achieved by dependence between $\boldsymbol{\mu}_1 \ldots \boldsymbol{\mu}_K$. Of course for functions of counts with aggregation then the rates should sum across units within each cell and so $\boldsymbol{\mu}_K = \sum_{l \in S_{i,k-1}} \boldsymbol{\mu}_l$. Louie and Kolaczyk (2006) give an example of multiple scale analysis for disease data.

8.1.2 Scaling Down

By scaling down, I mean trying to make inferences at a lower aggregation level than that used in the analysis. The classic situation where ecological bias arises, is an example of this scaling down: trying to make inference at the individual level from aggregate level analysis. Disaggregation is the reverse operation from aggregation and the parallel with the mathematical operation of integration carries over, so disaggregation is equivalent to differentiation. Hence the opposite of smoothing would be to add noise to an existing field. Plummer and Clayton (1996) essentially assume the knowledge of a distribution of noise at the lower level in an ecological application. In general, knowledge of the variation between the lower aggregation level units is important if attempting to make inference at a lower aggregation level.

8.1.3 Multiscale Analysis

By multiscale analysis, I mean where data is available at multiple resolution levels. For example, the focus of the analysis might be to include all data at levels of aggregation in an analysis of all the levels. This could be a joint analysis or could be separately carried out in the null case. Louie and Kolaczyk (2006) give an example of multiple scale analysis for disease data. More concretely, assume that we observe public health district data (in the US) and county level data. Figure 8.1 displays the public health districts (18) and counties (159) of the state of Georgia in the USA.

The county set is a unique subdivision of the district set i.e., each county falls uniquely within one PH district. Public health districts are administrative units within which certain health services are provided. It is therefore possible that some grouping effect based on health district could be found for counties that lie within a given district. For example, Dalton PH district includes the 6 counties of Whitfield, Murray, Fannin, Gilmer, Pickens, and Cherokee. These counties lie completely within Dalton and no other district. Hence in this case

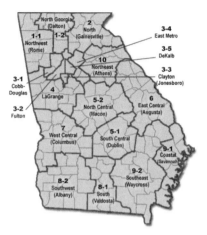

FIGURE 8.1
State of Georgia, United States: public health district boundary map (thick line) and county boundary map (thin line).

there is multiscale information which is completely aligned in the sense that the lower level county units fall completely and uniquely within the higher aggregation level units (districts).

In this particular case we could imagine the data defined with $K = 2$, with y_{i1} $i = 1, ..., 159$ where l_1 is the county level and y_{j2} $j = 1, ..., 18$ and l_2 is the district level. We could further assume a model of the form

$$y_{i1} \sim f_1(\mu_{i1}; l_1)$$
$$y_{j2} \sim f_2(\mu_{j2}; l_2).$$

Hence, under a null (separate) model we might have $\mu_{i1} = \exp\{\alpha_{10} + u_{i1}\}$, and $\mu_{j2} = \exp\{\alpha_{20} + u_{j2}\}$ where α_{10} and α_{20} intercepts and u_{i1} and u_{j2} are effects at level 1 and 2. These can be random effects or functions of measured predictors. Clearly to ensure linkage between levels and so to model the joint behavior μ_{i1} and μ_{j2} may be linked. One natural way to do this is to consider the contextual effect of district on county and so we could have:

$$\mu_{i1} = \exp\{\alpha_{10} + u_{i1} + u_{j2}\}_{i \in j}$$
$$\mu_{j2} = \exp\{\alpha_{20} + u_{j2}\}.$$

In effect, because there is dependence between the mean levels then joint estimation of the latent factors (u_{i1}, u_{j2}) must be considered. Additionally, it might be considered that the district level should have a contribution from effects at the county level and so a further possibility could be to consider $\mu_{j2} = \exp\{\alpha_{20} + u_{j2} + \sum_{i \in j} \mu_{i1}\}$. In the latter formulation it would be useful to

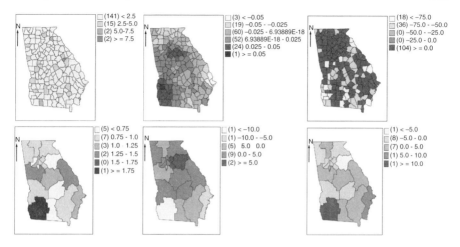

FIGURE 8.2
Multiscale model for the Georgia oral cancer data: posterior average effects: top row (left to right) county level $\widehat{\theta}, u_i, v_i,$; bottom row (left to right) PH level $\widehat{\theta}, u_i, v_i$.

keep the separate effect of level (u_{j2}). Further extensions or variants of these linkages are possible.

8.1.3.1 Georgia oral cancer 2004 example

As an example of analysis at multiple scales, I examine the Georgia PH district and county example. In this case, mortality counts from oral cancer were considered for the year 2004 in the US state of Georgia. The state-wide expected rate for oral cancer was obtained and applied to the local county populations. The count of oral cancer mortality within public health (PH) districts is also available for the same period (as a sum of constituent county counts). Expected rates can also be summed from counties or directly calculated from the district population. We have applied the different two level models to the Georgia PH-county data. A seprate model was fitted to each level as well as a joint model with contextual effect. Table 8.1 displays the results in terms of DIC and pD for the different fitted models. Figure 8.2 displays the posterior average maps for the joint model with Poisson data model the relative risks are defined as:

$$\theta_{i1} = \exp\{\alpha_{10} + v_{i1} + u_{i1} + \underset{i \in j}{v_{j2}} + \underset{i \in j}{u_{j2}}\}$$
$$\theta_{j2} = \exp\{\alpha_{20} + v_{j2} + u_{j2}\}$$

where each $v_{*1} + u_{*1}$ is a convolution of a UH and CH random effect.

TABLE 8.1
Goodness of fit results for separate and joint models for
Georgia oral cancer PH-county level data

Model	pD	DIC
County	97.68	507.01
PH district	18.72	124.96
Joint model: County	100.5	513.38
Joint model: PH district	18.07	123.77

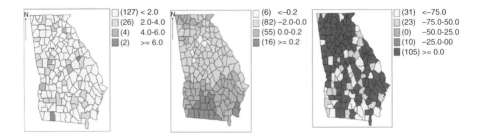

FIGURE 8.3
Georgia oral cancer model when a simple convolution model at county level is fitted: posterior average maps of (left to right) $\hat{\theta}, u_i, v_i$.

Figures 8.3 and 8.4 display the posterior avergaed maps for the $\hat{\theta}$, and u_i, v_i effects obtained when separate models are fitted to county level and PH level. Overall it appears that the joint model mainly benefits the PH analysis as the DIC is marginally lower under the joint model.

8.2 Misaligned Data Problem(MIDP)

While multiscale analysis can concern spatial units that are completely matched when aggregated there is also a situation where units are not matched and are termed *misaligned*. This often occurs when sampling at different spatial scales are not linked. A classic example of this scenario is where residential address of disease cases are to be related to measurements of environmental pollution obtained from a network of sites. The locations of the cases do not match the pollution measurement sites. Another example of misalignment is where data on disease is available in different administrative units that are not matched spatially. For example, census tracts, are not matched to, e.g., postal codes in the United Kingdom or zip codes in the United States. Thus in both cases some mechanism must be used to provide data on the same spatial

Multiple Scale Analysis

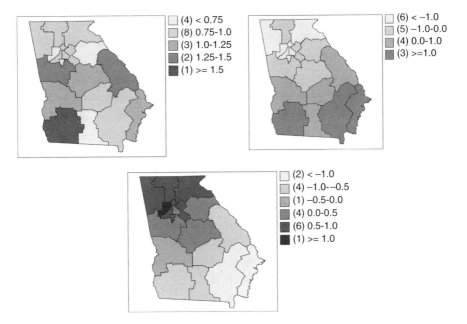

FIGURE 8.4
Georgia oral cancer PH district model only: posterior average maps of (left to right) $\widehat{\theta}, u_i, v_i$.

scale within or at the same spatial region or location. In the first example, it is usually the case that *interpolation* of pollution measurements to residential locations would be required. In the second case, it may be that disease outcome data (counts) need to be available in the same spatial units. The first situation is one where a predictor is misaligned, whereas in the second case, different spatial data observation levels are misaligned.

8.2.1 Predictor Misalignment

Predictor misalignment can take various forms. Here I will discuss two basic situations: misalignment which requires interpolation to a point location and misalignment where interpolation must be made over an area. In both cases, interpolation or measurement error is involved. Define s_i $i = 1, ..., N$ to be the locations of cases and controls, where $N = m + n$ with m cases and n controls and y_i is the corresponding binary case/control label. Also define the measured level of a predictor at a set of sites as $z(s_l)$, $l = 1, ..., L$ and $z_l \equiv z(s_l)$ for short. Usually we would assume that the predictor data is noisy and so even at s_l we would want a smoothed value. In addition, however, we would usually want to have $z(s_i)$, at the residential locations and this also involves interpolation. One approach to this situation is to assume a spatial Gaussian process for the measured predictor and to make a conditional prediction of the level of the predictor.

Define a Gaussian process model for the sets of sites, and define also the parameter vector $\theta = (\tau, \psi)^T$, and let $\mathbf{z}_s^T = (z(s_1), ..., z(s_L))$:

$$\mathbf{z}_s | \alpha, \theta \sim N(\boldsymbol{\mu}_s, \Gamma)$$

where $\boldsymbol{\mu}_{s_l} = \mu(s_l, \alpha)$, a predictor at the l th site, and Γ is a spatial covariance matrix. Often $\boldsymbol{\mu}_s$ will consist of trend surface components and $\Gamma_{ll'} = \tau \rho(s_l - s_{l'}; \psi)$ where τ is a variance and $\rho(.)$ is a correlation function measuring the relation between values of z at separation distance $s_l - s_{l'}$. Choices of $\rho(s_l - s_{l'}; \psi)$ are many, (see e.g., Cressie, 1993, Diggle and Ribeiro Jr., 2007) and a simple choice could be an exponential form such as $\rho(s_l - s_{l'}; \psi) = \exp\{-\psi||s_l - s_{l'}||\}$. More generally the powered exponential family defined by $\rho(s_l - s_{l'}; \psi) = \exp\{-(\psi||s_l - s_{l'}||)^k\}$ can be assumed with $0 < k < 2$, the extra parameter allowing for a slower distance decline at short separation for $k > 1$, when $\psi << 1$.

For predicting a new set of locations $(z(s_i), i = 1, ..., N)$ the predictive distribution is

$$f(\mathbf{z}(s_i)|\mathbf{z}(s_l)) = \int f(\mathbf{z}(s_i)|\mathbf{z}(s_l), \alpha, \theta) f(\alpha, \theta|\mathbf{z}(s_l)) d\alpha d\theta.$$

Under a Gaussian process assumption, and denoting the set of original sites as s and the interpolant sites as $s\prime$ then this distribution is just

$$N(\boldsymbol{\mu}_{s'} + \Gamma_{s,s'}^T \Gamma_s^{-1}(\mathbf{z}_s - \boldsymbol{\mu}_s), \Gamma^*) \tag{8.1}$$
$$\Gamma^* = \tau[\Gamma_{s'} - \Gamma_{s,s'}^T \Gamma_s^{-1} \Gamma_{s,s'}].$$

This distribution can be sampled from directly assuming that the inverse Γ_s^{-1} can be calculated. It is therefore possible to assume a joint model for the binary outcome y_i and the interpolated (latent) predictor $z(s_i)$:

$$y_i \sim Bern(p_i) \; i = 1, ..., N \tag{8.2}$$
$$\log it\, p_i = \beta_0 + \beta_1 z(s_i)$$
$$\begin{pmatrix} \mathbf{z}_s \\ \mathbf{z}_{s'} \end{pmatrix} \sim N\left(\begin{pmatrix} \boldsymbol{\mu}_s \\ \boldsymbol{\mu}_{s'} \end{pmatrix}, \tau \begin{pmatrix} \Gamma_s & \Gamma_{s,s'} \\ \Gamma_{s,s'}^T & \Gamma_{s'} \end{pmatrix}\right).$$

Essentially, $z(s_i)$ is a latent variable and must be estimated via the joint model. One issue that remains is whether $z(s_i)$ should be estimated in the same model as p_i or should be estimated from $\mathbf{z}_s$ separately. Note that if a joint model is assumed then both β_1 and $z(s_i)$ are unknown and so there is likely to be some difference in the estimated $z(s_i)$ compared to an estimate simply obtained by Bayesian Kriging. There is some support for the idea that Bayesian Kriging via (8.1) should be used separately, as there seems a priori little reason to assume that the disease outcome should influence the estimate of the latent predictor.

Predictor misalignment can also occur where a value of the predictor is needed over an area rather than at a point location. For example, count outcome data are commonly available for small areas and we might want to interpolate measured covariates over small areas. This process is equivalent to block Kriging in Geostatistics and can be specified as follows. Define the counts in m small areas as y_i, $i = 1, ..., m$. We assume the usual Poisson relative risk model with $y_i \sim Pois(e_i \theta_i) \quad \forall i$. Also assume a log linear model relating the relative risk to a block level covariate: $\log(\theta_i) = \beta_0 + \beta_1 z_i$. Here z_i denotes the value of z averaged over the i th small area. Assume a set of measurement site values as before: $\mathbf{z}_s^T = (z(s_l), ..., z(s_L))$ and also a set of small area average values: $\mathbf{z}_A^T = (z_1, ..., z_m)$ for areas $A_1, ...A_m$. We need the predictive distribution, which is now

$$f(\mathbf{z}_A | \mathbf{z}_s) = \int f(\mathbf{z}_A | \mathbf{z}_s; \boldsymbol{\alpha}, \boldsymbol{\theta}) f(\boldsymbol{\alpha}, \boldsymbol{\theta} | \mathbf{z}_s) d\boldsymbol{\alpha} d\boldsymbol{\theta}.$$

As in the point location case, there is a conditional distribution for these averages of the form:

$$N(\boldsymbol{\mu}_A + \Gamma_{s,A}^T \Gamma_s^{-1}(\mathbf{z}_s - \boldsymbol{\mu}_s), \Gamma_A^*) \qquad (8.3)$$

$$\Gamma_A^* = \tau[\Gamma_A - \Gamma_{s,A}^T \Gamma_s^{-1} \Gamma_{s,A}]$$

where Γ_s^{-1} is the inverse of the site covariance and

$$(\Gamma_{s,A})_{li} = |A_i|^{-1} \int_{A_i} \rho(s_l - u; \psi) du \qquad (8.4)$$

$$(\Gamma_A)_{ii'} = |A_i|^{-1} |A_{i'}|^{-1} \int_{A_i} \int_{A_{i'}} \rho(v - u; \psi) du dv$$

$$(\boldsymbol{\mu}_A)_i = |A_i|^{-1} \int_{A_i} \mu(s) ds.$$

The integrals in (8.4) must be evaluated. Banerjee et al. (2004) suggest a Monte Carlo integration where the random locations are generated and the integrals replaced by sums. Finally the joint model could be specified as

$$\mathbf{y} \sim Pois(\boldsymbol{\mu}(z_A)) \qquad (8.5)$$

$$\begin{pmatrix} \mathbf{z}_s \\ \mathbf{z}_A \end{pmatrix} \sim N\left(\begin{pmatrix} \boldsymbol{\mu}_s \\ \boldsymbol{\mu}_A \end{pmatrix}, \tau \begin{pmatrix} \Gamma_s & \Gamma_{s,A} \\ \Gamma_{s,A}^T & \Gamma_A \end{pmatrix}\right). \qquad (8.6)$$

Here, it could be assumed that $\boldsymbol{\mu}(z_A)$ is simply a linear function of the latent interpolated predictor, assuming no confounders are present. Hence it is often assumed that $\boldsymbol{\mu}(z_A)_i = e_i \exp(\beta_0 + \beta_1 z_i)$.

Note that, in general, integrating the first order intensity we get $\mu(z_A)_i = \int_{A_i} \lambda_0(u)\lambda_1(u)du$ where, if we assume $\lambda_0(u)$ to be constant over A_i, so $\lambda_0(u) \equiv e_i$. Denote the exposure of the j th individual within the i th area as $z_{j(i)}$, and an individual has an exposure $\lambda_1(u)_i = \exp(\beta_0 + \beta_1 z_{j(i)})$. Hence,

$$\mu(z_A)_i = e_i \int_{A_i} \exp(\beta_0 + \beta_1 z_{j(i)})ds = e_i \exp(\beta_0) \int_{A_i} \exp(\beta_1 z_{j(i)})ds \approx$$

$e_i e^{\beta_0} \sum \exp(\beta_1 z_{j(i)})$, where the sum is over the whole at-risk population within the small area.

Now usually individual level exposure is unavailable and indeed without knowledge of locations, this is not possible to directly estimate. However it is possible to use the region average (z_i) with some further assumptions to allow a closer approximation. For example we could assume that $z_{j(i)} \sim N(z_i, f(z_i))$ where the variance is assumed to be a function of the mean level. Wakefield and Shaddick (2006) suggest a function such as $f(z_i) = a + bz_i$. Hence it may be possible to consider a model whereby (8.3) could be used to predict block effects but within the Poisson model the individual effects could be used via

$$\mu(z_A)_i = e_i e^{\beta_0} \sum \exp(\beta_1 z_{j(i)})$$
$$z_{j(i)} \sim N(z_i, a + bz_i).$$

Other alternatives could be considered where point interpolation could be made to random points within each small area, instead of block Kriging.

It can be important to consider how individuals within areas are related to exposures, at least from the view of making inference at the individual level. If individual inference is to be made from aggregate data then ecological bias should be considered. Of course, it is still remains valid to make inference at aggregate levels from aggregate level models.

Some simple alternatives are often suggested to the above when using interpolated predictors. One simple approach is to use plug-in estimates z_i directly within the first level aggregated data model. For example,

$$y_i \sim Pois(e_i \theta_i) \ \forall i,$$
$$\log \theta_i = \beta_0 + \beta_1 z_i,$$

can be assumed where z_i is estimated separately via a Bayesian Kriging model. However in such a model, the error in the interpolated value is ignored. As discussed in Chapter 7, it is possible to make allowance for some biases by inclusion of random effects in models for example: $\log \theta_i = \beta_0 + \beta_1 z_i + \epsilon_i$ where ϵ_i is assumed to be unstructured with distribution $\epsilon_i \sim N(0, \tau_\epsilon)$. In effect we are assuming that error in the interpolation is simply random noise. While this will indeed make some allowance for the interpolation error, it is unlikely to properly correct for any correlated error and will also be associated with the error in the relation between y_i and z_i. Alternatives involving forms of Berkson measurement error, such as $\log \theta_i = \beta_0 + \beta_1[z_i + \epsilon_i]$, could be envisaged where either unstructured error with distribution $\epsilon_i \sim N(0, \tau_\epsilon)$, or spatially-structured error could be assumed. Given that z_i has already been Kriged it is less justified to assume spatially-correlated error here.

Another alternative would be to employ a simplified model for the Gaussian field that is to be interpolated. For example, it would be possible to assume a proper CAR model for the variation in the $\mathbf{z}_s^T$ so that at a location j, say, where z_j is observed, then assume

$$z_j \sim N(z_j^*, \tau)$$
$$z_j^* | z_{-j}^* \sim N(\overline{z_j}^*, \tau^*/n_{\delta_j})$$
$$\overline{z_j}^* = t_j + \phi \sum_{l \in \delta_j} (z_l^* - t_l)/n_{\delta_j},$$

where t_j is a trend component. The neighborhood (δ_j) defined for this example could be obtained as Dirichlet tesselation neighbors of the set of sample sites and outcome locations (if point-to-point interpolation is needed). The interpolated values are treated simply as missing. However this ignores some parts of the variation in the field due to the Markov neighborhood dependence that is assumed by the model.

Another simplifying assumption, in the case of point-to-block interpolation, is to consider the centroids of small areas as sampling points which augment the measurement at the sampling sites. In this case, the small area centroids could be considered to have missing data. More specifically, assume an augmented set $\mathbf{z}_{s'} : (\mathbf{z}_A, \mathbf{z}_s)$ where the z_A are missing:

$$y_i \sim Pois(e_i \theta_i)$$
$$\log = \beta_0 + \beta_1 z_i$$
$$\mathbf{z}_{s'} \sim N(\boldsymbol{\mu}_{s'}, \tau \Lambda)$$

where $(\boldsymbol{\mu}_{s'})_i = \sum_{i \in \delta_i} z_i/n_{\delta_i}$ and $(\Lambda)_{ii} = n_{\delta_i}^{-1}$. This CAR-Voronoi model was proposed by Greco et al. (2005) in an application to interpolation of groundwater uranium to the centroids of zip codes in South Carolina.

8.2.1.1 A binary logistic spatial example

In a study of a birth-defect disease outcome the residential addresses of births and births with defects were recorded within a US state. For an anonymized study area of 2 x 2 kms the locations of this binary outcome were recorded (see Figure 8.5). Soil chemical measurements were also made on a range of chemicals. In this example, I examine just one of these: Arsenic (As). Arsenic in soil is commonly found in agricultural areas where heavy pesticide use has occurred over many years. It is important to consider whether exposure to As (via a possible groundwater exposure pathway) during pregnancy could lead to adverse birth defect outcomes. Many residents of certain rural areas in the US have a groundwater domestic supply. In this example we know the residential location of outcomes (births with or without defects) during a particular month of pregnancy, but have a mismatched set of soil chemical sampling sites. Figure 8.5 displays the distribution of the sites (left panel) and the distribution of the births with their binary marks (right panel). The example displays a point-to-point misalignment problem. It is appropriate here to interpolate the soil chemical values from the measurement sites to the residential locations so as to provide a local exposure estimate. Of course, issues relating to appropriateness of the residential location as exposure "location" do arise, but here we assume that this location is the most relevant for the given month of exposure. The first model assumed is closely related to that given in (8.2) except that the model is not jointly estimated. Instead, the interpolation is carried out separately via Bayesian Kriging and a plug-in estimator is used ($\hat{z}_{KB}$). The model assumed for this estimator is a Gaussian spatial process with an overall mean parameter (β_z) and a power exponential correlation function so that the covariance matrix is given by

$$\sigma^2 \Sigma_{ij} = \sigma^2 \exp\{-(d_{ij}/\phi)^\kappa\} \tag{8.7}$$

with smoothing parameter $\kappa = 1$ which is equivalent to a Matérn covariance with exponential form. The prior distribution for β_z is assumed to zero-mean Gaussian with variance 3.0, whereas the inverse variance prior distribution is assumed for σ^2. A discrete prior distribution on a fixed range $(0, 2*max(d_{ij}))$ is used for ϕ. The parameters within the Kriging model were estimated using krige.bayes in the R package geoR. The posterior expected estimates under this model were $\overline{\beta}_z = 1.9903$, $\overline{\sigma}^2 = 5.00629$, and $\overline{\phi} = 0.0001$. This suggests a very small covariance distance effect. The overall logistic model used was

$$y_i \sim Bern(p_i)$$
$$\log = \beta_0 + \beta_1 \hat{z}_{KB}$$

with $\beta_* \sim N(0, \tau_*)$, $\tau_*^{-1/2} \sim U(0, 100)$. This is a simple Bayesian logistic regression model with zero-mean Gaussian prior distributions for the regression parameters. This is referred to as Model 1 in Table 8.2. Models 2 and 3 are variants of this model which are intended to account for the measurement error in using the Bayesian Kriged values in the model (rather than a joint model). Model 2 assumes an additive unit level uncorrelated error (UH) which has zero-mean Gaussian prior distribution. Model 3 assumes that instead of a UH term it assumes that the classical measurement error model is reversed and that the Kriged estimate is the true value observed with error. This is termed Berkson error. The final models utilize the original data and model the spatial covariation of that data directly. Model 4 assumes that the constant mean estimated from the As data has an additive uncorrelated error term ($v_i \sim N(0, \tau_v)$, $\tau_*^{-1/2} \sim U(0, 100)$) and an additive spatially-correlated error including the term $\mathbf{W} \sim \mathbf{MVN}(\mathbf{0}, \sigma^2 \mathbf{\Sigma})$ with covariance term

TABLE 8.2
Model fit results for the point misalignment example

Model	Components	DIC
1	$\beta_0 + \beta_1 \widehat{z}_{KB_i}, \beta_* \sim N(0, \tau_*), \tau_*^{-1/2} \sim U(0, 100)$	178.33
2	$\beta_0 + \beta_1 \widehat{z}_{KB_i} + v_i, v_i \sim N(0, \tau_v), \tau_*^{-1/2} \sim U(0, 100)$	116.22
3	Berkson error: $\beta_0 + \beta_1 (\widehat{z}_{KB_i} + v_i)$, priors as in (2)	155.47
4	$\beta_0 + \beta_1 \overline{(As + W_i)} + v_i, \mathbf{W} \sim MVN(\mathbf{0}, \sigma^2 \mathbf{\Sigma})$	177.50
5	as in (4) but $\beta_0 + \beta_1 \overline{(As + W_i)}$ only	176.10
6	$\beta_0 + \beta_1 \widehat{z}_i$	140.90

$\mathbf{\Sigma}_{ij} = \exp\{-(d_{ij}/\phi)\}$, with prior distributions $\sigma^2 \sim Exp(0.1)$, $\phi \sim Exp(0.1)$ which are reasonably overdispersed while favoring small values. Finally, model 6 consists of a joint model for the birth outcome and arsenic levels where the true arsenic level is a latent variable ($\widehat{z}_i$) interpolated via the point to point Bayesian model in (8.2). Again a power exponential covariance was assumed with $\kappa = 1$ as in (8.7). The assumed prior distributions were $\sigma^2 \sim Exp(0.1)$, $\phi \sim Exp(0.1)$ as in the previous model, and a random level β_z was assumed to have a zero-mean Gaussian prior distribution with variance 3.0. The resulting goodness-of-fit DIC measures for these models were computed and they are reported on Table 8.2. While these are only a subset of overall potential models and prior distribution choices which could be made, they represent a range of approaches often advocated for this type of modeling exercise. It is notable that the general random effect model (Model 2) provides the lowest DIC among the plug-in ME models and by this measure performs better than the potentially more plausible Berkson model. Among models making direct use of the data, the joint model (Model 6) yields the lowest DIC (140.90) and, with the exception of Model 2, this yields the lowest DIC. However for this particular example the overall best model judging by DIC alone is the Model 2. This might suggest that there is overall more noise in the relation between the outcome and the covariate than within the covariate itself. There are of course various criticisms that can be leveled at such models. First, approximate models, such as those using plug-in estimators do not have a priori justification based on Gaussian model or process theory. However from a pragmatic viewpoint, if a parsimonious description is achieved by such a model in a regression context then it is supported. The model is less justified when treating out-of-data prediction on a priori grounds. However joint modeling of the latent spatial effect could also be challenged in the sense that the true value of the spatial covariate is forced to depend on the outcome model parameters. This may also seem to be a criticism of that model. Of course prior distributional assumptions can be challenged for all models fitted and this also can impact model selection.

For the lowest DIC model the posterior mean estimates of the parameters $\beta_0, \beta_1, \tau_0, \tau_1, \tau_v$ are available after a burn-in of 60000 and a sample of

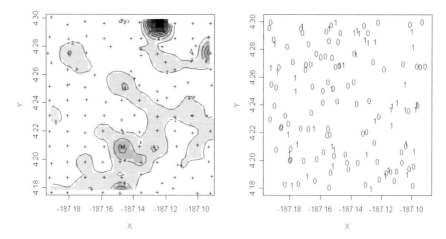

FIGURE 8.5
Spatial distribution of sampling sites (+) and multivariate B-spline smooth of Arsenic (image and contourplot from R package MBA) for the synthetic data example (left panel); spatial distribution of births with binary mark (1: birth defect, 0: otherwise) in same window.

10000 taken. This yielded estimates of mean (sd) of $\widehat{\beta}_0 = -1.715(3.832)$, $\widehat{\beta}_1 = -0.6846(1.951), \widehat{\tau}_0 = 106.1(4233.0), \widehat{\tau}_1 = 76.16(3419.0), \widehat{\tau}_v = 0.05634(0.1135)$. The overall conclusion of this analysis is that the Arsenic does not have a significant (linear) relationship with birth defect under a logit model. Further model variants could be conceived where changes to both the linear model and also to the interpolation model could be considered. Often environmental predictors have non-linear relations with health outcomes and so polynomial or spline models could provide better fits. In our experience, with other soil chemicals it is indeed the case that a nonlinear model yields a significant fit when a linear model does not. Further alternative interpolants such as spline models could be considered (see e.g., French and Wand, 2004).

Finally, Zhu et al. (2003) describe the analysis of misaligned data when spatiotemporal analysis is to be considered. They also consider (log) linear relations between outcome and predictor within a small area setting (zip code units).

8.2.2 Outcome Misalignment

Outcome misalignment occurs when spatial outcome units are not matched. One basic and commonly occurring form is where counts are observed within small areas but expected rates are only available in other misaligned areas. For example, in the US, zip codes don't align with counties but are a smaller-sized spatial (postal) unit; in the UK postcodes don't align with census enumeration districts. When comparing diseases within study areas it can be the case that different aggregation units are found and matching must take place. When multiple diseases are concerned then multivariate misalignment could occur. This latter is not considered in this volume, but could be of considerable practical importance when multiple disease analysis is considered. Multiple disease analysis is discussed in Chapter 9.

Outcome misalignment leads to the consideration of the overlay of different spatial aggregations to yield mismatched outcomes. In cases where data are needed to be matched on their spatial units then some form of allocation model must be considered. Banerjee et al. (2004), Ch 6, discuss in detail the issues related to interpolation of Gaussian processes to point locations and to areas. Note that in general, outcome misalignment can be treated as a problem in block to block interpolation in that areas of different sizes are to be matched, at least in terms of parameter estimation.

A common assignment rule is simply to use proportional allocation. Define a simple misalignment with a single large region (A) and a set of misaligned smaller regions that overlap $A : (B : \{b_1, ..., b_p\})$. Parts of the smaller regions overlap areas of A. It is possible to consider the problem first of the aggregation of count data from smaller to larger regions. Define M larger scale regions with counts $\{y_{A_i}\}$, $i = 1, ..., M$ and P smaller region counts $\{y_{b_l}\}$, $l = 1, ..., P$; also the corresponding expected rates: $\{e_{A_i}\}, \{e_{b_l}\}$, and area proportions $\{p_{b_{li}}\}$, where $p_{b_{li}}$ is the proportion of region b_l within the i th region.

First for aggregation to larger region counts, it is simple to assume proportional allocation and to derive the larger region counts from smaller regions:

$$y_i = \sum_l p_{b_{li}} y_{b_l}. \tag{8.8}$$

One consequence of proportionate allocation is that the resulting larger region y_i could be non-integer. Special rules could be devised to make sure the allocation resulted in integer allocation, but these can lead to edge effect problems. Note that many packages allow noninteger Poisson data (e.g., WinBUGS). Given two levels of count data a consideration might be given to employing a joint model for the different levels where :

$$y_i \sim Pois(\mu_i); \; \mu_i = \sum_l p_{b_{li}} e_{b_l} \theta_{b_l}. \tag{8.9}$$
$$y_{b_l} \sim Pois(\mu_{b_l}); \; \mu_{b_l} = e_{b_{li}} \theta_{b_l}.$$

Occasionally aggregation is required only for expected rates. The situation envisaged is where expected rates are available in smaller regions only and must be aggregated to match the counts in larger regions. For example, if a

Poison model for the larger region counts were assumed then

$$y_i \sim Pois(e_i.\theta_i) \tag{8.10}$$
$$e_i = \sum_l p_{b_l i} e_{b_l}.$$

In consideration of disaggregation then smaller regions can be allocated counts by proportionate allocation downwards. For example, assuming the same complete misalignment as above, then the derived count in a smaller region y_{b_l} could be derived via $y_{b_l} = \sum_i p_{il} y_i$ where the proportion of the larger region overlapping the smaller region is denoted by p_{il}.

8.2.3 Misalignment and Edge Effects

In Section 5.8, various issues were discussed relating to the importance of edge effects in disease mapping studies. These can be even more important in consideration of misaligned data at different aggregation levels. For example assume a simple proportionate aggregation as in (8.8) above. Close to the study area boundary, the neighborhoods of regions are censored. Assume that smaller regions misalign to a study area where the hull of the larger regions forms the outer boundary of the study area. This implies that the smaller regions can overlap the study boundary. To estimate the proportionate allocation the area of these smaller regions must be known as well as the counts within the regions. Hence, knowledge of an area larger than the study area is important. Further when modeling small area counts where aggregation to larger regions has been assumed, then the effects of local neighborhoods on the spatially structured risk could be considerable. In Model (8.9) above, any spatially-structured risk involving adjacencies will depend on knowledge of both the neighbors of the larger regions and the neighbors of the smaller regions. In Model 8.10 the relative risk in larger regions is not a direct aggregation of the lower level risk as in (8.9). In (8.9) the smaller regions could have neighborhood effects and these aggregate to the larger level. In (8.10) only the expected rate is aggregated. The effect of neighborhood censoring within CAR random effect models has been highlighted by Vidal-Rodiero and Lawson (2005).

If the study area does not correspond to the hull of the larger regions then some regions will overlap the boundary. This could also cause edge effect problems due to the need to assign data from the larger region to the area within the study area. Clearly misalignment can have significant implications for inference about areas at or close to the edge of study boundaries.

9

Multivariate Disease Analysis

Often it is appropriate to consider the analysis of the geo-referenced distribution of more than one disease. For example, the focus may be on a group of diseases with similar etiology in an epidemiological study. Another example, in the context of public health, could be the examination of the general health status of a region (possibly following a cluster alarm signal). In the latter case a range of disease types might be considered to find out if any show signs of unusual risk variation. In some cases, the focus is on the spatial distribution of the vector of diseases (for example relative risk estimation for the vector). In other cases the diseases are to be contrasted or correlation between their spatial distributions are to be considered. In this chapter the focus will be on relative risk estimation and modeling of risk both in terms of correlation between diseases and in terms of comparison. The simplest situation where such analysis can be considered is where two diseases are to be examined.

9.1 Notation for Multivariate Analysis

Both case event and count data will be considered in this chapter.

9.1.1 Case Event Data

Assume there are L diseases. Assume a fixed study area T common to all diseases. Misaligned study areas are not considered here. Assume that the residential addresses of each disease type are known and given by $\{s_{l_i}\}$, $l = 1, ..., L$ and $i = 1, ..., m_l$. It is assumed that there are m_l cases of the l th type within the study region. In addition, there may be observed a control realization for each of the diseases. The number of these are denoted by m_{c_l} and the total number of cases and controls for a l th disease is $N_l = m_l + m_{c_l}$. Hence the controls will be denoted $\{s_{l_i}\}$, $l = 1, ..., L$ and $i = m_l + 1, ..., N_l$. Often a common control for all diseases is used but this can be discussed later.

9.1.2 Count Data

Here, the simplest situation is considered where fixed spatial units (small areas) have a range of L disease counts observed in a fixed time period. Define these counts as $\{y_{l_i}\}$, $l = 1, ..., L$, $i = 1, ..., m$. hence, within each spatial unit there are L counts of disease. The corresponding expected rates and relative risks are defined as $\{y_{l_i}\}$, $\{\theta_{l_i}\}$.

9.2 Two Diseases

Often the simplest situation arises where two diseases are to be modeled where $L = 2$.

9.2.1 Case Event Data

For case events the observed data is now $\{s_{l_i}\}$, $l = 1, 2$ and $i = 1, ..., N_l$. Analysis will depend on the object of the study. If the object is to locally (in space) compare the probability of getting disease 1 or disease 2, then a *competing risk* formulation might be useful. This might condition on the observed total of cases, $N_T = m_1 + m_2$. On the other hand it might be important to jointly model each disease (perhaps with linking parameters) without conditioning.

First define the first order intensity for the l th disease as $\lambda_l(s|\psi_l) = \lambda_{l0}(s|\psi_{l0}).\lambda_{l1}(s|\psi_{l1})$, $l = 1, 2$. If the competing risk scenario is important then it is possible to simply consider the joint realization of the cases conditional on N_T. Assume that the diseases can be considered to be independent heterogeneous Poisson processes (PPs) governed by $\lambda_1(s|\psi_1)$, and $\lambda_2(s|\psi_2)$. This independence assumption would be at least conditional given parameters $(\psi_{l0}, \psi_{l1}, l = 1, 2)$. Hence we can assume that the joint realization is a PP and so conditioning on the joint realization it is possible to consider the probability that given a case at s it is of type 1 or 2. This is given by

$$\Pr(l(s) = 1 | case(s)) = \frac{\lambda_1(s|\psi_1)}{\lambda_1(s|\psi_1) + \lambda_2(s|\psi_2)}. \tag{9.1}$$

where $l(s)$ is the type label of the location and $case(s)$ means case at s. Note that the locations are now conditioned on and it is now possible to make inference about the labeling of the points. Substituting the full intensities into (9.1), gives

$$\frac{\lambda_{10}(s|\psi_{10}).\lambda_{11}(s|\psi_{11})}{\lambda_{10}(s|\psi_{10}).\lambda_{11}(s|\psi_{11}) + \lambda_{20}(s|\psi_{10}).\lambda_{21}(s|\psi_{11})}.$$

Note that in this case it is not possible to remove the background intensities (λ_{l0}) as these are different for each disease and the conditioning is on the case distribution. If it can be assumed that $\lambda_{10}(s|\psi_{10}) = \lambda_{20}(s|\psi_{10})$ then these cancel in (9.1) and so the type probability is just $\lambda_{11}(s|\psi_{11})/(\lambda_{11}(s|\psi_{11}) + \lambda_{21}(s|\psi_{11}))$. Hence there is some advantage in seeking a common control realization. Note that other conditional probabilities can be derived (such as $\Pr(case(s))$ given the joint case-control realization). Suitable likelihoods can be constructed from these probabilities and inference can proceed as usual. Lawson and Williams (2000) provide an early likelihood-based analysis of a competing risk scenario. As far as this author is aware, there are no published examples of the application of Bayesian approaches to these spatial competing risk scenarios.

If an unconditional analysis is to be considered, albeit conditional in the case-control realizations, then a joint likelihood could be derived for the joint realization. Define the binary case-control label as

$$y_{l_i} = \begin{cases} 1 & \text{if the event is a case} \\ 0 & \text{otherwise} \end{cases}$$

and the joint likelihood as

$$L(s|\psi_0, \psi_1) = \prod_{l=1}^{2} \prod_{i=1}^{N_l} p_{l_i}^{y_{l_i}} (1 - p_{l_i})^{(1-y_{l_i})}$$

$$p_{l_i} = \frac{\lambda_l(s_{l_i}|\psi_l)}{\lambda_{l0}(s_{l_i}|\psi_{l0}) + \lambda_l(s_{l_i}|\psi_l)}$$

$$= \frac{\lambda_{l1}(s_{l_i}|\psi_{l1})}{1 + \lambda_{l1}(s_{l_i}|\psi_{l1})}.$$

9.2.1.1 Specification of λ_{l1}

The specification of λ_{l1} $l = 1, 2$ can now be designed to model the relation between the two diseases. For example, simple functional dependence between the two could be modeled by a relation such as $\lambda_{11}(s|\psi_{11}) = \rho\lambda_{21}(s|\psi_{21})$. Often a log-linear specification is assumed for the intensity so that $\lambda_{l1}(s_{l_i}|\psi_{l1}) = \exp\{f(s_{l_i}, \psi_{l1})\}$. For descriptive models, often the linear component will consist of covariate terms and/or random effect terms. For example

$$f(s_{1_i}, \psi_{11}) = \beta_0 + \mathbf{x}'_i\boldsymbol{\beta}_1 + v_{1_i}$$
$$f(s_{2_i}, \psi_{21}) = \beta_0 + \mathbf{x}'_i\boldsymbol{\beta}_2 + v_{2_i},$$

could be assumed, where $\mathbf{x}'_i\boldsymbol{\beta}_*$ is a covariate predictor with common covariates and separate parameters, and v_{1_i}, v_{2_i} are random effects for each disease. Often these random effects would be uncorrelated, but more complex structuring could be assumed. A common random effect could be added to this model so that

$$f(s_{1_i}, \psi_{11}) = \beta_0 + \mathbf{x}'_i\boldsymbol{\beta}_1 + v_{1_i} + \delta w_i \qquad (9.2)$$
$$f(s_{2_i}, \psi_{21}) = \beta_0 + \mathbf{x}'_i\boldsymbol{\beta}_2 + v_{2_i} + w_i/\delta$$

and then a common random component is included. This could be spatially-structured (as a zero-mean Gaussian process) or could be uncorrelated. A linked component model where a common component has been included for two diseases (for count data) has been proposed by Knorr-Held and Best (2001) and further, Held et al. (2005) suggested a more general model. The components suggested were included to mimic an ecological regression on an unobserved component. While this was proposed for count data there is no reason why this idea could not be applied with case event data. Of course other forms of common parameterization can be considered and what is described here in (9.2) is a very basic proposal.

An alternative possibility is to consider that a bivariate effect is included in the models so that a defined correlation parameter is included. For example, the joint model

$$f(s_{1_i}, \psi_{11}) = \beta_{10} + \mathbf{x}'_i\boldsymbol{\beta}_1 + W_{1i}$$
$$f(s_{2_i}, \psi_{21}) = \beta_{20} + \mathbf{x}'_i\boldsymbol{\beta}_2 + W_{2i}$$
$$\begin{pmatrix} W_{1i} \\ W_{2i} \end{pmatrix} \sim \mathbf{N}(\mathbf{0}, \boldsymbol{\Gamma})$$
$$\text{with } \boldsymbol{\Gamma} = \begin{pmatrix} \kappa_1 & \sqrt{\kappa_1}\sqrt{\kappa_2}\rho_W \\ \sqrt{\kappa_1}\sqrt{\kappa_2}\rho_W & \kappa_2 \end{pmatrix}$$

with no spatial correlation could be considered, where ρ_W is a correlation parameter and κ_1, and κ_2 are variances. Further extension to cross-correlated spatially-structured prior distributions could also be considered.

9.2.2 Count Data

For counts the observed data is now $\{y_{l_i}\}$, $l = 1, 2$ and $i = 1, ..., m$. Analysis will depend on the object of the study. If the object is to locally (in space) compare the risk of getting disease 1 or disease 2, then a *competing risk* formulation might be useful. This might condition on the observed total of cases within spatial units $N_i = y_{1_i} + y_{2_i}$. On the other hand it might be important to jointly model each disease (perhaps with linking parameters) without conditioning.

In the simple case where we condition on a small area total then conditioning on the sum of two (conditionally independent) Poisson variates leads straightforwardly to a binomial model for either type count. For example we could model the disease 1 count as

$$y_{1_i} \sim bin(p_{1_i}, N_i);$$

this has sometimes been called the proportional mortality model (see e.g., Dabney and Wakefield, 2005). In this formulation the probability can be formulated to define the relation between the two diseases. For example the logit of the probability can be modeled with differences in risk terms:

$$\log it(p_{1_i}) = \log(\frac{e_{1_i}}{e_{2_i}}) + \beta_0 + W_i^*$$
$$W_i^* = (W_{1_i} - W_{2_i})$$

where $W_i^* = (W_{1_i} - W_{2_i})$ is the difference between the random effects for the two diseases. These could have UH and CH components also. In this formulation the relative expectation of the two diseases is also included and the random effect W_i^* (which is a difference) can be directly modeled with prior distributions for its components.

As mentioned above it is possible to consider

1. A joint model for the two diseases with common components:

$$y_{1_i} \sim Pois(\mu_{1_i}) \qquad (9.3)$$
$$y_{2_i} \sim Pois(\mu_{2_i})$$
$$\mu_{1_i} = e_{1_i} \cdot \exp\{\beta_{10} + \mathbf{x}'_i\boldsymbol{\beta}_1 + W_{1i}\}$$
$$\mu_{2_i} = e_{2_i} \cdot \exp\{\beta_{20} + \mathbf{x}'_i\boldsymbol{\beta}_2 + W_{1i}\}$$

where $\mathbf{x}'_i\boldsymbol{\beta}_1$, $\mathbf{x}'_i\boldsymbol{\beta}_2$ is the covariate predictor for disease 1 and 2 respectively, and W_{1i} is a common component. As before this component is a composite random effect term and could consist of UH and CH components. Instead, as in the case event scenario, separate random effects could be assumed which have a prior correlation so that

$$\mu_{1_i} = e_{1_i} \cdot \exp\{\beta_{10} + \mathbf{x}'_i\boldsymbol{\beta}_1 + W_{1i}\}$$
$$\mu_{2_i} = e_{2_i} \cdot \exp\{\beta_{20} + \mathbf{x}'_i\boldsymbol{\beta}_2 + W_{2i}\}$$
$$\begin{pmatrix} W_{1i} \\ W_{2i} \end{pmatrix} \sim \mathbf{N}(\mathbf{0}, \boldsymbol{\Gamma})$$
$$\text{with } \boldsymbol{\Gamma} = \begin{pmatrix} \kappa_1 & \sqrt{\kappa_1}\sqrt{\kappa_2}\rho_W \\ \sqrt{\kappa_1}\sqrt{\kappa_2}\rho_W & \kappa_2 \end{pmatrix}$$

where κ_1, and κ_2 are the prior variances of the two effects and as before the ρ_W is the prior correlation.

2. A shared component model which mimics an ecological regression on the unobserved shared component. In the example of Knorr-Held and Best (2001) this takes the form of

$$\mu_{1_i} = e_{1_i} . \exp\{\beta_{10} + u_{2_i} + \delta u_{1i}\} \quad (9.4)$$
$$\mu_{2_i} = e_{2_i} . \exp\{\beta_{20} + u_{1i}/\delta\},$$

where u_{2_i} is a separate random component for the first disease and u_{1i} is the shared component and δ is a scaling component that is necessary because of identifiability. Usually δ has the prior distribution $\log(\delta) \sim N(0, \kappa_\delta)$ with a small fixed variance of $\kappa_\delta = 0.17$ in Knorr-Held and Best (2001). In Knorr-Held and Best (2001), both the shared term u_{1i} and u_{2_i} were assumed to have a CAR prior distribution with

$$p(\mathbf{u}|\kappa) \propto \kappa^{(m-1)/2} \exp\left\{-\frac{\kappa}{2} \sum_{i \sim j} (u_i - u_j)^2\right\}$$

Dabney and Wakefield (2005) give examples of comparison of the proportional mortality/incidence model with shared component models in a particular application.

9.2.3 Georgia County Level Example (3 Diseases)

As an example of the application of the above models, we have examined a group of diseases which could a priori be considered to display the same or similar distributional features. These are ambulatory care-sensitive conditions and counts of these conditions were recorded. These are chronic diseases which are to some measure exacerbated by adverse environmental air pollution: asthma, chronic obstructive pulmonary disease (COPD), and angina. While both asthma and COPD are respiratory in nature, whereas angina is a coronary outcome, all the diseases could be affected by poor environmental air quality. Therefore, in some part, there may be some evidence of this feature in the spatial distribution. A common component could emerge from a joint analysis or at least some evidence of correlation between the spatial distributions could be apparent. In the following, a range of models have been fitted to parts of these data. A full multivariate analysis is examined in a later section (9.3). The models demonstrated here are the individual convolution models for each disease (Model 1), the joint model with common CAR component with separate UH components (Model 2) and the shared component model (Model 3). The diseases compared in Models 2) and 3) are asthma and COPD. Figure 9.1 displays the results for separate analysis of the three diseases (Model 1). Table 9.1 displays the results for these models in terms of DIC. It is clear that a common component as in Model 2 leads to a improved goodness-of-fit as judged by lower DIC. The improvement seems to be mostly to the asthma (DIC from 1144.7 to 1137.3). The form of the common component an be seen in Figure 9.3 and it appears that there is a considerable positive concentration of common risk in the southeastern portion of

TABLE 9.1
Model comparisons for the three disease examples: joint, common, and shared

model	DIC	pD
1 (COPD, angina, asthma)	1239.8, 840.9,1144.7	145.9, 121.1, 127.8
1 overall	3225.46	394.8
2 common (COPD, asthma)	1238.3, 1137.3	144.6, 121.5
3 shared (COPD, asthma)	1244.9, 1131.97	141.7, 119.0

the state. The separate posterior mean uncorrelated components appear to be quite different in distribution (Figure 9.2). Of course it should be borne in mind that there is some lack of identifiability in these components and so more reliance should be placed on the shared component model (Model 3 as in (9.4)). The results of the shared component model (Model 3 in Table 9.1) with prior distribution $\log(\delta) \sim N(0, \kappa_\delta)$ with $\kappa_\delta = 0.17$, seems to suggest that the shared component does reduce the DIC close to that of the common component model (overall DIC: 2375.6 and 2376.9 for common and shared component, respectively). However the shared component model yields a lower pD, as there is only one separate component, instead of two components. The posterior expected estimate (sd) of the δ parameter is 0.6855 (0.03712). The model fitted was

$$\mu_{AS_i} = e_{AS_i} \cdot \exp\{\beta_{10} + u_{2_i} + \delta u_{1i}\}$$
$$\mu_{COPD_i} = e_{COPD_i} \cdot \exp\{\beta_{20} + u_{1i}/\delta\}$$

with both u_{2_i} and u_{1i} having CAR prior distributions with $\tau_{u^*}^{-1/2} \sim U(0, 100)$. It is debatable whether there is an advantage in fitting the shared component model here as there is no separate component for COPD.

In Figure 9.4 the left panel displays the separate CAR component for asthma while the common shared component is displayed in the right panel. These components are remarkably similar in spatial distribution although the shared component had a greater concentration of common risk in the far south of the state. This is slightly different in distribution from the common component in Model 2.

9.3 Multiple Diseases

Many of the considerations that apply to two diseases extend straightforwardly to more than two diseases. It is often the case that a range of diseases might be of interest in both public health applications and in epidemiological

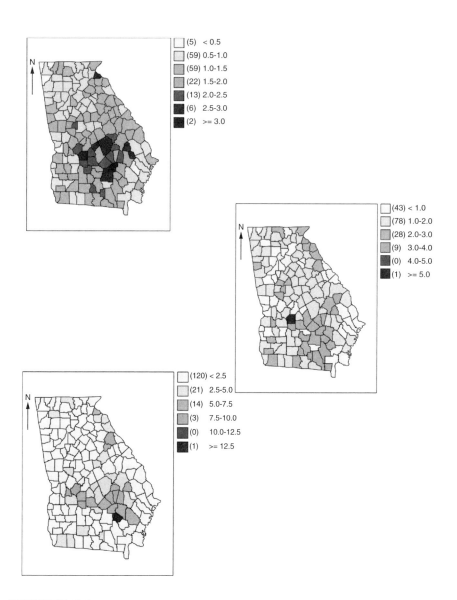

FIGURE 9.1
Georgia county level data, 2005: asthma, angina, and COPD (all ages standardised by the state rate).

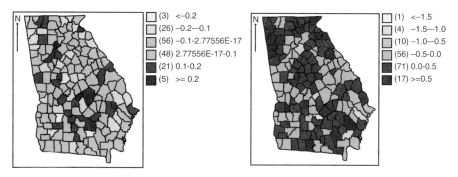

FIGURE 9.2
Georgia county level three diseases: posterior expected UH components for asthma (left) and COPD (right) when fitted with a common CAR component.

studies. However there are also more difficulties, as the possible comparisons increase with L. Multivariate disease modeling is now the focus.

9.3.1 Case Event Data

Assume that there are L diseases with $\{s_{l_i}\}$, $l = 1, ..., L$ and $i = 1, ..., N_l$ where $L > 2$. Here $N_l = m_l + m_{c_l}$ so that there are m_l cases and m_{c_l} controls for the l th disease. Here I will only consider models that are conditional on the realization of the case and control events. Hence the assumption is made that m_l, m_{c_l} and the locations of these cases and controls $\{s_{l_i}\}$, $l = 1, ..., L$ are fixed at the likelihood level. Often groups of disease are the focus. For example, it might be that a range of respiratory (asthma and COPD) and other chronic diseases (angina) are to be examined in relation to an air pollution source or to general pollution levels measured at sites. We are now interested in the case event intensities $\lambda_l(s|\psi_l) = \lambda_{l0}(s|\psi_{l0}).\lambda_{l1}(s|\psi_{l1})$, $l = 1, .., L$. Consideration similar to Section (9.2.1) lead to different forms of conditioning. First it is clear that conditional on an event at s, the probability it is of disease type k is a normalization: $\lambda_k(s|\psi_l)/\sum_{l=1}^{L}\lambda_l(s|\psi_l)$. Other normalizations can be derived.

It is important to focus on particular forms of inference. For example, if we are interested in competing risks of one disease over another and want to make inference about the distribution of case types then

$$\Pr(l(s) = k \text{ and } case(s)) = [\lambda_k(s|\psi_l)/\sum_{l=1}^{L}\lambda_l(s|\psi_l)].\Pr(case(s))$$

where $\Pr(case(s)) = \sum_l \lambda_l(s|\psi_l)/[\sum_l \lambda_l(s|\psi_l) + \sum_l \lambda_{l0}(s|\psi_{l0})]$.

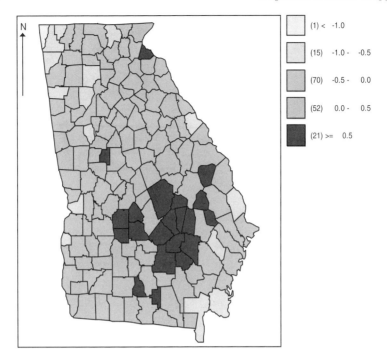

FIGURE 9.3
Georgia county level three diseases. Analysis with common component: posterior expectation of the common CAR component (Model 2).

In general, it is possible to define a likelihood for particular situations. Lawson and Williams (2000) proposed conditional independence likelihoods for a putative hazard example.

As far as this author is aware, there are few published examples of Bayesian analysis of multitype spatial disease realizations. A multivariate analysis of the residential locations of death certificates for respiratory disease (bronchitis) and air-way cancers (respiratory, gastric, and oesophageal) was proposed by Lawson and Williams (2000). Data were obtained for the years 1966–1976 for a small industrial town in the United Kingdom. These diseases were chosen as a set of diseases potentially related to adverse air pollution. Control diseases examined were coronary heart disease mortality (which is age-related but not usually affected directly by air pollution), and a composite control of lowerbody cancers (prostate, penis, testes, breast, cervix, uterus, colon, and rectum). These latter cancers were useful as they are less affected by respiratory inhalation insult and so can be regarded as a reasonable control which is matched on age to the risk profile of the case diseases. Figures 9.5 and 9.6 display the location maps of the residences. It is notable how the composite

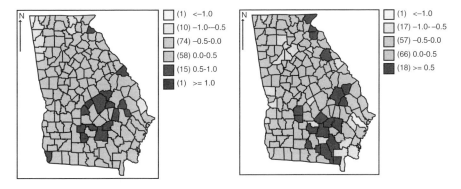

FIGURE 9.4
Georgia county level asthma and COPD: posterior expected estimates of the asthma CH component (left) and the shared component (right).

control follows closely the CHD spatial distribution. In the example of Lawson and Williams (2000), the first order intensity was related to a fixed putative pollution source via a distance measure, so that $\lambda_{l1}(s|\psi_{l1}) = 1 + f_l(s|\psi_{l1})$ and for each disease the link was defined as $f_l(s|\psi_{l1}) = \alpha_{1l}\exp[-\alpha_{2l}d(s) + \alpha_{3l}\log d(s)]$ where $d(s)$ is the distance from the location s to a fixed point (putative source). A joint likelihood was derived and estimation of parameters proceeded via McMC applied to the posterior distribution with uniform prior distributions for all parameters. The likelihood used conditional probabilities for different case types and a common control disease. Further, exploration of the possibility of weighting of different diseases was considered via the total intensity specification: $\lambda(s|\psi) = \sum_{l} w_l \lambda_l(s|\psi_l)$. Subsequent development of a weighted likelihood was considered in the context of prior expert opinion about what weight each disease should have in defining evidence for an effect.

Further non-Bayesian analysis of multiple diseases has been developed within a non-parametric smoothing approach by Diggle et al. (2005), where estimation of the conditional disease probability:

$$\widehat{p}(s) = \lambda_k(s|\psi_l) / \sum_{l=1}^{L} \lambda_l(s|\psi_l)$$

is carried out non-parametrically to produce surfaces of these probabilities.

The basic likelihood derived for the multitype situation can be seen as a special case of an ordinal logistic formulation where a probability of a disease type is to be modeled. Different formulations of ordinal logistic regression can be considered, but the commonest for nominal categories is the multinomial logit model. If a single common control is assumed it would be very convenient

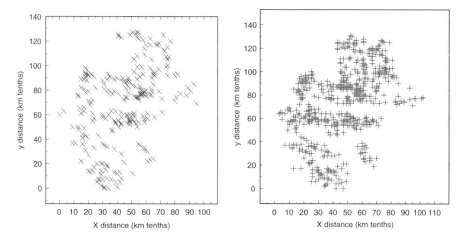

FIGURE 9.5
Arbroath mortality study: control disease realisations: composite cancer control (left panel); CHD control (right panel).

to consider that as a baseline category in the comparison. This is denoted as $L+1$ below. Hence, a possible multinomial logit model would be

$$\log\left(\frac{p_k(s_{k_i})}{p_{L+1}(s_{k_i})}\right) = \alpha_k + f_l(s_{k_i}|\psi_{l1})$$

$$\text{where } p_k(s_{k_i}) = \lambda_k(s_{k_i}|\psi_k)/\sum_{l=1}^{L+1}\lambda_l(s_{k_i}|\psi_l).$$

Again it would be straightforward to define a Bayesian hierarchical model around this formulation. For example, $\log\left(\frac{p_k(s_{k_i})}{p_{L+1}(s_{k_i})}\right) = \alpha_k + f_l(s_{k_i}|\psi_{l1})$ where $f_l(s_{k_i}|\psi_{l1}) = w_{l_i} + v_{l_i}$ where w_{l_i}, v_{l_i} are spatially-correlated and uncorrelated random effects. This would allow separate random effects for each disease. Zhou et al. (2007) gives an example of Bayesian formulation where spatial correlated effects are modeled with categorical ordinal outcomes, and these models could be modified for the simpler nominal case.

As an example of the application of this multinomial logit model, the Arbroath case event data has been examined. Only one control is assumed and it is regarded as the comparison group here. In the following, the composite lower body cancer is the control disease (label 1), followed by gastric and oesophageal cancer (label 2), respiratory cancer (label 3), and bronchitis (label 4). The multinomial logit model was fitted assuming a simple random effect model with y_{il} denoting a sparse indicator variable of dimension $N_T \times L$,

Multivariate Disease Analysis 213

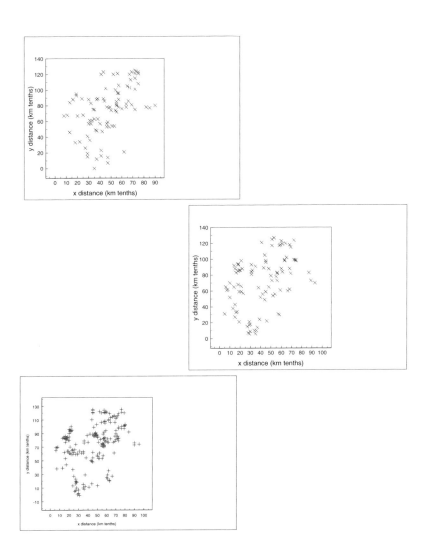

FIGURE 9.6
Arbroath mortality study: gastric and oesophageal cancer (top); respiratory cancer (middle); bronchitis (bottom).

where $N_T = \sum_l N_l$, which takes values as follows

$$y_{il} = \begin{cases} 1 \text{ if } l = 1 \\ 0 \text{ otherwise} \end{cases} i = 1, 250, l = 1, .., L$$

.
.

$$y_{il} = \begin{cases} L \text{ if } l = L \\ 0 \text{ otherwise} \end{cases} i = 437, 630, l = 1, .., L$$

In the Arbroath example, $L = 4$, and $N_T = 630$ with 250 control cases, 90 gastric and oesophageal cancer, 97 respiratory cancer, and 193 bronchitis case events. Assume that

$$\mathbf{y}_i \sim Mult(\mathbf{p}_i, 1)$$

where

$$\Pr(y_{il} = 1) = p_{il} = \lambda_l(s_{l_i}|\psi_l)/\{1 + \sum_{k=2}^{L} \lambda_k(s_{l_i}|\psi_k)\} \ l > 1$$

$$\Pr(y_{i1} = 1) = p_{i1} = 1/\{1 + \sum_{k=2}^{L} \lambda_k(s_{l_i}|\psi_k)\}$$

with $\lambda_l(s_{l_i}|\psi_l) = \exp(\alpha_l + w_{il})$ where

$$\alpha_l \sim N(0, \tau_l)$$
$$w_{il} \sim N(0, \tau_{w_l})$$

with a separate intercept and a simple uncorrelated effect for each disease. Suitably dispersed prior distributions were assumed for the variance parameters. More complex random effect structures could be envisaged of course. However this formulation serves to demonstrate the modeling approach. Following convergence the DIC for this model was 3054.97 with pD = 579.14. The model provides relative estimates of disease probabilities as well as scalar parameters. The posterior expected estimates of α_l, $l = 2, 3, 4$ with sd in brackets: -2.455(0.2331), -2.078(0.2208), -0.234(0.1921). It appears from this that gastric and oesophageal cancer and respiratory cancer have significantly different overall levels compared to the combined control whereas the bronchitis level is not significant. Figures 9.7 and 9.8 display the posterior expected estimates for the p_{il} for the control ($l = 1$), gastric and oesophageal cancer ($l = 2$), respiratory cancer ($l = 3$), and bronchitis ($l = 4$). It is noticeable that the control has a highly variable distribution, while, in relation to control, the gastric and oesophageal cancer appears with peaks in markedly different locations (the respiratory cancer has a similar patterning). The bronchitis distribution also differs from control but seems to have larger areas of elevated risk.

It is of course possible to extend this approach to models with more sophisticated random components (such as CH components based on full MVN prior

Multivariate Disease Analysis 215

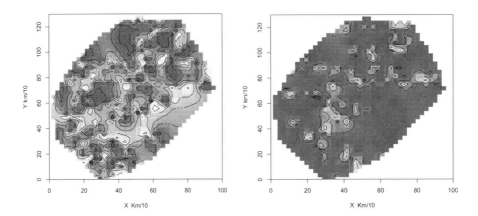

FIGURE 9.7
Arbroath study: posterior expected probability surface ($\bar{p}_{il}$) for the composite control (left), and gastric and oesophageal cancer (right).

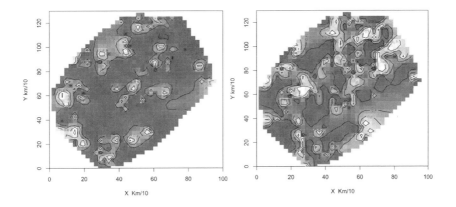

FIGURE 9.8
Arbroath study: posterior expected probability surface ($\bar{p}_{il}$) for the respiratory cancer (left), and bronchitis (right).

specification or approximate MRF models based on Voronoi neighborhoods). Indeed models including correlation between diseases, with for example share components or cross-correlation could be specified. This is largely unexplored in this application area.

9.3.2 Count Data

In the case of count data various possibilities exist. Assume that there are $\{y_{il}\}$, $l = 1,..,L$ and $i = 1,...,m$, with $L > 2$. Hence in each area there is a vector of counts representing the L different diseases. Various approaches can be adopted depending on the focus. First, by conditioning on the total count within the small area: $y_{T_i} = \sum_{l=1}^{L} y_{il}$, it is possible to consider the multinomial distribution for the count probability vector $\mathbf{y}_i$. On the other hand, it is also possible to examine the unconditional distribution of the counts assuming conditional independence and a Poisson count distribution. In the first case, assume that

$$\mathbf{y}_i \sim Mult(\mathbf{p}_i, y_{T_i})$$

and, because of the constraint, the probability vector is defined to be $0 < p_{il} < 1$, and $\sum_{l=1}^{L} p_{il} = 1 \; \forall i$. The log-likelihood is then considered to be

$$l(\mathbf{y}|\mathbf{p}) = \sum_{i=1}^{m} \sum_{l=1}^{L} y_{il} \log p_{il}.$$

To model the probabilities, it is useful to assume that they arise from a normalization such as:

$$p_{il} = \frac{\lambda_{il}}{\sum_k \lambda_{ik}}$$

where the constant term in the rate (λ_{il}) cancels out. Here it would also be convenient to assume that rate terms consist of a log linear function of covariates or random effects. A typical general example could be $\lambda_{il} = e_{il}\theta_{il}$ where $\theta_{il} = \exp\{\alpha_l + \mathbf{x}'_i\boldsymbol{\beta} + u_{il} + w_{il}\}$ where α_l is a disease specific intercept, $\mathbf{x}'_i\boldsymbol{\beta}$ a linear predictor, and $\mathbf{u}_l$, and $\mathbf{w}_l$ are disease specific random effects. Other forms are possible in specific applications. Clearly by normalization, the conditioning on the total disease count in each area yields relative inference concerning the disease distribution. An example of the application of such a model was made to the chronic three-disease example for county-level data for Georgia. In this case a description of the three diseases is sought and no covariates are included. Hence the form $\theta_{il} = \exp\{\alpha_l + u_{il} + w_{il}\}$ is assumed where u_{il} has a CAR prior distribution specification for each disease and w_{il} has a zero-mean Gaussian specification. In that analysis the converged sampler

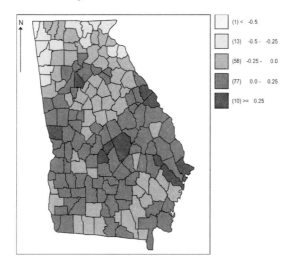

FIGURE 9.9
Georgia county level three chronic diseases 2005: asthma: spatially-correlated random effect (u_1).

yielded a DIC of 1879.9 with pD = 240.112. The resulting spatially-correlated random effects for the three diseases are shown in Figures 9.9, 9.10, 9.11. It is clear that under this multinomial model the spatially-structured risk is quite different for each of the three cases. In fact the distribution of high risk areas of COPD seems to be inversely related spatially to those of angina.

Alternative formulations of multivariate risk can be envisaged. In fact the shared component models discussed in Section 9.2.2 have been extended to multiple diseases by Held et al. (2005). In their formulation a Poisson likelihood is assumed:

$$y_{il} \sim Poisson(e_{il} \exp[\eta_{il}])$$

and

$$\eta_{il} \sim N(\alpha_l + \sum_k \delta_{k,l} u_{ki}, \tau_l)$$

and

$$\sum_{l=1}^{n_k} \delta_{k,l} = 0$$

and the terms $\log \delta_{k,1}, \ldots, \log \delta_{k,n_k}$ have multivariate normal distribution with mean zero and given marginal variance. Once more than two diseases are examined however the interpretation of a shared component is more difficult.

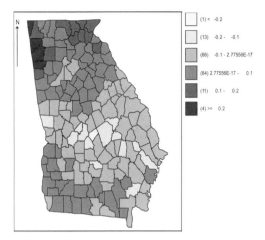

FIGURE 9.10
Georgia county level three chronic diseases 2005: COPD: spatially-correlated random effect (u_2).

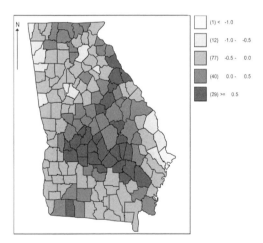

FIGURE 9.11
Georgia county level three chronic diseases 2005: angina: spatially-correlated random effect (u_3).

9.3.3 Multivariate Spatial Correlation and MCAR Models

9.3.3.1 Multivariate Gaussian models

In general, once multiple diseases are admitted into an analysis there is a need to consider relations between the diseases. This can be done in a variety of ways. A basic approach to this is to consider cross-correlation between the diseases. There is considerable literature on the specification of cross-correlation models for Gaussian processes (see e.g., Banerjee et al., 2004). In general, define an L-dimensional vector $\mathbf{Y}_i$ $i = 1, ..., m$ observed at a set of sites. For a multivariate Gaussian process, a common assumption would be that

$$\mathbf{Y} \sim \mathbf{MVN}_{mL}(\boldsymbol{\mu}, \mathbf{A_Y})$$

where $\boldsymbol{\mu}$ is $m \times L$ and $\mathbf{A_Y}$ has dimension $mL \times mL$. It is convenient to consider a block representation of $\mathbf{A_Y}$ which stresses the covariance in cross-covariance form:

$$\mathbf{A_Y} = \left\{ \begin{array}{cccc} A_{11} & A_{12} & . & A_{1L} \\ A_{21} & A_{22} & . & . \\ . & . & . & . \\ A_{L1} & & & A_{LL} \end{array} \right\}.$$

Here each of the diagonal block matrices are internal covariances within the given field whereas the off-diagonal block matrices define the cross-correlations between components. The dimension of the block matrices is $m \times m$ if all the fields are observed at the same m locations (sites), whereas if the different fields are measured at different numbers of sites then each diagonal matrix will be square and will have different dimension and the off-diagonals will not necessarily be square either.

Various models can be assumed for the overall covariance structure of a set of Gaussian fields. Often simple assumptions are made to allow for computation. Banerjee et al. (2004) discuss various examples of separable models and asymmetric cases (mainly for simple situations where each field is measured on the same grid). They also extend the analysis by considering the linear model for coregionalization (LMC) which specified that a multivariate process is a linear function of iid spatial processes with zero mean, variance 1 and spatial covariance function $\rho(h)$ for distance h. More generally separate covariance functions $\rho_l(h)$ can be assumed so that the cross-covariance is defined as $A_{ll'} = \sum_{j=1}^{L} \rho_j(s - s')T_j$ for locations s and s', where T_j is the covariance matrix for the j th component. An alternative, computationally attractive, conditional specification was also proposed by Royle and Berliner (1999).

While in general full multivariate cross-correlation models could be employed for modeling continuous multivariate spatial processes, their implementation is not straightforward and in particular their computational demands often force the consideration of simpler formulations. Note that within a disease mapping context these models could form joint prior distributions for spatial random effects (rather than models for observed Gaussian fields), especially for case event models where continuous spatial effects are naturally favored. Hence for the i th case event we might be interested in the vector of intensities:

$$\boldsymbol{\lambda}(s_i|\boldsymbol{\psi}) = \exp\left(\boldsymbol{\Delta}_i + \mathbf{Y}_i\right)$$

where $\boldsymbol{\Delta}$ includes fixed and uncorrelated random effects and $\mathbf{Y}$ is a multivariate spatial Gaussian process. For count data this might take the form, for the i th small area with area denoted as a_i:

$$\boldsymbol{\theta}_i = \exp(\boldsymbol{\Delta}_i + \boldsymbol{\mu}_i)$$
$$\boldsymbol{\mu}_i = \int_{a_i} \mathbf{Y}(u) du$$

Often for count data and approximately for case event data a Markov random field (MRF) specification is adopted at least for simplicity of implementation. In the next section these multivariate CAR models are discussed.

9.3.3.2 MVCAR models

The MVCAR model of Gelfand and Vounatsou (2003) specifies that the $m \times L$ matrix of random effects $\boldsymbol{\phi}$ in the model

$$y_{il} \sim Poisson(e_{il} \exp[x'_{il}\boldsymbol{\beta}_l + \phi_{il}])$$

is defined with a constraint that the spatial effects separate into non-spatial and spatially structured effects:

$$\boldsymbol{\phi} \sim \mathbf{N}_{mL}(\mathbf{0}, \mathbf{H}_1)$$

where $\mathbf{H}_1 = [\boldsymbol{\Lambda} \otimes (D - \alpha W)]^{-1}$ with $\otimes$ denoting Kronecker product, and D is a $m \times m$ diagonal matrix with elements which are the number of neighbors of the i th region and W is an adjacency matrix where $W_{ii} = 0$ and $W_{ij} = 1$ if the areas i, j are adjacent (i.e., $i \sim j$) and 0 otherwise. Here $\boldsymbol{\Lambda}$ is a $L \times L$ positive definite matrix of non-spatial precisions, defining the relation between diseases and α is a common spatial autocorrelation parameter. This is denoted as the $MCAR(\alpha, \boldsymbol{\Lambda})$ model. This model can be extended to allow for separate autocorrelation (smoothing) for each disease:

$$\boldsymbol{\phi} \sim \mathbf{N}_{mL}(\mathbf{0}, \mathbf{H}_2)$$

where $\mathbf{H}_2 = [Q(\boldsymbol{\Lambda} \otimes I_{m \times m})Q']^{-1}$ and $Q = diag(R_1, ..., R_L)$ and $R_l = chol(D - \alpha_l W)$, $l = 1, ..., L$, where $chol()$ denotes the cholesky decomposition. This has been termed the $MCAR(\boldsymbol{\alpha}, \boldsymbol{\Lambda})$. Extensions and variants to these models have been proposed by Kim et al. (2001) and Jin et al. (2005). Restriction to the conditional ordering of the effects in the GMCAR model of Jin et al. (2005), have led to a different approach.

9.3.3.3 Linear model of coregionalization

A classic approach to modeling cross-correlation between spatial fields is to adopt a simple model for the relation between selected fields. Within Geostatistics, the linear model of coregionalization (LMC) is commonly assumed for this purpose Wackernagel, 2003. In that model a set of random spatial functions $\{z_l(s); l = 1, ..., L\}$ are modeled via a linear combination of uncorrelated factors $(Y_u^l(s))$ and $u = 1, .., S$ components:

$$z_l(s) = \sum_{u=0}^{S} \sum_{l=1}^{L} a_u^l Y_u^l(s).$$

This idea has been used by Jin et al. (2008) in extending the multivariate models for disease mapping to allow order-free modeling. In their formulation the model at the likelihood level is

$$y_{il} \sim Poisson(e_{il} \exp[x'_{il}\beta_l + \phi_{il}])$$

where ϕ_{il} are random effects for each unit and disease. The joint distribution of ϕ is defined to be

$$\phi \sim \mathbf{N}_{mL}(\mathbf{0}, \mathbf{G})$$

where $\mathbf{G} = (A \otimes I_{m \times m})(I_{L \times L} \otimes D - B \otimes W)^{-1}(A \otimes I_{m \times m})'$ with $\otimes$ denoting Kronecker product and B includes smoothing parameters in the cross-covariances of the field, D is a $m \times m$ diagonal matrix with elements which are the number of neighbors of the i th region, and W is an adjacency matrix where $W_{ii} = 0$ and $W_{ij} = 1$ if the areas i, j are adjacent (i.e., $i \sim j$) and 0 otherwise.

This is defined as a $MCAR(B,\Sigma)$ distribution, where the diagonal elements of B are a correlation within a spatial process and the off-diagonals are the cross-correlations between any two processes. These are scalar quantities. Essentially, the difference with the $MCAR(\alpha, \Lambda)$ lies in the the elements of B: if the $b_{jl} = 0$ and $b_{jj} = \alpha_j$ then the $MCAR(\alpha, \Lambda)$ results. Special prior distribution constructions must be examined to ensure that the eigenvalues of B lie in the correct range. Jin et al. (2008) give examples of its use and compare different formulations.

9.3.3.4 Model fitting on WinBUGS

Currently, only the intrinsic (improper) version of the $MCAR(\alpha, \Lambda)$ model is available automatically on WinBUGS. This version forces the value of $\alpha = 1$, and this implies that this can be used as a prior distribution only, assuming that propriety of the posterior distribution can be assured. The command in WinBUGS for this is the mv.car distribution. It is also possible to fit a proper $MCAR(\alpha, \Lambda)$ if it is assumed, via the LMC, that

$$\phi = (A \otimes I_{m \times m})\mathbf{u}$$

where $\mathbf{u}_l$ $l = 1,...,L$ are assumed to have proper univariate CAR prior distributions (on WinBUGS: car.proper distribution) with common smoothing parameter α. This fixes A as it is the Cholesky decomposition of Λ although it may be preferred to allow a separate prior specification for Λ. Again by the LMC it is possible to extend this idea to fitting $MCAR(\boldsymbol{\alpha}, \boldsymbol{\Lambda})$ models. In that case, as before assign proper univariate CAR prior distributions to $\mathbf{u}_l$ $l = 1,...,L$ but with separate smoothing parameters: $\alpha_l, l = 1,...,L$. Assuming an inverse Wishart prior distribution for $\boldsymbol{\Lambda}$ determines A.

9.3.4 Georgia Chronic Ambulatory Care-Sensitive Example

In the example above concerning chronic ambulatory care sensitive diseases, three diseases were examined: asthma, COPD, and angina. These were examined as counts for the year 2005 in Georgia counties. In Section (9.2.3), the analysis of these diseases was limited to two diseases only. Here all three are considered together in a multivariate framework. Here, it is assumed that each disease has a log-linear link to a linear predictor which consists of random effect components. In particular it is assumed that two additive random effects are included in the form

$$\log(\mu_{l_i}) = \log(e_{l_i}) + \alpha_l + W_{l_i} + U_{l_i}$$

where

$$U_{l_i} \sim MVN(\mathbf{0}, \Sigma)$$
$$W_{l_i} \sim MCAR(1, \Omega).$$

The first effect is an uncorrelated effect with zero mean and diagonal covariance matrix where $\Sigma = diag(\tau_1,...,\tau_L)$. For the second term an intrinsic CAR model was assumed using the mv.car distribution. The 3×3 precision matrix has assigned to it a Wishart prior distribution with parameter matrix R, and the covariance matrix defined as Ω^{-1}. Additional assumptions about the model components were made. These include a Wishart prior distribution of the precisions of the uncorrelated effects (Σ^{-1}). Flat (uniform) priors for the intercept terms (α_l). The following code was used to specify the covariance priors and resulting standard deviations:

```
omega[1:3, 1:3] ~dwish(R[ , ], 3)
sigma2[1:3, 1:3] <- inverse(omega[ , ])
sigma[1] <- sqrt(sigma2[1, 1])
sigma[2] <- sqrt(sigma2[2, 2])
sigma[3]<-sqrt(sigma2[3,3])
```

The sigma[1:3] terms are conditional standard deviations of the disease components (1: asthma, 2: COPD, 3: angina). It is also possible to compute the correlations between spatial components in this formulation:

> corr12<-sigma2[1,2]/(sigma[1]*sigma[2]) # asthma and COPD
> corr13<-sigma2[1,3]/(sigma[1]*sigma[3]) # asthma and angina
> corr23<-sigma2[2,3]/(sigma[2]*sigma[3]) # COPD and angina

The above analysis allows the computation of the correlations between the spatial random effects for each disease. This is useful but it is sometimes useful to also consider the correlation of the relative risks directly. To compute the correlation between the relative risks it is possible to examine functionals of the posterior distribution. This can be easily implemented in WinBUGS with the commands below. RR1, RR2, RR3 are the relative risk for the asthma, COPD and angina where within a for() loop with index i they are assigned as

RR1[i] <- exp(alpha[1] + S[1,i]+U[i,1])
RR2[i] <- exp(alpha[2] + S[2,i]+U[i,2])
RR3[i] <- exp(alpha[3] + S[3,i]+U[i,3])

where S is the MCAR and U is the (spatially) uncorrelated random effects. Then the empirical posterior correlation can be computed as

```
mu1<-mean(RR1[])
mu2<-mean(RR2[])
mu3<-mean(RR3[])
sd1<-sd(RR1[])
sd2<-sd(RR2[])
sd3<-sd(RR3[])
mu12<-inprod(RR1[],RR2[])/N
mu13<-inprod(RR1[],RR3[])/N
mu23<-inprod(RR2[],RR3[])/N
CRR12<-(mu12-mu1*mu2)/(sd1*sd2)
CRR13<-(mu13-mu1*mu3)/(sd1*sd3)
CRR23<-(mu23-mu2*mu3)/(sd2*sd3)
```

The results of fitting this model to the 3 diseases yielded the following results. After convergence, the DIC was 3226.61 with pD as 392.92. This is slightly higher than the DIC for the separate analyses (DIC = 3225.5 and pD = 394.8). However this is not a large difference and so there appears to be little differentiating the models except that the MCAR model provided estimates of correlation and a lower pD of 392.9. From the point of view of parsimony then this might be preferred. The correlation estimates are given in Table 9.2. In that table the top right hand triangle is the posterior expected correlation estimates from the posterior distribution for the spatially-structured random effects, while the lower triangle has the empirical correlations calculated from the posterior sample for the estimated relative risks. It is interesting to note

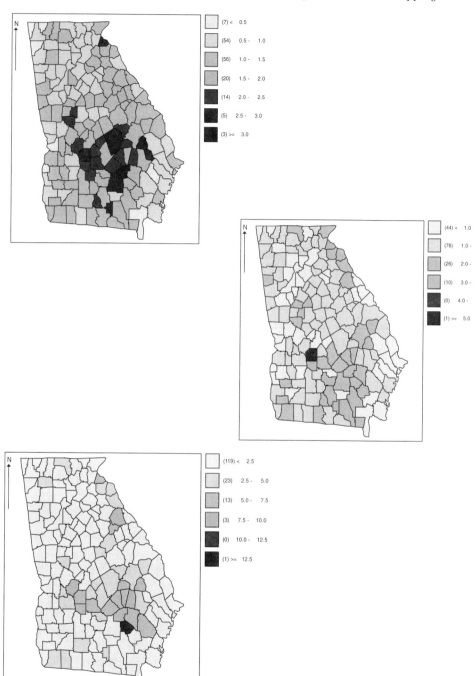

FIGURE 9.12
Georgia county level three disease analysis 2005: posterior expected relative risks under an additive UH and MVCAR model: top left: asthma, right: COPD, bottom left: angina.

TABLE 9.2
Correlation of spatial random effects under the MCAR model for the Georgia chronic disease example: upper triangle: correlation of the spatially structured effects; lower triangle: correlation of the relative risks

Disease	Asthma	COPD	Angina
asthma		0.6617(0.128)	0.7536(0.105)
COPD	0.6004(0.033)		0.7825(0.099)
angina	0.5417(0.044)	0.516(0.042)	

that the correlations between the spatially-structured effects are higher for angina versus COPD and asthma than those correlations for the relative risks. This effect might suggest that there is more uncorrelated noise in the angina relation than in the asthma versus COPD relation, which remains relatively stable. In general, there does appear to be a strong correlation between the risk for these chronic diseases.

10
Spatial Survival and Longitudinal Analysis

10.1 General Issues

In many biostatistical applications there is a need to consider temporal variation. The commonest examples are often found in clinical or behavioral intervention trials where a state can be reached by a patient. The time at which the state is reached could be of primary interest. The end point could be a vital outcome (death) or disease remission, cure or cessation of a behavior. In all cases the time of the event is the important random variable. This is the typical scenario where *survival analysis* is employed. On the other hand, in some clinical or intervention studies, the variation of response over time is to be monitored. For example, cholesterol concentrations in blood might be monitored under different treatments, and the effect of these treatments over time is to be examined. In an intervention trial for diet change, food intake might be repeatedly measured via self-report questionnaire. In these cases, the time of measurement is usually fixed and the measurement itself is the random variable.

In clinical trial applications there is not often a need, or interest, in examining the residential address of the patient or the even the neighborhood or county of residence. As clinical trials are designed experiments, and usually randomized, then there should be less need to consider location of residence or area of residence as a factor affecting outcome. Two factors should be considered, however, that could impact a decision to ignore spatial effects. First, at the design stage there could be a geographical bias in the areas displaying risk for the disease. This bias could manifest itself by differential recruitment. Any strong geographical differences might need to be represented in the study design if the study is to be representative of the population concerned. Even after design, in the analysis phase, there could be strong reasons to examine spatial effects. First, it could be important to know if there were any spatial effects such as confounding between space and intervention groups, in relation to outcome. Second, all studies whether designed or observational, have confounders which could be unknown or known but not included in the study. Hence if the researcher wants a) to explore factors affecting the study and b) to make sure that confounding is allowed for in the analysis, then the use of spatial information can help with both these tasks. Clayton et al. (1993)

have stressed the usefulness of including spatial correlation terms in models to make allowance for confounder and ecological biases and so there is reasonably strong arguments for always including contextual terms that have spatial structure.

10.2 Spatial Survival Analysis

In survival analysis the time to endpoint (T) is the random variable. Assume that a sample of individuals have associated endpoint times. Denote this sample of times as $\{t_i\}$ $i = 1, ..., m$. For now assume these are all observed exactly. In addition to an endpoint time a geo-reference is also available. The geo-reference could be an address location, in which case it is denoted as s_i, or a contextual spatial effect, denoted as w_i. The contextual effect is simply a factor that may have spatial correlation so that areas close together have similar risk. In addition to these ingredients, each observation unit can have covariates associated and, for the i th unit/person these are denoted by the vector $\mathbf{x}_i$. These covariates could be individual or contextual/ecological (see e.g., Chapter 7). Reviews of Bayesian survival methods can be found in Gustafson (1998) and Ibrahaim et al. (2000). For Bayesian proportional hazard modeling see Carlin and Hodges (1999).

10.2.1 Endpoint Distributions

Often a failure time or endpoint distribution is specified for the time to endpoint and this is often chosen from distributions on the R^+ line. Common choices are among the Weibull or extreme value, lognormal, or gamma families.

Hence for a parametric survival model, at the 1st level of the hierarchy, the data model consists of an endpoint distribution. For flexibility, we will assume a Weibull distribution in what follows. The probability of an endpoint at time t_i under a Weibull distribution is specified by

$$f_T(t_i) \equiv f(t_i) = \rho \mu t_i^{\rho-1} \exp(-\mu t_i^\rho). \tag{10.1}$$

The survival and hazard functions derived from this specification are

$$S(t_i) = 1 - \int_0^{t_i} f(u)du = \exp(-\mu t_i^\rho)$$

$$h(t_i) = f(t_i)/S(t_i) = \rho \mu t_i^{\rho-1}.$$

The parameterization emphasizes the modeling of a function of the mean of the distribution via μ. Note that this allows a straightforward interpretation of the model component for this distribution: covariates and contextual effects can be included within μ and the parameter ρ provides the shape of the distribution. Often modeling proceeds via the hazard function, rather than the density, and for the Weibull this decomposes into two components:

$$h(t_i) = h_0(t_i).h_1(t_i) = \rho t_i^{\rho-1}.\mu_i.$$

Here $h_0(t_i)$ is regarded as a baseline hazard, while $h_1(t_i)$ is a non-baseline component which is usually the focus of modeling. Usually it is assumed that a predictor term is linked to the parameter μ_i and each unit will have a different μ_i depending on covariates. A log-linear specification is often assumed. A non-spatial example could be

$$\log(\mu_i) = \mathbf{x}'_i \boldsymbol{\beta} + v_i \qquad (10.2)$$

where $\mathbf{x}'_i$ is a row vector of fixed covariates, $\boldsymbol{\beta}$ the corresponding parameter vector, and v_i is an uncorrelated random effect. For the Weibull distribution, Carlin and Hodges (1999) suggested the formulation (10.2) in a non-spatial setting. This can be extended easily to include a spatial contextual effect. For example, we could specify

$$\log(\mu_i) = x'_i \beta + \Omega_i \qquad (10.3)$$
$$\Omega_i = w_i + v_i$$

where Ω_i contains the unit specific random effect terms. These effects could be individual unit level or they could be contextual.

While the Weibull is a flexible distribution there are many alternatives, some of which do not impose the constraints of proportionality of hazard. A wider class of models is the accelerated failure time (AFT) models. These models replace the t within the survival and hazard functions with a modulated function of covariates: $t \exp\{x'\beta\}$. This leads to a covariate acceleration/deceleration of risk. The Weibull is a special case of this general class. This lead to a linear model in the log of time:

$$\log T = \alpha + x'\beta + \epsilon$$

where ϵ is an error term independent of x (not necessarily zero centered) and α is an intercept. This model could also be extended with addition of random effects which are individual or contextual in the form of

$$\log T = \alpha + x'\beta + \Omega + \epsilon$$

where Ω is a random effect term (as in (10.3) above).

10.2.2 Censoring

Censoring is always an important issue in survival analysis as it is often the case that times are not observed exactly. For parametric models this can be treated via a survival function term product in the likelihood. For example for right censoring we can assume

$$L = \prod_u f(t_u) \prod_c S(t_c)$$

where u denotes uncensored and c denotes right censored. Note that this likelihood simplifies if you assume a censoring indicator γ which takes 0 for censored and 1 for uncensored, as

$$L = \prod_{\text{all } t} h(t)^\gamma S(t).$$

Other likelihood forms can similarly be derived for alternative censoring mechanisms.

10.2.3 Random Effect Specification

As most survival data is observed at the individual unit level there could be either individual covariates or random effects or contextual effects relating to the individual. For example, the age of an individual could be a personal covariate and the location coordinates of the individual's address could be regarded as personal covariates also. In addition, there could be an individual level random effect which allows for frailty among individuals. This could be correlated spatially or uncorrelated. If the residential address of the individual (unit) were known then this could be used directly. An uncorrelated frailty can easily be specified for each individual via

$$v_i \overset{iid}{\sim} N(0, \tau_v)$$

For a spatially correlated term, w_i could be specified to have a full multivariate normal prior distribution with spatial correlation included within the covariance matrix, i.e.,

$$\mathbf{w} \sim \mathbf{N}(\mathbf{0}, \boldsymbol{\Sigma}) \tag{10.4}$$

where the elements of the covariance are a function of distance between locations: $\boldsymbol{\Sigma}_{ij} = f(d_{ij})$ where d_{ij} is the distance between the i th and j th location. Essentially this is a zero-mean spatial Gaussian process prior distribution. Note that it would also be possible to specify an intrinsic Gaussian prior distribution for $\mathbf{w}$ as long as a suitable neighborhood structure could be specified. A distance-threshold or Dirichlet tesselation neighborhood metric could be used to define neighbors. While it is commonplace to model uncorrelated frailty at the individual level, it is less common to find modeling of spatially correlated effects at this level.

As contextual effects, it is relatively straightforward to include spatial effects. For example, with cancer registry data, often individual outcomes have associated county, postal district, or zip code of residence. They often don't have address location for confidentiality reasons. Hence contextual spatial information is often available for these data. This consists of aggregated spatial groupings or factors. For instance, in

$$\Omega_i = w_i + v_i$$

the two effects could be specified as spatially-contextual. For example, assume there are $j = 1, ..., J$ counties within which individuals can live. The county of residence for the i th individual is denoted $c_j \underset{i \in j}{}$ so that

$$w_i = w(\underset{i \in j}{c_j})$$

Hence, the effect for the i th person is assigned from the j th county that they live in. This also applies to any other contextual aggregate effects: for example, uncorrelated effects could also be defined in this way:

$$v_i = v(\underset{i \in j}{c_j}).$$

Henderson et al. (2002) modeled individual level frailty but resorted to modeling the spatial structure of leukemia survival data using district level (contextual) effects. In that work the covariance function $f(.)$ between the i th and j th *district* was modeled via the power exponential and Matérn covariance functions (as in Section 5.4.2). At this level of aggregation, it is natural to consider a CAR specification of the spatial effect at individual level:

$$w_i | w_{-i} \sim N(\overline{w}_{\delta_i}, \tau_w / n_{\delta_i})$$

where $w_i = w(\underset{i \in j}{c_j})$. Henderson et al. (2002) also compared their contextual models with fully specified covariance and CAR components and found that a Matérn covariance, specified with correlation function ($\boldsymbol{\rho}_{ij} = \boldsymbol{\Sigma}_{ij}/\tau$):

$$\boldsymbol{\rho}_{ij} = (d_{ij}/\phi)^\kappa K_\kappa(d_{ij}/\phi)/[2^{\kappa-1}\Gamma(\kappa)] \qquad (10.5)$$

where $K_\kappa(.)$ is a modified Bessel function of the third kind with $\kappa = 2$, gave the best fitting model based on the DIC goodness-of-fit criterion. Hence, in this case, the geostatistical model yielded a more appropriate model than the CAR model for district effects. This was not found to be the case in other comparisons of Gaussian prior distributions (e.g., see Best et al., 2005).

Banerjee et al. (2003) first reported the use of spatially-correlated frailty in application to linked birth-death individual level infant mortality data for the US state of Minnesota. In their formulation a Weibull parametric model was assumed and the hazard was assumed to be

$$h(t_{ij}|\mathbf{x}) = \rho t_{ij}^{\rho-1} \exp\{\mathbf{x}'_{ij}\boldsymbol{\beta} + w_i\}$$

where i th subject in the j th stratum $(i = 1, .., n_j)$ and $\mathbf{x}'_{ij}$ is a row vector of covariate values and $\boldsymbol{\beta}$ the corresponding parameter vector. The random effect w_i was assumed to have either a CAR prior specification, an uncorrelated zero-mean Gaussian prior distribution, or a fully-specified multivariate normal prior specification as in Henderson et al. (2002). In fact, the fully specified model yielded the lowest DIC in model comparison in this case also.

10.2.4 General Hazard Model

Banerjee and Carlin (2003) proposed a relaxation of the Weibull model to allow a semiparametric formulation whereby, with subject j, $(j = 1, ..., n_i)$ in the i th county:

$$h(t_{ij}|\mathbf{x}_{ij}) = h_{0i}(t_{ij}) \exp\{\mathbf{x}'_{ij}\boldsymbol{\beta} + w_i\}$$

where h_{0i} denotes the county-specific baseline hazard. For an individual with censoring indicator γ_{ij} (0 if alive, and 1 if dead) the likelihood contribution is then

$$h(t_{ij}; \mathbf{x}_{ij})^{\gamma_{ij}} \exp\{-H_{0i}(t_{ij}) \exp\{\mathbf{x}'_{ij}\boldsymbol{\beta} + w_i\}\}$$

where $H_{0i}(t_{ij}) = \int_0^{t_{ij}} h_{0i}(u)du$ a county-specific cumulative baseline hazard, and the covariates are assumed to be not time-dependent. The baseline hazard appears in this likelihood and so this must be estimated. Different approaches have been proposed for estimation of this baseline. One approach assumes a Gamma process which is a function of a parametric cumulative hazard (see e.g., Ibrahaim et al., 2000). Another is the use of Beta mixtures (Banerjee and Carlin, 2003). A related spatial model was developed by Bastos and Gamerman (2006) whereby they allowed time-dependent covariates which are fixed within small time periods and a CAR spatial frailty. They assume no separate baseline risk however.

10.2.5 Cox Model

The Cox proportional hazards model has been applied in a spatial context by Henderson et al. (2002). In the context of leukemia survival, the authors posited that the partial likelihood could be used without recourse to the estimation of the baseline. For ordered uncensored times $\{t_{(1)}, ..., t_{(m)}\}$, then the partial likelihood is given by

$$\prod_i [\exp\{\mathbf{x}'_{(i)}\boldsymbol{\beta}\} / \sum_{j \in R_i} \exp\{\mathbf{x}'_j \boldsymbol{\beta}\}]$$

where R_i is the set of those individuals at risk just before the i th event time. This can be used for the estimation of regression parameters. This is less parametric than models that include baseline components. Note that spatial effects can be included again as contextual effects by extending the specification of the intensity term $\exp\{\mathbf{x}'_{(i)}\boldsymbol{\beta}\}$ to include random effects:

$$\exp\{\mathbf{x}'_{(i)}\boldsymbol{\beta} + w_i + v_i\} \tag{10.6}$$

where w_i, v_i are random contextual effects which could be at an aggregate level such as district or county. In addition these effects could also be purely individual (in the sense of frailty rather than context).

10.2.6 Extensions

Extensions of the above approaches have been proposed into a variety of more complex applications. For example, spatial cure rate modeling has been proposed by Cooner et al. (2006) and joint spatial survival modeling of date of diagnosis and vital outcome from cancer registry data has been examined by Zhou et al. (2008). In that approach, prostate cancer registry data was available for the US state of South Carolina for the years 1997–2001. The endpoints are defined as T_{1ij} and T_{2ij} and their joint distribution is defined by conditioning as $f_{T_2}(t_{2ij}|t_{1ij})$ and $f_{T_1}(t_{1ij})$. In this formulation, flexible Weibull parametric models were assumed for both $f_{T_2}(t_{2ij}|t_{1ij})$ and $f_{T_1}(t_{1ij})$:

$$f_{T_1}(t_{1ij}) \sim Weib(\gamma_1, \lambda_{1ij})$$
$$f_{T_2}(t_{2ij}|t_{1ij}) \sim Weib(\gamma_2, \lambda_{2ij}|t_{1ij})$$

with log linear models for the scale parameters:

$$\log \lambda_{1ij} = \alpha_1 + \mathbf{x}'_{1ij}\boldsymbol{\beta}_1 + W_{1ij}$$
$$\log \lambda_{2ij} = \alpha_2 + \mathbf{x}'_{2ij}\boldsymbol{\beta}_2 + \beta_{2t}t_{1ij} + W_{2ij}.$$

In the second model, dependence on the the first endpoint is simply via a linear function dependent on t_{1ij}. Different forms of random effect combination were specified within a selection of models for W_{1ij} and W_{2ij}. Figure 10.1 displays the empirical Kaplan-Meyer curves for the date of diagnosis (left panel) and vital outcome (right panel) endpoints. The data-based K-M curves are shown in solid line. The posterior predicted survival is shown on these curves (as dashed lines) and seems to approximate well the empirical behavior at least for ages less than 80. A range of eight models were fitted with different combinations of separate and common random effect structures. Of the eight models, a model with an uncorrelated effect (UH) for age and separate correlated effect (CH) for survival time had the lowest DIC (DIC = 115722, pD = 57.0). The second lowest DIC was for a model with separate UH effects only in each component (DIC = 115725, pD = 55.7). These models were denoted Model 6 and Model 3, respectively. Figure 10.2 displays various posterior average random effect maps for different fitted models for these data. The top row displays results for the best-fitting model: Model 6. The left panel is the UH map for age-at-diagnosis while the right panel is the CH map for survival

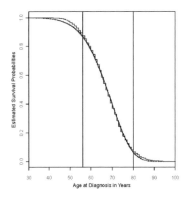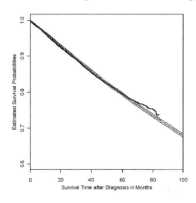

FIGURE 10.1
Nonparametric (Kaplan-Meier) survival curves compared to posterior predictive survival from the best fitting model based on DIC: left) age at diagnosis, b) survival time after diagnosis. Solid line is data-based while the dashed line (with 95% credible limits) is the posterior predictive survival under the best-fitting model.

time. It is noticeable that the lowered survival appears in the coastal and upstate areas of SC whereas the UH map seems to display a relatively random pattern. The bottom panel highlights the similarity of the UH effect map for other models fitted for survival after diagnosis. It would appear that the CH component is absorbed by a single UH component under these models and so the result is similar to the CH map for the best fitting model.

10.3 Spatial Longitudinal Analysis

There are many situations when variation in an individual response is time-dependent, and the focus is the monitoring over time of the associated outcome. The areas of application for these methods are manifold and in particular they are commonly applied in clinical trials and community-based behavioral intervention trials.

In many clinical settings, it is possible to conduct trials of new treatments. These trials are essentially designed experiments where patient outcomes are compared between treatment groups. Often these are monitored over time to establish whether a treatment has been effective or not. In the simplest case, when only two time points are employed, this leads to simple comparisons. However if more that two time points are monitored then considerable complication can arise. In what follows, I will emphasize a general framework for longitudinal analysis which incorporates spatial referencing. Recent gen-

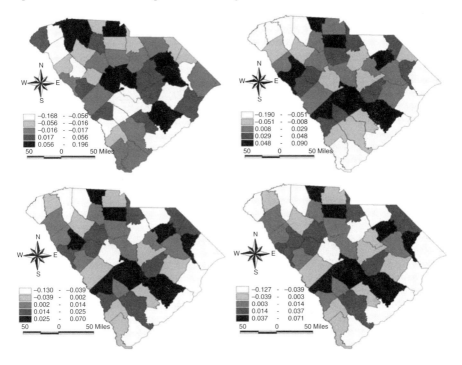

FIGURE 10.2
Posterior average maps of the random effects. top left: UH effect for age-at-diagnosis for best model, top right: CH effect for survival after diagnosis for the best model, bottom left: UH effect joint Model 3 and bottom right: UH effect joint Model 5, both for survival after diagnosis.

eral references to this area are Diggle et al. (2002), Verbeke and Molenberghs (2000), and Molenberghs and Verbeke (2005).

Define an outcome for the i th individual at a given time j as y_{ij}. Usually in designed studies the time period is fixed and measurements are made at these fixed times. Hence, the time label denotes a fixed time period. Further define there to be $i = 1, ..., m$ individuals or individual observation units and $j = 1, ..., J$ time periods. The periods could be units of time, such as minutes, hours, days, weeks, years, etc. depending on the study design. Usually recruitment to the study will be randomized in a clinical or intervention trial setting. However in observational studies, where such control is not possible, it is likely that there will be considerable extra variation and potential for imbalance. Hence there are strong reasons for inclusion of confounder or random effects when observational designs are used.

A simple general approach is to consider a linear model formulation whereby the outcome of interest is assumed to depend on underlying parameters, and these parameters can be time-dependent. For example, serum cholesterol (ldl) levels could be measured in patients on a trial for a new cholesterol drug at two time points (baseline and 6 weeks). There are two groups: old drug, fixed dose (ODFD), and new drug, fixed dose (NDFD). These groups are labeled $l = 1, 2$, respectively. The outcome of interest is y_{ilj} at given times $j = 1, 2$, and groups $l = 1, 2$. A statistical model for this situation will depend on the nature of the outcome. If the outcome is continuous then a Gaussian or gamma error model might be assumed. In the cholesterol trial, a continuous Gaussian variate could be assumed, as a first model, and so

$$y_{ilj} \sim N(\mu_{ilj}, \tau_y)$$

where μ_{ilj} is the mean for the i th person at time j in group l, and the τ_y is a variance.

The mean parameter would then be specified as a function of available covariates and random effects. For example, it might be important to allow for age of the individual. Define this as x_{1i}. One simple regression model might be

$$\mu_{ilj} = \alpha_l + \beta_1 x_{1i} + \beta_2 t_j$$

where t_j is the time of the j th period. Here each group has a separate intercept. The regression parameters would be commonly assumed to have zero-mean Gaussian prior distributions with small precisions. In this model a group effect is specified, with a covariate effect and constant linear effect over time. There is assumed to be no correlation in the error. More complex models could of course be envisaged. Addition of random effects can include frailty terms and a temporal dependence term (which can substitute for the regression on t_j):

$$\mu_{ilj} = \alpha_l + \beta_1 x_{1i} + v_i + \eta_j$$

with, possibly, a random walk prior dependence:

$$\eta_j \sim N(\eta_{j-1}, \tau_\eta), \ j > 1,$$

and a zero-mean Gaussian prior distribution with for the individual level frailty:

$$v_i \sim N(0, \tau_v).$$

If inference about spatial effects are important then a variety of possibilities exist. First, if individual level information is available then there are two approaches to the inclusion of space. First it may be possible to estimate an individual level; effect directly for residential location data. Denote a spatially-structured effect as w_i. Also assume that $w_i \equiv w(s_i)$ where s_i is the i th residential address. A model could be assumed of the form

$$\mu_{ilj} = \alpha_l + \beta_1 x_{1i} + w_i + \eta_j$$

where w_i can be assumed to have a fully-specified multivariate Gaussian prior distribution and so

$$\mathbf{w} \sim \mathbf{N}(\mathbf{0}, \mathbf{\Gamma})$$

where the elements of $\mathbf{\Gamma}$ are defined as $\mathbf{\Gamma}_{ij} = \tau_w \rho(d_{ij})$ and $d_{ij} = ||s_i - s_j||$. Suitable forms of the correlation function $\rho(.)$ have been discussed elsewhere (Sections 5.4.2, 10.2). Among these the power exponential or Matérn are common choices.

10.3.1 General Model

A general linear mixed model for a continuous outcome can be specified where the vector $\mathbf{y} : \{\mathbf{y}_1, ..., \mathbf{y}_m\}$ where $\mathbf{y}_i = (y_{i1}, ..., y_{iJ})$ is a realization of a random vector from a multivariate normal distribution of form:

$$\mathbf{y}_i = \mathbf{x}'_i \boldsymbol{\beta} + \mathbf{z}'_i \boldsymbol{\gamma} + \mathbf{e}_i \tag{10.7}$$

where $\mathbf{x}'_i \boldsymbol{\beta}$ is a linear predictor (which can include a group indicator), $\mathbf{z}_i$ is a vector of g random effects for the i th individual, $\boldsymbol{\gamma}$ is a $g \times J$ unit matrix and the model for $\mathbf{e}_i$ is $\mathbf{e} \sim \mathbf{N}(\mathbf{0}, \tau \boldsymbol{\Sigma})$ where $N = mJ$, $\mathbf{x}'$ is an $N \times p$ matrix of covariates, $\boldsymbol{\beta}$ is a parameter vector, τ is a variance and $\boldsymbol{\Sigma}$ is a block diagonal matrix with $J \times J$ blocks each representing the variance matrix of the vector measurements on a single subject. Various assumptions about the structure of the covariance can be made depending on the focus of the study. I do not pursue this here.

An extension of the approach can be easily made to generalized linear mixed models (GLMM) whereby (10.7) is modified to allow different distributional assumptions at the data level and a link function is introduced to connect the mean to the linear predictor:

$$\mathbf{y}_i \sim f(\boldsymbol{\mu}_i)$$
$$\text{where } E(\mathbf{y}_i) = \boldsymbol{\mu}_i$$
$$\text{and } g(\boldsymbol{\mu}_i) = \boldsymbol{\eta}_i = \mathbf{x}'_i \boldsymbol{\beta} + \mathbf{z}'_i \boldsymbol{\gamma}.$$

Suitable models for $f(\boldsymbol{\mu}_i)$ could be the Poisson, binomial, gamma, and others in the standard exponential family.

10.3.2 Seizure Data Example

As an example of the effect of spatial structure on the analysis of longitudinal data I have taken a famous data set which was examined by Breslow and Clayton (1993): the epileptic seizure data (see also Diggle et al., 2002). This consists of a randomized clinical trial of anticonvulsive therapy to inhibit seizures in epilepsy. The trial consists of data from 59 patients, treated in two groups (0,1: control, treatment) and the seizure count at four time points was monitored. There are three covariates: baseline seizure count (x_{1i}), treatment (x_{2i}), and age (x_{3i}) in years.

In Bayesian hierarchical modeling of longitudinal data there is no need to consider marginal models (Verbeke and Molenberghs, 2000) as a model for the full hierarchy is assumed and fitted. Hence our model will explicitly assume a full hierarchical model with random effects. The basic model is a GLMM with

$$y_{ij} \sim Poiss(\mu_{ij}) \qquad (10.8)$$
$$\log \mu_{ij} = \alpha_0 + \log(x_{1i}) + \alpha_1 x_{2i} + \alpha_2 x_{3i} + \gamma_j. \qquad (10.9)$$

Here the baseline count is treated as a log(offset) with no estimable parameter while the treatment group is associated with parameter α_1. Age is also treated linearly with parameter α_2, while an overall intercept is also fitted. Finally the temporal effect in this model is defined by a time dependence. There is no regression on time but a random walk prior specification is used to model the variation. Hence the following is specified in the prior distributions:

$$\alpha_0 \sim U(-1000, 1000)$$
$$\alpha_1 \sim N(0, \tau_{\alpha_1}); \alpha_2 \sim N(0, \tau_{\alpha_2})$$
$$\gamma_1 \sim N(0, \tau_\gamma)$$
$$\gamma_j \sim N(\gamma_{j-1}, \tau_\gamma) \quad j > 1.$$

Regression parameters are given dispersed zero-mean Gaussian prior distributions with $\tau_{\alpha_1}, \tau_{\alpha_2}, \tau_\gamma \sim Ga(0.001, 0.001)$. Table 10.1, line 1 displays the overall goodness-of-fit results for the basic model with no spatial random effects. Under this model, the DIC was found to be 1775.4, with pD = 5.311. To demonstrate the effect of spatial referencing on a clinical trial data set, I have made a random allocation of patients to the 46 counties of South Carolina. While this single realization does not represent the possible spatial variation across counties, it does represent the type of variation one might expect if there were a set of contextual regions (of any size) within which the patients resided. While this initial assumption is crude in that it does not allow for population variation, it should demonstrate less spatial structure than a typically significant spatial confounder. I would then expect that there would be little or no advantage found in adding a spatial contextual effect at the county level. However, as will be seen below, this is in fact not the case, even for a blanket randomization such as this.

To examine the effects of spatial context I have fitted a sequence of simple spatial context random effect models with different combinations of UH effects (v_i) and CH effects (w_i). These are contextual effects and so they are defined as

$$w_i = \underset{i \in j}{w(c_j)}$$
$$v_i = \underset{i \in j}{v(c_j)}.$$

TABLE 10.1
Comparison of four models for the seizure data: basic and Models 1–3

Model	pD	DIC
basic model	5.311	1775.4
model 1	33.93	1485.7
model 2	38.24	1494.2
model 3	36.0	1488.9

They are defined as a conventional zero-mean Gaussian prior distribution for the $\{v_i\}$ and an improper CAR prior distribution for the $\{w_i\}$. The models fitted were as follows:

Model 1 is as (10.9) but with v_i added, Model 2 is as (10.9) but with w_i added, and finally the Model 3 is a convolution model with v_i and w_i added to the basic model. The variance parameters for these distributions were assumed to have their distributions defined via standard deviations with $U(0, 100)$ distributions (Gelman, 2006). The results of fitting these different models is found on Table 10.1. It is interesting to note that *all* models including spatially-referenced random effects (Models 1, 2, 3) have a considerably lower DIC than the basic model. The best fitting model by the DIC criterion is Model 1 with only the uncorrelated effect. Although the convolution model yields a slightly higher DIC, it has lower DIC and pD than a model with only the correlated spatial effect. It is reassuring that the uncorrelated noise model has best fit as we did not present any overt spatial structure in the data. However the fact that any spatially-referenced model (whether correlated or not) leads to much reduced DIC is an important consideration. The reason for the closeness of the Model 1 and 2 is largely because the CH effect (w_i) will often mimic small scale variation found. Figure 10.3 displays the results of these different models. It is perhaps surprising how clustered even the UH component appears. Of course one has to be careful of over-interpreting these results as they are based on a *single* realization of synthetic contextual labeling. On the other hand, it is commonly the case that a single geo-referenced sample would be available if the true spatial referencing were known. Finally it is interesting to consider how the parameter estimates for the non-spatial parameters performed under the different models. Table 10.2 displays the estimates for the final converged sample for the basic model and the "best" model (Model 1). It is notable that under the random effect model the coefficient for the group effect (α_1) becomes significant while under the basic model it is not well estimated. It seems therefore that by reducing the noise in the model the group effect becomes more clearly apparent.

It is also interesting to compare standardized residuals for these models. I have computed the average standardized residual for each individual $\widehat{r}_i = \frac{1}{4}\sum_j (y_{ij} - \widehat{\mu}_{ij})/\sqrt{\widehat{\mu}_{ij}}$ where $\widehat{\mu}_{ij}$ is the posterior averaged value of μ_{ij}. Figures 10.4 and 10.5 display these residuals for the basic model and Model 1. There is

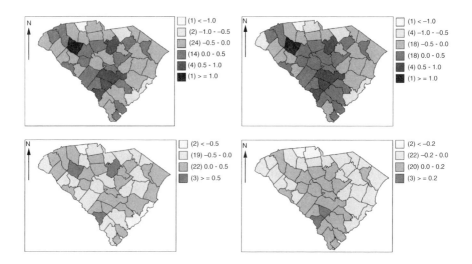

FIGURE 10.3
Seizure example: posterior average effect maps: correlated (w_i) and uncorrelated (v_i) random effects. Top left model 1 (v_i), top right model 2 (w_i); bottom row Model 3: Left v_i an right w_i.

TABLE 10.2
Posterior average estimates of the model parameters under the four different seizure models

Model	α_0	α_1	α_2
basic model	-6.188(0.623)	-0.079(0.046)	0.0101(0.003)
model 1	-2.368(0.751)	-0.5028(0.074)	0.0207(0.006)

TABLE 10.3
Parameter estimates for the basic model compared to the best fitting model (Model 1)

Model	γ_1	γ_2	γ_3	γ_4
basic model	4.681(0.614)	4.636(0.614)	4.619(0.616)	4.530(0.615)
best model (model 1)	0.607 (0.724)	0.559 (0.724)	0.543 (0.725)	0.455 (0.726)

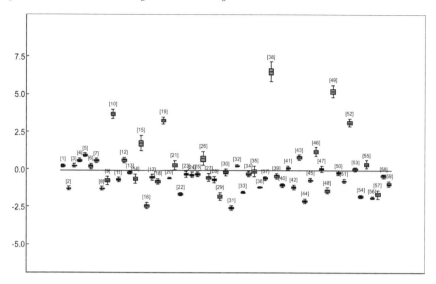

FIGURE 10.4
Standardized average residuals for the 59 individuals for the seziure data under the basic model.

a noticeable reduction in variation in the residuals and a much better fit with Model 1. Patients 16, 38, and 52 remain with large residuals but 10, 49, and 19 have been accommodated under Model 1. Finally it is clear that random effect models are important here, and of course it would be useful in this modeling to consider *individual* frailty (besides contextual UH effects). Indeed this would often be the first step of extending a model, before considering any spatial effects. In this example in fact, an uncorrelated frailty term (constant over time) with $v_i \sim N(0, \tau)$ $i = 1, ..., 59$ leads to a substantial reduction in DIC (DIC =1277.180, pD =59.8) and also a significant group effect is found. Overall, in the 'pretend' spatial data set it is clear that extra variation is present. Assuming the spatial referencing were correct, then spatial contextual effects appear to be important. It is also clear that individual frailty could be even more important to consider in the context longitudinal data analysis.

10.3.3 Missing Data

Missing data is a very important aspect of longitudinal studies. Apart from general forms of missing data found in all studies, missingness can often arise in longitudinal studies due to individuals failing to remain in the study. This *dropout*, as it is known, is common in clinical or intervention trials and often by the end of a study 20%–30% of the participants may have left. There are different forms of missingness mechanisms. Verbeke and Molenberghs (2000)

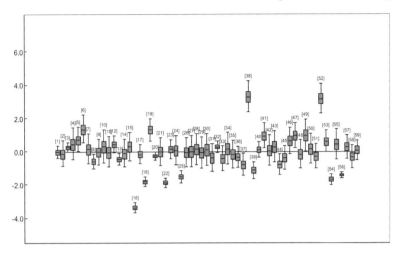

FIGURE 10.5
Standardised average residuals for the 59 individuals for the seziure data under the Model 1 with spatial contextual random effect.

define three basic forms. While it may be important in given applications to consider which if these apply, here I will focus simply on the implications of geo-referencing on missingness. In particular the effect of geo-referencing on dropout will be discussed. Usually a geographical delimitation of a study area is made in the design of a trial. During the course of a trial it is possible for participants to move their residence. This is more likely to occur when longer term studies, such as behavioral intervention trials are considered. These trials can last up to 2–3 years. Ultimately, the mobility of participants could lead to a move outside the study area. If there is a geographical limit on the study in terms of recruitment then those who remove from the study area would be regarded as dropout. With intention-to-treat approaches to trials (Lachin, 2000) it may be important to use all data for those entered into the trial and so it may be important to consider this geographical dropout. Denote the outcome for the i th participant at time j as y_{ij}. Consider residential history defined as $s_i(t)$ where t is continuous and s denotes a spatial location (residence address). A study lasts for $j = 1, ..., J$ time points. Outcome measurements are made at the J time points. Often the residential history will be discretized into the study time periods so that s_{ij} may be censored between time points. Besides location shifts, it would be useful to define a dropout indicator R whereby r_{ij} denotes presence (1) or absence (0) at the j measurement. While r_{ij} could be 0 for a range of reasons one such could be related to s_{ij}. Hence it might be important to consider (y_{ij}, r_{ij}, s_{ij}) jointly in modeling a given outcome. This area appears to have been little explored.

10.4 Extensions to Repeated Events

There are many other application areas where spatial context can be fruitfully introduced or allowed for. One such area that is closely related to longitudinal analysis is the analysis of repeated events. This is sometimes called repeated event analysis or event history analysis. For a recent review see, for example, Cook and Lawless (2007). This is an extension of longitudinal and survival analysis where instead of making one measurement at different times or observing a single time-to-endpoint, the time period is fixed and within that period a sequence of events are observed on an individual. In the simplest case, the sequence could consist of just a single type of event and its repeated occurrence is observed. Hence for a single person this is a point process in time, assuming the event does not have finite duration. An example of a simple sequence would be a sequence of doctor's visits where the time of visit is recorded. If we are only concerned about the times of the visits (and not their nature or duration) then this can be considered to be point process at the unit (patient) level.

At this point it is worth noting that there is an immediate connection between the linear or generalized linear modeling approach of longitudinal analysis and the time-based analysis for survival data. With repeated events, time is important but when fixed time periods are used and observations are collected within these time periods then the resulting counts of events can be considered within the framework of the conventional hierarchical model. For example for simple doctor visits the count of visits in time periods might be treated at the first level of hierarchy as a Poisson random variable whereas if the times are recorded directly then a (heterogeneous) Poisson process model might be appropriate. This, of course, mirrors the duality of the point process and count model when binning of events takes place in a spatial context.

10.4.1 Simple Repeated Events

Assume first that a patient resides at an address and makes repeated doctor visits. Denote the address as s_i and the sequence of visits as $\{t_{i,1}, ..., t_{i,n_i}\}$ where t denotes visit time within a study period (t_0, t_T) and n_i is the number of visits. If some basic assumptions are made concerning conditional independence (given knowledge of all confounding and event history) independence of the events might be a reasonable starting model. If a modulated (heterogeneous) Poisson process (PP) were assumed for the event times then a conditional PP likelihood could be assumed whereby the first order intensity for the i th individual could be defined as

$$\lambda_i(t) = \lambda_{i0}(t) \exp(\mathbf{x}'_i \boldsymbol{\beta} + g_i(t)).$$

Note that here the covariates are included via the fixed vector $\mathbf{x}'_i$ and the $g_i(t)$ function could be a smooth function of time. The baseline function $\lambda_{i0}(t)$ is also unit specific. The associated unconditional likelihood is given by

$$\prod_{i=1}^{m}\prod_{j=1}^{n_i} \lambda_i(t_{ij})\exp\{\Lambda_i\} \tag{10.10}$$

$$\text{where } \Lambda_i = \int_{t_0}^{t_T} \lambda_i(u)du. \tag{10.11}$$

For fixed covariates, then $\Lambda_i = \exp(\mathbf{x}'_i\boldsymbol{\beta})\int_{t_0}^{t_T}\lambda_{i0}(u)\exp(g_i(u))du$. Various methods can be used to include time-varying effects. Discretizing to allow piecewise linear terms in the baseline and $g_i(t)$ are possible, while semiparametric models could also be assumed. Of course gamma or Dirichlet processes could also be used (Ibrahim et al., 2000). Cook and Lawless (2007, Ch. 3), discuss various possibilities for Poisson process models.

10.4.2 More Complex Repeated Events

Two generalizations are immediate from the simple event case. First, multiple types of events could occur. Second, a feature associated with the event could be important and could vary with time or type of event. An example of the first situation could easily arise with the progression of a disease. Visits to doctors could be interspersed with hospital visits, or nurse visits. In fact, complex patterns of repeated events are more usual when making observational studies on disease progression, than in clinical settings. The observed data could then be of the form $\{t^s_{i,1},....,t^s_{i,n_i}\}$ where s denotes the event type. For a fixed number of event types (L) then s could be defined as $s = 1, ..., L$. The second situation arises when the event has attached a mark. For instance, a severity score or biomarker might be measured at a given visit. Alternatively the visit itself could have a duration. In the first case, the mark or measurement could be jointly modeled with the event time and the correlation between the mark and time could be directly modeled. In this case the observed data for the i th individual would be $\{t_{i,1}, x_{i,1}...., t_{i,n_i}, x_{i,n_i}\}$, where x is the mark value. When duration is associated with event time, there can be a complication, as this type of mark directly affects the subsequent event times (as duration is time-based). In this case the observed data would be $\{t_{i,1}, d_{i,1}...., t_{i,n_i}, d_{i,n_i}\}$ where d is the event duration. This might be appropriate where hospital visits involve stays of different lengths.

Ultimately we might have a mixture of these situations where multi-type events also have marks or durations and so, for the mark case, we would observe $\{t^s_{i,1}, x_{i,1}...., t^s_{i,n_i}, x_{i,n_i}\}$.

Spatial Survival and Longitudinal Analysis

Modeling approaches for these different situations depends on the observed data and also the study purpose. For example, if known times are observed it is often convenient to conditionally model the marks given the times so that we have the joint model

$$[x, t] = [x|t][t]. \tag{10.12}$$

In this case, we consider the times to be governed by point process model. In addition, the model for the marks could be a simple Gaussian distribution: $[x|t] \sim N(\mu(t), \tau)$. Here dependence on time could be made explicit in the mean parameterization ($\mu(t)$). This mean function could be specified to include covariates (individual or contextual) as well as time dependence: e.g., $\mu(t) = x'\beta + \gamma(t)$. Note that the point process model could also have covariate dependence and so it is debatable whether there should be multiple entry of the same covariate in each model. For example, patient age could affect both the mark (e.g., blood pressure) and doctor visit times.

On the other hand, if we do not observe directly the visit time but simply the number of visits within a fixed time period, then usually the mark would not be any longer associated directly with the individual (as the information has been averaged). Either the mark would be based on the time period, or is an average mark for all events within the period. As the exact times have been lost the resulting data would consist of counts of visits and an average mark or the value of (say) a contextual variable pertaining to the time period. In this case the observed data would be y_{ij} where there are $j = 1, ..., J$ time periods. We might be interested in the joint distribution $[x, y] = [x|y][y]$. In effect, the model for the mark is conditioned on the count and a separate count model is specified. If the focus were on the counts per se then the alternative formulation of $[x, y] = [y|x][x]$ could be considered and often the visit frequency is modeled conditionally, i.e., via $[y|x]$ treating the mark as a covariate.

These formulations do not include any explicit spatial dependence. In the next section some proposals for how this dependence could be incorporated are proposed. As in Sections 10.2 and 10.3 the spatial effects can be regarded as contextual.

10.4.2.1 Known times

10.4.2.1.1 Single events
As a first pass we could assume that the times follow a heterogeneous Poisson process (hPP) so that

$$f(t) = \lambda(t) \exp(-\int_{t_0}^{t_T} \lambda(u) du)$$

and conditional on n_i, then the likelihood element for the i th individual would be

$$\prod_{i=1}^{m}\prod_{j=1}^{n_i} \lambda_i(t_{ij})\exp\{-\Lambda_i\}, \qquad (10.13)$$

$$where\ \Lambda_i = \int_{t_0}^{t_T} \lambda_i(u)du.$$

and $\lambda_i(t) = \lambda_{i0}(t)\exp(\mathbf{x}_i'\boldsymbol{\beta} + g_i(t))$. Note that it is possible to include within a Bayesian hierarchy the spatial effect, especially if it is not time-dependent. For example, a contextual effect at the individual level could be included as $\Lambda_i = \exp(\mathbf{x}_i'\boldsymbol{\beta} + w_i) \int_{t_0}^{t_T} \lambda_{i0}(u)\exp(g_i(u))du$ where $w_i = \underset{i\in j}{w(c_j)}$ as defined in Section 10.2.3. More generally, the spatial location of the individual (s_i) could be incorporated in the PP model and a space-time formulation could be considered directly:

$$\lambda_i(s,t) = \lambda_{i0}(s,t)\exp(\mathbf{x}_i'\boldsymbol{\beta} + g_i(t) + h_i(s))$$

and the resulting likelihood would be

$$\prod_{i=1}^{m}\prod_{j=1}^{n_i} \lambda_i(s_i, t_{ij})\exp\{-\Lambda_i\}$$

$$where\ \Lambda_i = \int_{t_0}^{t_T}\int_W \lambda_i(v,u)dvdu$$

and W is the study region area. In addition, variants of the specification could allow for nonparametric specification of the $g_i(t)$, $h_i(s)$ functions with a simple alternative being piecewise discretization in time and space. For space it may be possible to define a common surface, $h(s)$ say, and to form piecewise constant components from a tiling of the distribution of individuals. However if the spatial component is zero-centered then further approximations may be available (see e.g., Chapter 7). Of course integral approximations such as proposed by Berman and Turner (1992), could also be considered.

10.4.2.1.2 Multiple event types Alternatively, simpler intensity based methods can be pursued. Denote the intensity of the i th individual and j th type as $\lambda_{ij}(t|H_i(t))$, where $H_i(t)$ is the event history. A different period of observation is allowed for each individual: $[0, \tau_i]$. Denote the history of the i th subject as $H_i(t) = \{N_i(s),\ 0 < s < t\}$ where $N_{ij}(t)$ is the number of event of j th type occurring on the i th individual in the interval $[0, t]$ and

$N_i(t) = \{N_{i1}(t), ..., N_{iJ}(t)\}'$. Hence the history of the process concerns the preceding count accumulation. The resulting likelihood is given for the times t_{ijk}, $k = 1, ..., N_{ij}(t)$ so that

$$\prod_{j=1}^{J}\left\{\prod_{k=1}^{n_{ij}}\lambda_{ij}(t_{ijk}|H_i(t_{ijk}))\exp(-\int_0^{T_i}\lambda_{ij}(u|H_i(u))du)\right\}. \tag{10.14}$$

This of course assumes that the distribution of events are functionally independent. In many situations this would of course not be reasonable. For example, a doctor's visit might precipitate a hospital visit. When multiple events can arise with dependence over time then it is often useful to consider transition models for the occurrence of events of different types. These types of models specify the probability that an event of a given type will occur in an interval of time given the preceding event. Hence, they are naturally specified conditioning on the preceding event, its type and time, or type of preceding events and their times. Alternatively a competing risk approach could be envisaged. I do not consider these further here.

Incorporation of random effects can be effected via redefinition of (10.14): $\lambda_{ij}(t_{ijk}|H_i(t_{ijk}), R_{ij}) = \lambda_{ij}(t_{ijk}|H_i(t_{ijk}), \exp\{W_{ij}\})$ where W_{ij} would represent the individual random effect for the j th event type. This could be decomposed into a number of individual specific or contextual random effects. For example, it would be possible to consider a set of L counties labeled c_l, $l = 1, ..., L$. Then we could consider as a first example, the hierarchical random effect model: $W_{ij} = (u_{ij} + v_i + w_i)$ where $w_i = w(c_l)_{i \in l}$ and $v_i = v(c_l)_{i \in l}$ and these are spatial contextual effects and $u_{ij} = u_{1i} + u_{2j}$ where u_{1i} is a general individual frailty and u_{2j} is a type specific effect. The prior distributions for the effects could be overdispersed zero-mean Gaussian for u_{1i}, u_{2j}, v_i while for w_i a CAR formulation would be possible. A variety of other formulations would be possible of course. As before it would be possible to consider the intensity as a function of both time and space and hence to include a specific spatial component in its definition. However it is probably simpler and more convenient to consider conditioning on the spatial component defined at a higher level of the hierarchy via a random effect.

10.4.3 Fixed Time Periods

When fixed time periods are observed, it is usual to collect events within the time periods into counts of events. Of course when this is done information about the time sequence of events is lost within the time period. Hence even if the residential location of individuals is known and hence known at a fine spatial resolution level, the resolution level in time is aggregated. In general, we denote the count within the j th time period as y_{ijl}, where i denotes the individual and $j = 1, ..., J$ denotes the time periods, and l denotes the event type ($l = 1, ..., L$).

10.4.3.1 Single events

In the case of a single outcome then we observe $\{y_{ij}\}$ for a sequence of J times and $i = 1, ..., m$. The simplest modeling approach is to assume a generalized linear model for the counts with some form of time dependence. For example it could be assumed that $y_{ij} \sim Pois(\mu_{ij})$ and a log linear model could be assumed for the mean:

$$\log \mu_{ij} = \mathbf{x}'_i \boldsymbol{\beta} + v_i + \gamma_j \tag{10.15}$$

where $\mathbf{x}'_i$ is a vector of individual fixed covariates and $\boldsymbol{\beta}$ is the corresponding parameter vector, and v_i is an individual frailty and γ_j is a common temporal effect. The temporal effect could have a variety of specifications :

1. An uncorrelated prior distribution (for example, $\gamma_j \sim N(0, \tau_\gamma)$)

2. A correlated prior distribution (for example, a random walk: $\gamma_1 \sim N(0, \tau_\gamma)$, $\gamma_j \sim N(\gamma_{j-1}, \tau_\gamma)$ $j > 1$)

3. Trend regression on time $\gamma_j = \beta t_j$ (or a higher order polynomial) where t_j is the time of the j th period (start or end or middle by convention)

If the focus is on a parsimonious description of the overall behavior then option 2 may be favored as it is relatively non-parametric. However if a specific linear or polynomial estimate of trend is required, then option 3 may be preferred. The individual frailty would usually be assumed to have an uncorrelated zero-mean Gaussian distribution:

$$v_i \sim N(0, \tau_v).$$

Incorporation of spatial contextual effects can follows as before by extending the model in (10.15) to include an individual contextual component:

$$\log(\mu_{ij}) = \mathbf{x}'_i \boldsymbol{\beta} + v_i + w_i + \gamma_j \tag{10.16}$$

where $w_i = w(c_l)$ and $w(c_l)|w(c_{-l}) \sim N(\overline{w}(c_{\delta_l}), \tau_w/n_{\delta_l})$ where $w(c_l)$ is the value of w for the l th county and $\overline{w}(c_{\delta_l})$ is the average value of w for the neighborhood (δ_l) of c_l. The number of counties in this neighborhood is given by n_{δ_l}. Note that v_i is an individual frailty effect here. An alternative specification could also assume an uncorrelated county effect, as for w_i except without correlation. Steele et al. (2004) describe essentially the model in (10.15) albeit with multiple events.

10.4.3.2 Multiple event types

In the case of multiple event types assume a count of the form y_{ijl} where l denotes the event type. Steele et al. (2004) describe a competing risk model where a multinomial form is assumed for the vector of events within a time period. The multinomial probability ratio (relative to the zero event case) was assumed to be defined by a logit link to a linear predictor with covariate random effect and trend components similar to (10.16) but without the spatial dependence. In general, one approach to these problems assumes that conditionally on $N_{ij} = \sum_l y_{ijl}$,

$$\mathbf{y}_{ij} \sim Mult(\mathbf{p}_{ij}, N_{ij}).$$

If it is important to consider the relative preference for visit types then this might be useful. Otherwise without conditioning it would be possible to consider

$$y_{ijl} \sim Pois(\mu_{ijl}).$$

A log linear link could be assumed whereby

$$\log(\mu_{ijl}) = \mathbf{x}'_i \boldsymbol{\beta} + v_i + w_i + u_{il} + \gamma_{jl} \qquad (10.17)$$

where v_i, w_i are contextual effects, as before, u_{il} is an individual level effect specific to the event type and γ_{jl} is a temporal effect specific to the event type. As before the contextual random effects can be defined to depend on small area level geographies, while the temporal effects could be regression based or random walk based. In the multinomial model of Steele et al. (2004) the temporal component γ_{jl} was assumed to be defined by a quadratic regression in time and the individual random effect u_{il} was assumed to have a multivariate normal distribution (between event types only).

10.4.3.2.1 Asthma-comorbidity Medicaid example
Sutton (2005) provides an example of the analysis of individual level outcomes with fixed time periods in a Bayesian setting. This work was based on Medicaid data on asthma (ICD-9 493) and CHF (ICD-9 428, 402, 518.4), comorbidities for recipients between age 50–64 in South Carolina for the period 1997–1999. Three groups were identified: asthma only, CHF only, and asthma and CHF. Of the 1857 individuals, 223 were in the comorbidity group. Recorded for each recipient were number of days between multiple dates of medical service; type of visit (inpatient, ER, outpatient, doctor's office), and recipients demographic information (age at first visit, gender, race, county of residence). As a brief guide, Figure 10.6 displays the event profiles summarized by time to second visit, for the asthma and CHF individuals separately.

An analysis was performed with the SC Medicaid data mentioned above where asthma and CHD were analyzed together. Here I only display the comorbidity group analysis. The analysis was carried out for 21 time periods (of 50 days each).

A model of the form

$$y_{ijl} \sim Poiss(\mu_{ijl})$$
$$\log(\mu_{ijl}) = \mathbf{x}'_i \boldsymbol{\beta} + v_i + w_i + \gamma_{jl}$$

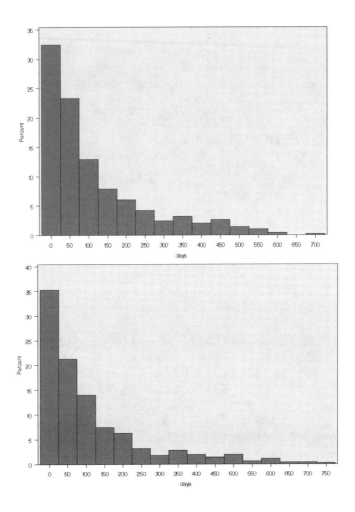

FIGURE 10.6
Distribution of number of days between first and second visit for recipients with: top panel: asthma, bottom panel: CHF.

was fitted to the data for the comorbidity group. Individual frailty was not required in this example, based on a variable selection criterion. Here the fixed covariates were β_0 (common intercept), β_1(age), factors for race and gender, and two spatial contextual effects at county level: $v_i = v(c_l)$ with $i \in l$
$v_i \sim N(0, \tau_v)$ and $w_i = w(c_l)$ and $w(c_l)|w(c_{-l}) \sim N(\overline{w}(c_{\delta_l}), \tau_w/n_{\delta_l})$ where $i \in l$
$w(c_l)$ is the value of w for the l th county and $\overline{w}(c_{\delta_l})$ is the average value of w for the neighborhood (δ_l) of c_l. All regression parameters had over-dispersed zero-mean Gaussian prior distributions while variances were assumed to have $Ga(0.05, 0.0005)$ distributions. The temporal effects for each type were assumed to have independent random walk prior distributions

$$\gamma_{jl} \sim N(\gamma_{j-1,l}, \tau_\gamma) \; \forall l.$$

The converged model fit yields the following results:

1. The posterior expected estimates (sds) of β_0, β_1 respectively were -4.6 (0.9982) and -0.008553 (0.01118).

2. The posterior expected estimates (sds) of the race and gender effects were not significant for gender but showed a significance for the white versus african-american racial groups

3. The temporal effects were estimated and are shown in Figure 10.7. The sequence of visit types are l = 1, l = 2, l = 3, l = 4—inpatient, ER, outpatient, doctor's office.

Finally the posterior average maps of the county specific random contextual effects are shown in Figure 10.8. It is clear that there is some spatial effect (w map) displayed in the NE of the state within largely rural areas, whereas the UH effect (v map) seems to be largely random. It should be noted here however that analysis of discrete time events lacks a considerable amount of information due to the grouping within time periods and information about sequencing is lost, hence there is an inevitable limitation to this form of analysis.

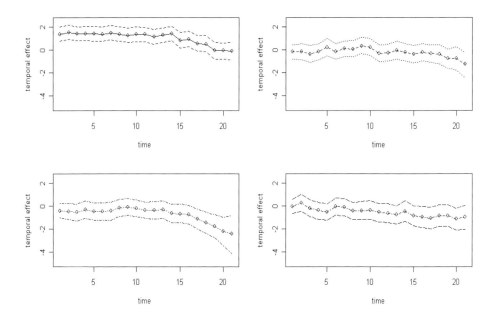

FIGURE 10.7
Posterior average temporal profiles for the effect γ_{jl}, for $l = 1,..,4$. Mean profiles and 95% credible interval shown. Top left $l = 1$, top right $l = 2$, bottom left $l = 3$, bottom right $l = 4$.

Spatial Survival and Longitudinal Analysis

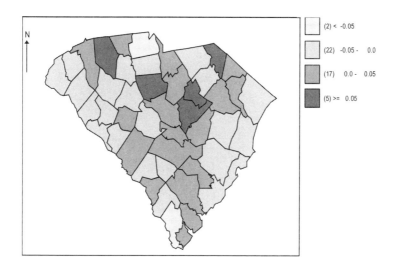

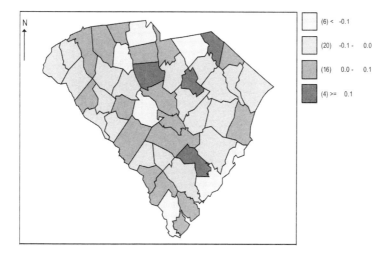

FIGURE 10.8
Posterior average maps of the county-level random effects (w_i, v_i) form the multiple event model for the Medicaid data.

11
Spatiotemporal Disease Mapping

As in other application areas, it is possible to consider the analysis of disease maps which have an associated temporal dimension. The two most common formats for observations are

(1) Geo-referenced case events which have associated a time of diagnosis or registration or onset, i.e., we observe within a fixed time period J and fixed spatial window W, m cases at locations $\{s_i, t_i\}, i = 1, \ldots, m$;

(2) Counts of cases of disease within tracts are available for a sequence of J time periods, i.e., we observe a binning of case events within $m \times J$ space-time units: $y_{ij}, i = 1, \ldots, m, j = 1, \ldots, J$.

The analysis found for spatial data (see Chapters 5 and 6) can be extended into the time domain without significant difficulty.

11.1 Case Event Data

In the case event situation, few examples exist of mapping analysis. However, it is possible to specify a model to describe the first-order intensity of the space-time process (as in the spatial case). The intensity at time t can be specified as:

$$\lambda(s,t) = \rho g(s,t).f_1(s;\theta_x).f_2(t;\theta_t).f_3(s,t;\theta_{xt}), \qquad (11.1)$$

where ρ is a constant background rate (in space $\times$ time units), $g(s,t)$ is a modulation function describing the spatiotemporal 'at-risk' population background in the study region, f_k are appropriately defined functions of space, time and space-time, and $\theta_x, \theta_t, \theta_{xt}$ are parameter vectors relating to the spatial, temporal, and spatiotemporal components of the model.

Here each component of the f_k can represent a *full* model for the component, i.e., f_1 can include spatial trend, covariate and covariance terms, and f_2 can contain similar terms for the temporal effects, while f_3 can contain *interaction* terms between the components in space and time. Note that this final term can include *separate* spatial structures relating to interactions which are not included in f_1 or f_2. The exact specification of each of these components will

255

depend on the application, but the separation of these three components is helpful in the formulation of components.

The above intensity specification can be used as a basis for the development of Bayesian models for case events. If it can be assumed that the events form a modulated Poisson process in space-time, then a likelihood can be specified, as in the spatial case. For example, a parsimonious model could be proposed where a regression component and a random effect component are assumed:

$$\lambda(s,t) = \rho g(s,t) \exp\{\mathbf{P}(s,t)'\boldsymbol{\beta} + T(s,t)\} \qquad (11.2)$$

where $\mathbf{P}(s,t)$ is a covariate vector, $\boldsymbol{\beta}$ a regression parameter vector, and $T(s,t)$ is a random component representing extra variation in risk. The term $T(s,t)$ could be decomposed in a number of ways. For example, it could represent a spatiotemporal Gaussian process (Brix and Diggle, 2001). However, a simpler approach might be to consider: $T(s,t) = a(s) + b(t) + c(s,t)$ where a discretized version of the random fields could be envisaged so that any realization of the field $\{s_i, t_i\}$ has separable correlation structure and

$$a(\mathbf{s}) \sim MVN(\mathbf{0}, K_a(\tau_x, \phi)),$$
$$b(t) \sim N(f(\Delta t), \tau_b),$$
$$\text{and } c(s,t) \sim N(0, \tau_c I), \qquad (11.3)$$

where $K_a(\tau_x, \phi)$ is a parameterized spatial covariance matrix and I is an identity matrix, with variances τ_b, and τ_c, and Δt is a distance measures in space and time. In this approach the likelihood remains that of a conditionally modulated Poisson process.

This type of model can be included within a likelihood specification and a full Bayesian analysis can proceed using extensions to the analysis for purely spatial data. In these extensions either the integrated intensity of the process

$$\Lambda(\boldsymbol{\theta}) = \int_W \int_0^J \lambda(u,v) dv du$$

where $\boldsymbol{\theta} = (\boldsymbol{\beta}, \tau_x, \phi, \tau_b, \tau_c)$ is the parameter vector, must be estimated or the background is concentrated out of the model by conditioning. In Lawson (2006b), an example of application of this model to a well known space-time case event dataset was given (Burkitt's Lymphoma in the Western Nile district of Uganda for the period 1960–1975). In that dataset the location of cases and a diagnosis date (days from January 1st 1960) are known as well as the age of the case. There is no background population information in this example. While it might be possible to consider an approach where a population effect in $g(s,t)$ were estimated from the case event data, this is particularly assumption dependent and this was not pursued. The following analysis essentially assumes that the population is homogeneously distributed over space. This is of course a strong assumption.

Spatiotemporal Disease Mapping

The estimation of the $\Lambda(\boldsymbol{\theta})$ was adopted in this case. The intensity was integrated over space-time using Dirichlet tile approximations (Berman and Turner, 1992). The model details were as follows. The basic form of the model assumed was

$$\lambda(s_i, t_i) = \exp\{\beta_0\} \cdot \exp\{a(s_i) + b(t_i) + c(s_i, t_i)\},$$

where the prior distributions are defined as in (11.3) for the main components $(a(s_i), b(t_i), c(s_i, t_i))$. A zero-mean spatial Gaussian process was assumed for the spatial component with exponential covariance function $\tau_x \exp(-\phi d)$, where d is the distance between any two locations and with variance τ_x and covariance range ϕ. The temporal component is defined by $b(t_i) \sim N(a_t b(t_{i-1}), \tau_b)$ where a_t could take a variety of forms. This parameter could be constant or could be dependent on time differences, for example, $a_t = 1/\Delta t_i$ where $\Delta t_i = t_i - t_{i-1}$. The space-time component is a residual effect, namely $c(s_i, t_i) \sim N(0, \tau_c)$. The parameter prior distributions were assumed to be

$$\exp\{\beta_0\} \sim N(0, 0.001^{-1})$$
$$\tau_x \sim Ga(0.1, 0.1)$$
$$\phi \sim U(0, 2)$$
$$\tau_b \sim Ga(0.1, 0.1)$$
$$\tau_c \sim Ga(0.1, 0.1).$$

For the converged sample (after 20,000 iterations based on two dispersed chains), the posterior estimate of ϕ was 0.0024 (sd: 0.0025) while that of $exp\{\beta_0\}$, τ_x, τ_b, τ_c were 11.77 (sd: 0.164), 8.481 (sd: 6.789), 22.68 (sd: 8.916), and 7.064 (sd: 5.087), respectively. All these parameters had positive lower and upper 95% credible limits. The Figure 11.1 suggests that there is a peak in the spatial component in the north and temporal variations with marked changes in the west of the area. However, the parameter estimates suggest that the overall rate and space-time component are well estimated but the spatial and temporal effects are not important in this example. Alternative formulations don't yield results of any great difference from this model. For example, a model including a covariate (age) was examined but the parameter for this covariate was found to have a credible interval crossing zero and so we have not reported this model here.

11.2 Count Data

Note that the above case event intensity specification can be applied in the space-time case where small-area counts are observed within fixed time periods

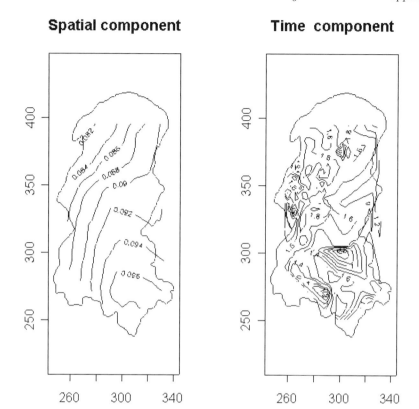

FIGURE 11.1
Burkitt's lymphoma: two displays of the components of a Bayesian model with a spatial Gaussian random field component, a temporal correlation component, and an uncorrelated space–time component. The spatial and temporal components are displayed here.

$\{l_j\}$, $j = 1, \ldots, J$, by noting that

$$E\{y_{ij}\} = \int_{l_j} \int_{a_i} \lambda(u, v) \, du \, dv,$$

where y_{ij} is the count in the $i - j$ th unit, under the usual assumption of Poisson process regionalization. In addition, the counts are independent conditional on the intensity given, and this expectation can be used within a likelihood modeling framework or within Bayesian model extensions. In previous published work in this area, cited above, the expected count is assumed to have constant risk within a given small-area/time unit, which is an approximation to the continuous intensity defined for the underlying case events. The appropriateness of such an approximation should be considered in any given application (see also Chapter 5) If such an approximation is valid, then it is straightforward to derive the minimal and maximal relative risk estimates under the Poisson likelihood model assuming $E\{y_{ij}\} = \lambda_{ij} = e_{ij}\theta_{ij}$, where e_{ij} is the expected rate in the required region/period. The maximal model estimate is $\widehat{\theta}_{ij} = y_{ij}/e_{ij}$, the space-time equivalent of the SMR, while the minimal model estimate is

$$\widehat{\theta} = \frac{\sum_i \sum_j y_{ij}}{\sum_i \sum_j e_{ij}}.$$

Smooth space-time maps, e.g., empirical Bayes or full Bayes relative risk estimates, will usually lie between these two extremes. If the full integral intensity is used, then these estimates have the sums in their denominators replaced by integrals over space-time units.

Development of count data modeling based on tract/period data has seen considerable development. In the context of a typical Poisson likelihood model where $y_{ij} \sim Pois(e_{ij}\theta_{ij})$, then the log relative risk ($\log(\theta_{ij})$) is usually the focus of modeling. The first example of such modeling was by Bernardinelli et al. (1995). In their approach, they assumed a model for the log relative risk of the form

$$\log(\theta_{ij}) = \mu + \phi_i + \beta t_j + \delta_i t_j, \tag{11.4}$$

where μ is an intercept (overall rate), ϕ_i is an area (tract) random effect, βt_j is a linear trend term in time t_j, δ_i an interaction random effect between area and time. Suitable prior distributions were assumed for the parameters in this model and posterior sampling of the relevant parameters was performed via Gibbs Sampling. Note in this formulation there is no spatial trend, only a simple linear time trend and no temporal random effect. The model in (11.4) above consists of spatial, temporal, and interaction terms. These could be extended in a number of ways. In general, we could consider three groups of components for $\log(\theta_{ij})$: $\log(\theta_{ij}) = \mu_0 + A_i + B_j + C_{ij}$ where A_i is the spatial group, B_j is the temporal group, and C_{ij} is the space-time interaction group. In (11.4) above, $A_i = \phi_i$, $B_j = \beta t_j$, and $C_{ij} = \delta_i t_j$.

Waller et al. (1997) and Xia and Carlin (1998) (see also Carlin and Louis, 2000) subsequently proposed a different model where the log relative risk is parameterized as

$$\log(\theta_{ijkl}) = \phi_i^{(j)} + \delta_i^{(j)} + \text{fixed covariate terms } (kl),$$

where $\phi_i^{(j)}$ and $\delta_i^{(j)}$ are uncorrelated and correlated heterogeneity terms which can vary in time. This model was further developed and simplified by Xia and Carlin (1998), who also examined a smoking covariate which has associated sampling error and spatial correlation. Their model was defined as

$$\log(\theta_{ijkl}) = \mu + \zeta t_j + \phi_{ij} + \rho p_i + \text{fixed covariate terms } (kl),$$

where an intercept term μ is included with a spatial random effect nested within time $\{\phi_{ij}\}$, a linear time trend ζt_j, and p_i is a smoking variable measured within the tract unit. In these model formulations no spatial trend is admitted and all time-based random effects are assumed to be subsumed within the ϕ_{ij} terms. In this formulation, $A_i = \rho p_i$, $B_j = \zeta t_j$, and $C_{ij} = \phi_{ij}$, and other covariate effects.

To allow for the possibility of time-dependent effects in the covariates included (race and age), Knorr-Held and Besag (1998) formulated a different model for the same data set (88 county Ohio lung cancer mortality, 1968–1988). Employing a binomial likelihood for the number at risk $\{n_{ijkl}\}$ with probability π_{ijkl}, for the counts, and using a logit link to the linear predictor, they proposed

$$\eta_{ijkl} = \ln\{\pi_{ijkl}/(1 - \pi_{ijkl})\},$$

where

$$\eta_{ijkl} = \alpha_j + \beta_{kj} + \gamma_{lj} + \delta z_i + \theta_i + \phi_i. \quad (11.5)$$

The terms defined are α_j, a time-based random intercept; β_{kj}, a kth age group effect at time j; γ_{lj}, a gender × race effect for combination l at the jth time; a fixed covariate effect term δz_i, where the z_i is an urbanization index; and θ_i, ϕ_i are correlated and uncorrelated heterogeneity terms which are not time-dependent. No time trend or spatial trend terms are used, and these effects will (partly) be subsumed within the heterogeneity terms and the $\alpha_j + \beta_{kj} + \gamma_{lj}$ terms. In this formulation, $A_i = \delta z_i + \theta_i + \phi_i$, $B_j = \alpha_j + \gamma_{lj} + \beta_{kj}$, including the covariate-time interactions and $C_{ij} = 0$.

More recent examples of spatiotemporal modeling include extensions of mixture models (Boehning et al., 2000), which examine time periods separately without interaction, and the use of a variant of a full multivariate normal spatial prior distribution for the spatial random effects (Sun et al., 2000), and the extension of the Knorr-Held and Besag model to include different forms of random interaction terms (Knorr-Held, 2000). Although the more complex interaction terms proposed in the latter work did not fit the data example well, the simpler formulations seem to provide a parsimonious representation of space-time behavior in risk. For example, a log relative risk can be defined purely in terms of random effects via:

$$\log \theta_{ij} = \beta_0 + u_i + v_i + \tau_j + \psi_{ij}$$

where the correlated and uncorrelated spatial components (CH, UH) are defined to be constant in time (u_i, v_i). In addition, there is a separate temporal random effect (τ_j), and finally a space-time interaction term (ψ_{ij}). In this case, $A_i = u_i + v_i, B_j = \tau_j, C_{ij} = \psi_{ij}$. Often an autoregressive prior distribution can be used for τ_j: $\tau_j \sim N(\gamma \tau_{j-1}, \kappa_\tau)$. This allows for a type of non-parametric temporal effect (random walk when $\gamma = 1$). The prior distribution for the interaction term can be simply zero mean normal (i.e., $\psi_{ij} \sim N(0, \tau_\psi)$), but more complex prior distributions could be used. This model has also been applied recently within a surveillance context (Lawson, 2004). Extensions to non-separable space-time interaction can be made by different prior distribution specifications for ψ_{ij}. Denote $\boldsymbol{\psi}$ as the matrix of interaction terms $\{\psi_{ij}\}$. For example, Type II random walk interaction (Knorr-Held, 2000) is defined by the prior distribution

$$[\boldsymbol{\psi}|\tau_\psi] \propto \exp(-\frac{\tau_\psi}{2} \sum_{i=1}^{m} \sum_{j=2}^{J} (\psi_{ij} - \psi_{i,j-1})^2),$$

whereas Type III interaction consists of time-averaged spatial correlation where

$$[\boldsymbol{\psi}|\tau_\psi] \propto \exp(-\frac{\tau_\psi}{2} \sum_{j=1}^{J} \sum_{i \sim l} (\psi_{ij} - \psi_{lj})^2),$$

and finally Type IV interaction is fully space-time dependent and is defined as

$$[\boldsymbol{\psi}|\tau_\psi] \propto \exp(-\frac{\tau_\psi}{2} \sum_{j=2}^{J} \sum_{i \sim l} (\psi_{ij} - \psi_{lj} - \psi_{i,j-1} + \psi_{l,j-1})^2).$$

These different prior distributions were fitted by Knorr-Held (2000) to the 21 year Ohio respiratory cancer dataset (for white male counts only) but he found that Type II interaction was favored with Type I also offering a lower deviance than Type III or IV. However given that the interaction term could be regarded as a form of residual (made up of unobserved confounding unaccounted for by the main effects) and also the fact that identifiability of highly structured interaction (as in Type III and IV) from main effects may be doubtful, there may be a need to consider less structured priors in applications. While it is certainly true that in other spatial statistical applications non-separable space-time interaction could be important in making a parsimonious description of the variation (Gneiting et al., 2007), it may be the case that for epidemiological data where expected rates and covariates are often available that parsimonious description is possible without such assumptions.

In other developments for space-time count data, Zhu and Carlin (2000) have examined the use of covariates at different levels of aggregation within misaligned spatial regions. Misalignment is discussed in Section 8.2. In addition, there has been development of both descriptive and mechanistic models for space-time infectious disease modeling (Cressie and Mugglin, 2000; Knorr-Held and Richardson, 2003).

Overall, there are a variety of forms which can be adopted for spatiotemporal parametrization of the log relative risk, and it is not clear as yet which of the models so far proposed will be most generally useful. Many of the above examples exclude spatial and/or temporal trend modeling, although some examples absorb these effects within more general random effects. Allowing for temporal trend via random walk intercept prior distributions provides a relatively non-parametric approach to temporal shifting, while it is clear that covariate interactions with time should also be incorporated. Interactions between purely spatial and temporal components of the models have not been examined to any extent, and this may provide a fruitful avenue for further developments. If the goal of the analysis of spatiotemporal disease variation is to provide a parsimonious *description* of the relative risk variation, then it would seem to be reasonable to include spatial and temporal trend components in any analysis (besides those defined via random effects).

Finally, it is relevant to note that there are many possible variants of the two basic data formats which may arise, partly due to mixtures of spatial aggregation levels, but also to changes in the temporal measurement units. For example, it may be possible that the spatial distribution of case event data is only available within fixed time periods, and so a hybrid form of analysis may be required where the evolution of case event maps is to be modeled. Equally, it may be the case that repeated measurements are made on case events over time so that attached to each case location is a covariate (possibly time-dependent) which is available over different time periods.

11.2.1 Georgia Low Birth Weight Example

An example of the application of a variety of space-time models to very low birth weight count data for the counties of Georgia for the years 1994–2004 are available. In this example the observed data is the count of births with very low birth weight (<1500 grams) in counties of Georgia, United States for each year during 1994–2004. The total birth count for the same period and county is also available. These data are publicly available from the Georgia Department of Health OASIS Web site (http://oasis.state.ga.us/). Figure 11.2 displays the crude rate ratios for this example. It is clear that some areas in rural Georgia have particularly high crude rates (>>0.0175).

In an analysis of very low birth weight, it may be important to provide a parsimonious description of the relative risk variation in space and time, and also to examine the data for possible risk anomalies. To this end I have examined a set of seven models that a priori may provide a parsimonious model. The choice is from a range of those discussed above and the models have been chosen to represent the different modeling approaches and also demonstrate features of the modeling process. This is in no way a comprehensive model fitting exercise, and alternative models could be hypothesized. The basic likelihood is assumed to be $y_{ij} \sim bin(p_{ij}, n_{ij})$ and the logit of the probability of very low birth weight is directly modeled. The overall crude rate ratio ($\sum \sum y_{ij} / \sum \sum n_{ij}$) for the counties of Georgia for this 11-year period is 0.0175. The first three models are simple separable models with no ST interaction: a model with a random UH spatial and temporal term and an autoregressive temporal effect (g_j) (Model 1); a model with no temporal

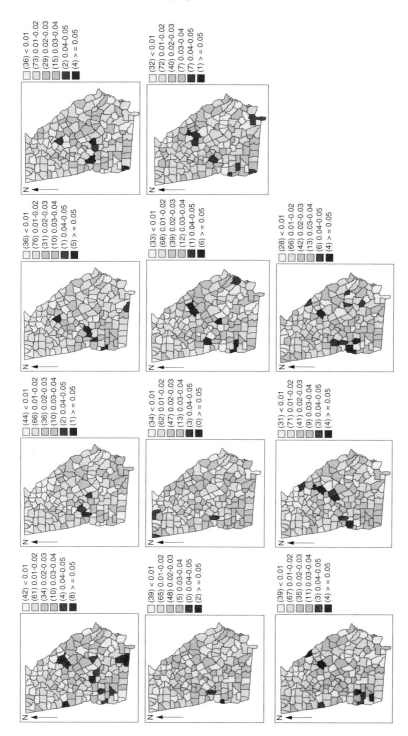

FIGURE 11.2
Georgia county level crude rate ratios for very low birth weight in relation to births 1994–2004: row-wise from 1994 to 2004.

dependence (Model 2); and a model with only temporal trend and spatial UH.

1) $\log it(p_{ij}) = \alpha_0 + a_{1j} + v_i + g_j$
   with $\alpha_0 \sim N(0, 0.0001)$, $v_i \sim N(0, \tau_v)$
   $a_{1j} \sim N(0, \tau_{a1})$
   $g_j \sim N(g_{j-1}, \tau_g)$
2) $\log it(p_{ij}) = \alpha_0 + a_{1j} + v_i$
   with $a_{1j} \sim N(0, \tau_{a1})$, $v_i \sim N(0, \tau_v)$
3) $\log it(p_{ij}) = \alpha_0 + a_1 t_j + v_i$
   with $v_i \sim N(0, \tau_v)$, $a_1 \sim N(0, \tau_{a1})$

Table 11.1 displays the DIC results for Models 1–7. It is clear that model 1–3, while parsimonious, is far from the best model. The product interaction Models (4,5) are more parsimonious, and the lowest among these is the original Bernardinelli et al model with an added spatial UH component (Model 5). It is also clear, however, that amongst the models fitted, the models proposed by Knorr-Held yield the lowest DIC model. This is Model 6 which is the model with Type I ST interaction and spatial CH and UH and temporal dependence. This model is lower than the Type II interaction (Model 7), although it is less parsimonious. Of course these results depend on prior specifications and in any particular applications sensitivity to prior specification should be examined.

For the model with lowest DIC, various posterior summaries are available. Figure 11.3 displays the sequence of 11 years of exceedence probabilities for the lowest DIC model fitted to these data (Model 6). These probabilities were estimated from $\widehat{\Pr}(p_{ij} > 0.0175) = \sum_{g=1}^{G} I(p_{ij}^g > 0.0175)/G$, where p_{ij}^g is the sampled value of p_{ij} from a posterior sample of size G. Given the caveats mentioned in Chapter 6 concerning the use of exceedence probabilities with inappropriate models, with the current 'best' model we would expect there to be reasonable reliability and stability in these estimates. It is notable that

TABLE 11.1
Space–time models for the Georgia oral cancer dataset; models are explained in text.

Model	$\overline{D}$	pD	DIC
1	8162.8	168.67	8331.68
2	8162.2	168.25	8330.43
3	8161.5	159.13	8320.64
4	8122.12	127.02	8249.14
5	8112.63	128.08	8240.71
6	7966.9	252.33	8219.23
7	8072.8	162.57	8235.37

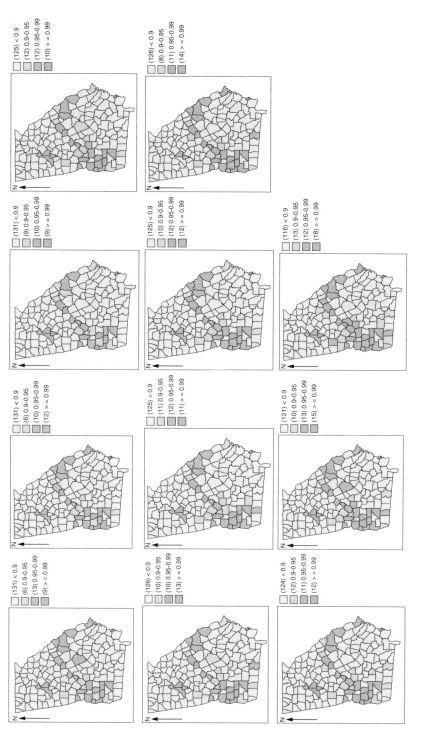

FIGURE 11.3
Georgia county level exceedence probability from a ST interaction model with Type I interaction: $\Pr(p_{ij} > 0.0175)$ is estimated as an average of posterior sample values of $I(p_{ij} > 0.0175)$. Row-wise 1994 to 2004.

most counties where high exceedences are found are rural counties (Dougherty, Terrell, Marion, Baldwin, Handcock, Richmond, and Burke) although Richmond county includes Augusta. Periodically the counties within Atlanta also signal (DeKalb and Fulton). In general there appears to be a stable patterning of the very low birth weight in that the spatial clusters seem to persist over time, whereas space-time clusters appear periodically in Atlanta.

11.3 Alternative Models

As in the case of spatial disease modeling, there are a wide variety of model variants available in the space–time extension. For example, semi-parametric models may be favored and it is straightforward to extend the spatial spline models discussed in 5.7.2, to the spatiotemporal situation. A recent example of a form of ST semi-parametric modeling is found in Cai and Lawson (2008). This is not pursued here.

11.3.1 Autologistic Models

Another important variant that was examined in the spatial case, in Chapter 5, was the autologistic model. For binary data this is an attractive likelihood variant. In Chapter 5, the ability of this model to capture some of the spatial correlation effects was noted (see Section 5.7.1). Besag and Tantrum (2003) proposed the use of autologistic models in a spatiotemporal setting. The use of pseudolikelihood allows conditioning on the neighborhood counts which are now time labeled. Define the binary outcome variable y_{ij} and assume that $y_{ij} \sim Bern(p_{ij})$. A model for p_{ij} could be constructed as

$$\frac{\exp(y_{ij}.A_{ij}))}{1+\exp(A_{ij})}$$

where A_{ij} is a function of the sum of neighboring areas and also a sum of neighboring areas at previous times. For example, define the current sum as $S_{\delta_i,j} = \sum_{l \in \delta_i} y_{lj}$, the sum over the neighborhood at a previous time as $S_{\delta_i,j-1} = \sum_{l \in \delta_i} y_{l,j-1}$. we can then consider a variety of models where space-time dependence can be captured by different forms of $S_{\delta_i,j}$ and $S_{\delta_i,j-1}$. Table 11.2 displays the results of fitting a range of autologistic models to the 21-year Ohio respiratory cancer county level dataset. In Chapter 5, an analysis of one year (1968) of this data was described. Here we examine the 21-year sequence of data from 1968–1988. Once again, for the sake of exposition, we threshold

TABLE 11.2
Autologistic space–time models: models 1–4; convolution model (Model 5)

Model	DIC	pD	MSPE	DIC (added v_i, ψ_{ij})	MSPE
1	2488.49	42.78	0.4582	1833.71 (pD: 559.9)	0.2151
2	2386.66	61.5	0.4544	1765.17 (pD:548.6)	0.2266
3	2511.94	64.6	0.4534	1824.39 (pD: 565.5)	0.2107
4	2542.65	105.29	0.4407	2539.82 (pD: 129.3)	0.4348
5	1936.00	90.34	0.3344	-	-

the $i-j$ th value at 2:

$$y_{ij} = \begin{cases} 1 \text{ if } smr_{ij} > 2 \\ 0 \text{ otherwise} \end{cases}.$$

Then we consider $y_{ij} \sim Bern(p_{ij})$ with

$$p_{ij} = \frac{\exp(y_{ij}.A_{ij}))}{1+\exp(A_{ij})}$$

with A_{ij} parameterized with a variety of covariates based on neighborhood sums. We define two sets of neighbors. The simplest model is defined to be a function of the sum of first order spatial neighbors, i.e., the neighbors defined as the adjacent small areas (in this case, I define adjacency as having a common boundary). I also examined an extended neighborhood (2nd order) where counties adjacent to the neighbors (excluding those already in the 1 st order neighborhood) are included. Hence the current sum of 1st order neighbors is $S_{\delta_{1i},j} = \sum_{l \in \delta_{1i}} y_{lj}$, while the second order is $S_{\delta_{2i},j} = \sum_{l \in \delta_{2i}} y_{lj}$. the sums at previous times are $S_{\delta_{1i},j-1}$ and $S_{\delta_{2i},j-1}$. The main autologistic models considered here are defined for the predictor A_{ij}:

1) $A_{ij} = \alpha_{1j} + \alpha_{2j} S_{\delta_{1i},j}$
2) $A_{ij} = \alpha_{1j} + \alpha_{2j} S_{\delta_{1i},j} + \alpha_{3j} S_{\delta_{1i},j-1}$
3) $A_{ij} = \alpha_{1j} + \alpha_{2j} S_{\delta_{1i},j} + \alpha_{3j} S_{\delta_{2i},j}$
4) $A_{ij} = \alpha_{1j} + \alpha_{2j} S_{\delta_{1i},j} + \alpha_{3j} S_{\delta_{1i},j-1} + \alpha_{4j} S_{\delta_{2i},j} + \alpha_{5j} S_{\delta_{2i},j-1}$.

Note that the regression parameters can be allowed to vary with time: there are no other random components in the model. For these models, α_{1j}, α_{2j}, α_{3j}, α_{4j}, α_{5j} have been assumed to vary with time but there is no prior dependence, i.e., $\alpha_{*j} \sim N(0, \tau_{\alpha_*})$ Here we also compare a conventional convolution model with components $A_{ij} = \alpha_0 + \alpha_{1j} + u_i + v_i + \psi_{ij}$ and type I interaction (Model 5) with $\alpha_0 \sim U(-a,a)$ with a large, $\alpha_{1j} \sim N(\alpha_{1j-1}, \tau_{\alpha_1})$, $u_i|u_{-i} \sim N(\overline{u}_{\delta_{1i}}, \tau_u/n_{\delta_{1i}})$, $v_i \sim N(0, \tau_v)$, $\psi_{ij} \sim N(0, \tau_\psi)$.

In this case, the random effect convolution model with Type I interaction appears to yield a relatively good model, based on DIC, compared to the autologistic model using a first order neighborhood and a single first order lagged neighborhood. To compare models with additional random effects it is reasonable to extend the autologistic models to include uncorrelated effects (v_i and ψ_{ij} where the same prior distributions are assumed as in the convolution model). For the four autologistic models this was carried out, and the resulting DICs are listed in the fifth column of Table 11.2. It is clear that the DICs are considerably lower than the convolution model for the first three autologistic models. While it is difficult to generalize from one data example, this result does suggest that autologistic models could be useful when modeling binary space-time health data, especially when added random effects are included. The added random effects included here are uncorrelated (v_i, ψ_{ij} Type I) and so are relatively simple to implement. Note that other GOF measures can be examined, such as MSPE (see Section 4.1) and these may be useful when other features of the model are important such as predictive capabilities. The MSPE for each model was also calculated and in Table 11.2 they are shown in columns four and six. While the convolution model yields the lowest DIC compared to simple autologistic models, the autologistic models with added random effects yield lower MSPEs and DICs for most models. However, the models with lowest DIC is not the lowest MSPE model. In this case, Model 2, with a lagged neighborhood effect has lowest DIC, whereas Model 3 has the lowest MSPE.

11.3.2 Latent Structure ST Models

In Chapter 5 some approaches to spatial latent structure modeling were examined. Space-time data often provides a greater latitude for the examination of latent features, as inherently there is likely to be more possibility of complexity when dealing with three dimensions instead of two. Again count data is the focus although many of the proposals here could be applied in the case event situation. As in the spatial case, it is possible to extend the fixed convolution model to include a random mixture of effects. For example, one could propose a log-linear model where

$$\log(\theta_{ij}) = \alpha_0 + \sum_{k=1}^{K} w_{ij}\lambda_{jk}$$

where K is fixed and $\sum_{k=1}^{K} \lambda_{jk} = 1, \sum_{i} w_{ij} = 1$ and $w_{ij} > 0$ $\forall i, j$, with α_0 as overall intercept. In this formulation there is a weight for each space-time unit, which is normalized as a probability over space for a given time, while for identifiability a sum to unity constraint is placed on the temporal profiles. The weights here could be regarded as loadings but are not assigned to a particular component. The temporal profiles are labeled by component. An extension to this idea could be made where the number of components are allowed to be random and then the joint posterior distribution of $(K, \{\boldsymbol{\lambda}_k\})$ would have to be sampled. This could be done via reversible jump McMC (Green, 1995) or via variable selection approaches (Kuo and Mallick, 1998; Dellaportas et al., 2002).

Alternative formulations are possible and two of these are mentioned here. First it may be possible to use a principal component decomposition in this context. Bishop (2006) discusses how a special prior specification on the loading matrix W in a Gaussian latent variable formulation leads to parsimonious description. The columns of the loading matrix W span a linear subspace within the data space that corresponds to the principal subspace. Of course it might also be useful to consider time-dependence in the specification of the component model. Extending the proposal of Wang and Wall (2003) it would be possible to consider a form such as

$$\log \theta_{ij} = \alpha_0 + \log(e_{ij}) + \lambda_j f_i$$

where f_i is the spatially-referenced risk factor, with $\sum_i f_i = 0$ and λ_j is the temporally referenced loading. Further it might be useful to consider an autoregressive prior distribution for the loading vector so that

$$\lambda_j \sim N(\lambda_{j-1}, \tau).$$

Further extension could be imagined.

An alternative to these approaches is to consider a mixture model extension of the spatial mixture models of Section (5.7.5.1). In this case the log of the relative risk is modeled via mixture product of separable components:

$$\log \theta_{ij} = \alpha_0 + \sum_{k=1}^{K} w_{i,k} \chi_{k,j}.$$

Here both $w_{i,k}$ and $\chi_{k,j}$ are unobserved but each are separate functions of space and time. Identification is supported by the separation of the spatial loading weights and temporal profiles, although further conditions can be specified. There are K components, and constraints given by $\sum_i w_{i,k} = 1$ $\forall k$, $0 < w_{i,k} < 1$ $\forall i, k$. In disease mapping studies it is reasonable to assume that underlying groupings of temporal risk profiles occur and these are region-based. Hence, we could easily be interested in finding spatial groupings of risk which are associated with specific temporal profiles.

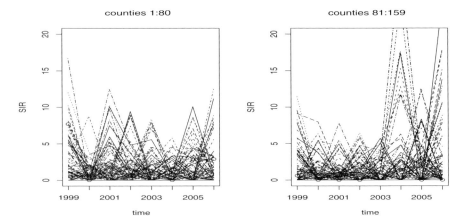

FIGURE 11.4
Georgia, United States, county level asthma ambulatory incidence for <1 year for the period of 1999–2006. Standardized incidence ratios with expected rates computed from the total period area rate adjusted for county × year population.

Suitable prior distributions depend on the application. Some prescriptions are

$$w_{i,k} = w^*_{i,k} / \sum_l w^*_{l,k} \qquad (11.6)$$

$$w^*_{i,k} \sim lN(\alpha_{i,k}, \tau_w), \ \forall i, k$$

$$\boldsymbol{\alpha} \sim MCAR(1, \Lambda)$$

$$\boldsymbol{\chi}_j \sim N(\boldsymbol{\chi}_{j-1}, \Sigma)$$

$$\Sigma = \boldsymbol{\tau}'_\chi I_0$$

where $\boldsymbol{\tau}_\chi = \{\tau_{\chi_1}, ..., \tau_{\chi_K}\}$, $\Lambda \sim Wish(\Phi, K)$, and fixed Φ. In the following, I have examined asthma ambulatory sensitive cases reported per year for <1 year age group in counties of Georgia over an 8-year period (1999–2006). These data are publicly available from the state of Georgia Department of Health OASIS online health data system. The expected rates were computed from the overall rate (= total rate/total population <1year) times the local population count <1 year. Figure 11.4 displays the SIR profiles for the 159 counties for the 8-year period. These incidence rates displayed considerable differences in their temporal behavior. It is of interest to examine whether there is any spatial clustering or aggregation in the temporal variation.

Figures 11.5 and 11.6 display the posterior averaged temporal profiles and weight maps for a converged run of Model (11.6)(500,000 burn-in and 2000 sample size). It is quite notable that component 1 and 3 are quite well defined (and have decreasing and increasing trend, respectively) while component 2

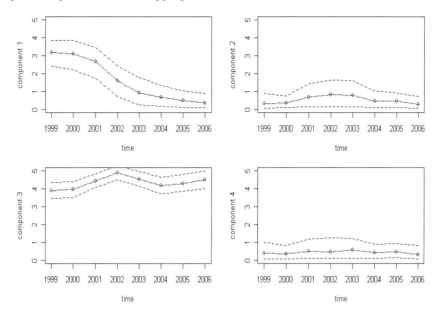

FIGURE 11.5
Space-time latent component mixture model with four fixed temporal components: row-wise from top left posterior expected temporal component profiles $(\chi_{j,l})$: components $l = 1, 2, 3, 4$.

and 4 are negligible. This is borne out by the probability weightings for the component 1 and 3 compared to 2 and 4.

Finally, it is worth noting that the Dirichlet process (DP) mentioned briefly in Section (5.7.5), can also be applied in space-time. Kottas et al. (2007) present an approach to disease modeling in space-time using Dirichlet process extensions.

11.4 Infectious Diseases

Often space-time health data concern infectious disease and its spread. This is a special case where consideration of the spread and how to model that effect could be important. Purely descriptive models can be defined of course that seek to mimic the behavior of the spread and such have been proposed by Mugglin et al. (2002) for flu epidemics, and Knorr-Held and Richardson (2003) for meningococcal disease.

FIGURE 11.6
Space-time latent component mixture model: posterior expected weight maps for the four temporal components. Row-wise from top left component 1, 2, 3, 4.

11.4.1 Case Event Data

In general it is possible to model the space-time labeling of infectious disease cases as in the non-infectious case. However, there are advantages to considering infectious disease case event modeling from a survival perspective.

11.4.1.1 Partial likelihood formulation in space–time

An alternative approach is to assume that the observed process has only a time-dependent baseline i.e., $\lambda_0(s,t) \equiv \lambda_0(t)$. This may be reasonable where the temporal progression of a disease is the main focus (such as in survival analysis). The set of observed space and time coordinates $\{s_i, t_i\}$ are conditioned upon, and a risk set (R_i) can be considered at any given time t_i. In the absence of censoring then $R_i = \{i,...,n\}$. Then the probability that an event at (s_i, t_i) out of the current risk set is a case is just

$$P_i = \lambda(s_i, t_i) / \sum_{k \in R_i} \lambda(s_k, t_i).$$

This is just an extension to the Cox proportional hazard model. Importantly in this formulation, when $\lambda_0(s,t) \equiv \lambda_0(t)$ the background hazard cancels from the model and the partial likelihood is given as

$$L = \sum_{i=1}^{n} [\log \lambda(s_i, t_i) - \log \sum_{k \in R_i} \lambda(s_k, t_i)].$$

Hence, this form enables relatively simple modeling of space-time progression of events. Lawson and Zhou (2005) use this approach to modeling progression of a foot-and-mouth epidemic, while it has also been used for a measles epidemic in a non-Bayesian context by Lawson and Leimich (2000) (see also Neal and Roberts, 2004, for another measles modeling approach; and Diggle, 2005).

11.4.2 Count Data

Often a descriptive approach would be considered first in the modeling of infection spread. By descriptive I mean using model elements to mimic the spread (without directly modeling the infection process). Mugglin et al. (2002) suggested a descriptive approach to flu space-time modeling in Scotland. In their case they applied the model to weekly ER admissions for influenza in Scottish local government districts for the period 1989–1990. The model proposed for ER admission count y_{ij} in the ith district and jth time period was of the form

$$y_{ij} \sim Poisson(e_{ij} \exp(z_{ij}))$$

where e_{ij} is the number of cases expected under non-epidemic conditions, and z_{ij} is the log relative risk. Here z_{ij} is modeled as

$$z_{ij} = d_i' \alpha + s_{ij}$$

where $d_i' \alpha$ is a linear predictor including site dependent covariates, with d_i' the ith row of the $n \times p$ covariate design matrix and α a p-length parameter vector, and s_{ij} is defined by a vector autoregressive model $(\mathbf{s}_j : (s_{1j}, ... s_{mj})')$

$$\mathbf{s}_j = H\mathbf{s}_{j-1} + \boldsymbol{\epsilon}_j.$$

Here, H is an $m \times m$ autoregressive coefficient matrix and ϵ_j is an epidemic forcing term. Spatial structure appears in both H and ϵ_j. The form of the epidemic curve is modeled by the Gaussian Markov random field prior distribution for ϵ_j:

$$\epsilon_j \sim MVN(\beta_{\rho(j)} 1, \Sigma)$$

where β determines the type of behavior, $\rho(j)$ indicates the stage of the disease, and Σ is a variance-covariance matrix. The model was completed with prior distributions specified for all parameters within a Bayesian model hierarchy. An alternative but somewhat simpler approach to descriptive modeling has been proposed by Knorr-Held and Richardson (2003). In their example, monthly counts of meningococcal disease cases in the departments of France were examined for 1985–1997. The model assumes the same likelihood as Mugglin et al. (2002) such that

$$y_{ij} \sim Poisson(e_{ij}\exp(z_{ij})).$$

At the second level they assume for the endemic disease process

$$z_{ij} = r_j + s_j + u_i$$

where r_j denotes temporal trend, s_j denotes a seasonal effect of period 12 months and a CAR prior distribution for **u**. They assume no space-time interaction for the endemic disease. For the epidemic period an extra term is included:

$$z_{ij} = r_j + s_j + u_i + x_{ij}r_{ij}^T\beta$$

where x_{ij} is an unobserved temporal indicator (0/1) which is dependent in time (but not in space) and r_{ij} is a $p \times 1$ vector (a function of the vector of observed number of cases in period $j-1$) and β is a p-dimensional parameter vector. The authors propose six different models to describe the epidemic period depending on the specification of $r_{ij}^T\beta$. Whether an epidemic period is present completely depends on the value of x_{ij}. In this formulation the x_{ij} are essentially unobserved binary time series, one for each small area. Unlike the Mugglin et al. (2002) formulation, these have to be estimated.

Both these approaches seem to have been successful in describing the retrospective epidemic data examined. It will be instructive to see whether these different approaches will be successful in the prospective surveillance of infectious disease.

Mechanistic count models that address the infection mechanism have been proposed for the temporal spread of measles (Morton and Finkenstädt, 2005). These were based on susceptible-infected-removed (SIR) models where account is made of the numbers at each time point in each class. They also account for underascertainment in their model. For daily measles case reporting in London, Morton and Finkenstädt (2005) defined the true infective count for period j as I_j and the reported count as y_j linked by a binomial distribution to allow for underascertainment: $y_j \sim bin(\rho, I_j)$ where ρ is a reporting probability. The susceptible population at the $j+1$ th period is S_{j+1} while removal is D_j and additions B_{j+1}. In that model infectives and susceptibles are modeled as

$$I_{j+1} \sim f_1(r_j I_j^\alpha S_j, K_{j+1})$$
$$S_{j+1} \sim f_2(S_j + B_{j+1} - I_{j+1} - vD_{j+1})$$

where K_{j+1} is some underlying latent series of events, and f_1 and f_2 are suitable distributions. The distribution f_1 is called the transmission distribution. The term r_j is a proportionality constant that modulates the interaction term $I_j^\alpha S_j$ and can be regarded as an infection rate. The α term can also be estimated. For populations where the susceptible population is large compared to the infectives at each time period then the effect of $B_{j+1} - I_{j+1} - vD_{j+1}$ may be small and so simpler models could be conceived where $S_{j+1} \sim f_2(S_j)$. Of course, for finite small populations this could be a bad approximation.

Extending this to the spatial situation within a Bayesian Hierarchical modeling framework is straightforward (see Lawson, 2006b, Ch. 10). A space–time infection model could be proposed where there are $i = 1, ..., m$ small areas and $j = 1, ..., J$ time periods. A simple form could be

$$y_{ij} \sim bin(\rho, I_{ij}) \quad (11.7)$$
$$I_{ij} \sim Pois(\mu_{ij})$$
$$S_{ij+1} = S_{ij} - I_{ij} - R_{ij}$$
$$R_{ij} = \beta I_{ij}$$

where $\mu_{ij} = S_{ij} I_{ij-1} \exp\{\beta_0 + b_i\}$. The term $\exp\{\beta_0\}$ describes the overall rate of the infection process while a spatially correlated term b_i is included and it is assumed to have a CAR prior distribution. The susceptible model is deterministic and with a fixed β the removal proportion is fixed. Many variants of these specifications could be considered. For example, we could specify $\mu_{ij} = S_{ij} f_*(I_{ij-1}, \{I\}_{\delta_{ij-1}}, \exp\{\psi_{ij}\})$ where dependence in f_* is on the previous count I_{ij-1}, on the counts in a predefined neighborhood δ_{ij-1}, $\{I\}_{\delta_{ij-1}}$ say, and a linear predictor including both covariates and random effects which could be spatially or temporally correlated.

An example of the application of this model to publicly available flu culture positives (C+) from the 2004–2005 flu season reported for bi-weekly periods for the counties of the state of South Carolina in the United States is given in Lawson (2006b, Ch. 10), for the model specified in (11.7). In that case it was assumed that $\beta = 0.001$. Figure 11.7* displays the flu season count variations for a selection of four counties in South Carolina (Beaufort, Richland, Charleston and Horry). Beaufort and Horry both have high older age group populations, while the main urban centers in the state are in Richland (city of Columbia) and Charleston (city of Charleston). Figure 11.8 displays the thematic maps of the counts for a selection of three time periods during the season.

Figure 11.9† displays the posterior average infection rates ($\widehat{\mu}_{ij}$) for a selection of four counties in the state, along with their 95% credible intervals. Interestingly, Horry county peaks much earlier than other areas (time period

---
*Wiley permission Figure 1.25 ch 1 of Lawson 2006
†Wiley permission Figure 10.3 Lawson 2006

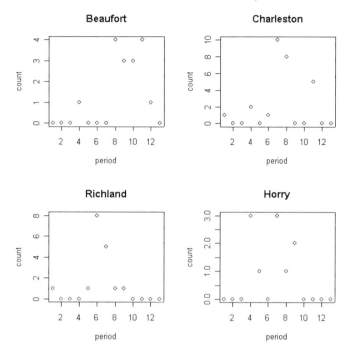

FIGURE 11.7
South Carolina influenza confirmed C positive notifications: count profiles for the period December 18, 2004–April 10, 2005 for a selection of four counties.

4–6) while Beaufort seems to display lag effects into period 10–12. In fact there appears to be considerable spatial and temporal variation in the mean infection level.

11.4.3 Special Case: Veterinary Disease Mapping

While most work in disease mapping has been targeted towards human health, there is a growing literature now in the application of disease mapping to veterinary health. Veterinary health covers the analysis managed animal populations, but may also cover wild populations, in particular where zoonosis is possible (disease transmission between species). This has arisen partly because of recent outbreaks of BSE in cattle, foot and mouth disease among sheep and cattle (FMD) and the spread of Sars or avian flu and its potential for evolution within the human population. Often in these examples, space-time variation is the most important feature to be modeled and so it is justified to include discussion of this topic here.

The basic descriptive disease mapping techniques, such as the commonly used convolution models (Chapter 5) can of course be applied to veterinary

Spatiotemporal Disease Mapping

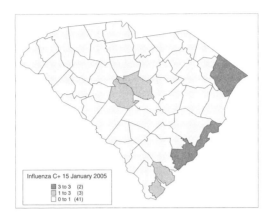
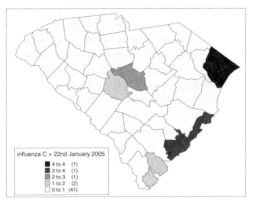
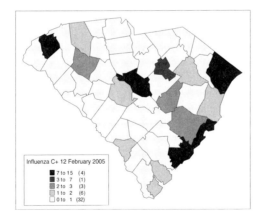

FIGURE 11.8
South Carolina influenza confirmed positive notifications: count thematic maps for a selection of three time periods in the 2004–2005 season.

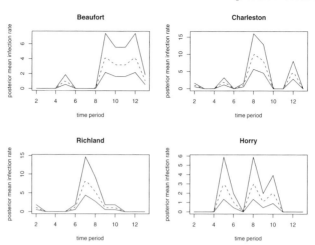

FIGURE 11.9
South Carolina influenza confirmed positive notifications: posterior mean infection rate estimates for 13 time periods with credible 95% intervals for a selection of four urban counties: Beaufort, Charleston, Richland, and Horry.

data and there are now various examples of this in the literature (e.g., Stevenson et al. 2000, 2001, 2005; Durr et al., 2005). Competing risk multivariate analysis has also been proposed (Diggle et al., 2005). An overview of GIS-based applications is found in Durr and Gatrell (2004).

Often the data that arises in veterinary applications is akin to human health data. It is usually discrete and could be in the form of a marked point process in space and time (animal locations and their disease state and a date of observation), could be counts of animals with a disease (such as within farms) or counts of infected farms within parishes or counties. Within smaller spatial units (such as farm buildings) it is also possible to model individual animal outcomes over time.

Figure 11.10[‡] displays the bi-weekly standardized incidence ratios for FMD over parishes within northwest England (Cumbria), United Kingdom during 2001. The space-time spread of the disease is clearly shown. The denominators for the SIR were calculated from the overall rate for the whole space-time window. In a retrospective analysis it is reasonable to standardize within such an overall rate. However in a surveillance context this would not be possible. In that case one option would be to use a historical rate. The spread of infection can be described via models that attempt to summarize the spatial and temporal effects. For example, the count of FMD premises (farms) within the i th parish at a given time period (j) in Cumbria (y_{ij}) could be modeled

---

[‡]Preventive vet med permission Lawson and Zhou (2005) Figure 4.

as a binomial random variable with various hierarchical elements. Define the number of farms within the i th parish as n_i, then assume

$$y_{ij} \sim bin(p_{ij}, n_i)$$
$$\log it(p_{ij}) = A_i + v_i + \xi_j$$
$$\text{where } A_i = \beta_0 + \beta_1 x_i + \beta_2 y_i + \beta_3 x_i y_i.$$

The term A_i is purely a trend component in the spatial coordinates (x_i, y_i) of the parish, while v_i, ξ_j are random effects that are meant to capture the spatial and temporal random variation. This is the descriptive model reported by Lawson and Zhou (2005). The random terms are assumed to have prior distributions given by:

$$v_i \sim N(0, \tau_v)$$
$$\xi_j \sim N(\xi_{j-1}, \tau_\xi).$$

Hence temporal dependence is assumed to be modeled by an autoregressive term in the logit link. Suitable parameter prior distributions were assumed for $\boldsymbol{\beta}, \tau_v, \tau_\xi$. There is no spatial correlation term, as it was felt that the dynamic nature of the risk would be better described via uncorrelated spatial risk and correlated temporal risk. This descriptive model was only partially successful in describing the variation. In this model the spatial trend component is fixed in time. Not considered by the authors was the possibility of making the regression parameters in the trend component time-dependent. This might be an attractive option in some cases as it would allow the spatial model to have a dynamic element. For example, it could be assumed that $\boldsymbol{\beta}_j \sim \mathbf{MVN}(\boldsymbol{\beta}_{j-1}, \tau_\beta I_n)$, where I_n is a unit matrix where $n = 4$.

11.4.3.1 Infection modeling

Infectious disease spread is particularly important in veterinary applications. There are few examples of mechanistic Bayesian modeling of such spread. Höhle et al. (2005) gives an example where swine fever within pig units is modeled spatially via a survival model where the hazard function is a function of the count of infected animals within the unit and also the count in neighboring units. Bayesian models for the UK FMD outbreak at farm level were also proposed where a survival model was assumed at the farm level (Weibull in this case) for the risk of infection and then a count model for the number infected conditional on the infection of the farm (Lawson and Zhou, 2005). This also had spatial dependence included. These models are adequate where a relatively slow epidemic is apparent, but are likely to be inadequate when a full epidemic curve with peaking and recession is to be modeled.

11.4.3.2 Some complicating factors

There are a number of complicating factors that appear in veterinary examples that should be highlighted. First of all, it is often the case, that for

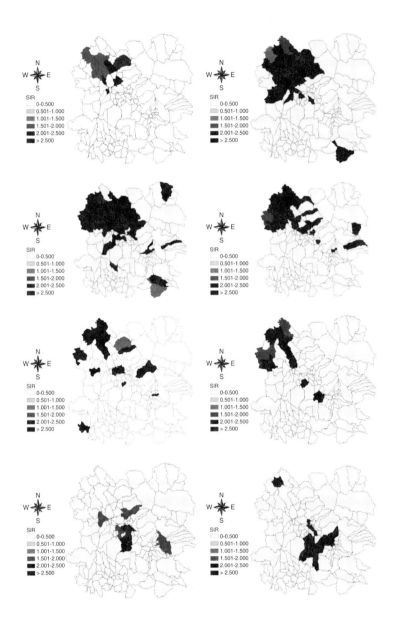

FIGURE 11.10
Foot and mouth disease (FMD) epidemic northwest England, United Kingdom 2001: bi-weekly maps of standardised incidence ratios for February 2nd two-week period until June 1st two-week period (row-wise).

important infection epidemics, *intervention* by veterinary agencies will dramatically alter the progression of the disease (and also the ability to observe the progression). In the FMD outbreak in the United Kingdom on 2001, ring culling was introduced. This entailed slaughtering all farm animals within a fixed radius of a newly found case of FMD. This culling is an attempt to intervene in the spread of the disease. The effect of this is to introduce a particular form of spatiotemporal censoring during the epidemic and this can lead to considerable missing information that could affect model predictions. This also leads to the other important aspect of modeling veterinary disease and that is the surveillance or predictive capability of models. In the FMD outbreak in the United Kingdom in 2001, statistical models were used on a daily basis to predict the progression of the disease. The online surveillance of disease spread is very important and predictive capability is of course an important and natural ingredient of Bayesian modeling with recursive Bayesian updating an essential ingredient.

Finally, it is also important to note that there can be a major difference in data acquisition within veterinary health compared to human health. Animals usually do not report disease to vets! Hence they have to be sampled, and, unless registries of disease are set up with mandatory reporting, there is the possibility that *under-reporting* or *underascertainment* of cases could become a major problem. While this is less important for managed herds (such as on farms), it could be very important for wild populations.

For wild populations, animals are free to move and their mobility can make sampling very problematic. For wild populations, access to animals and the fact that they move around in space-time leads to extra complications. Distance sampling (Buckland et al., 2001) is one approach to assessing mobile population density. Remote sensing could be used for density estimation also. However this does not usually allow the health of animals to be assessed. Hunter surveys and special culling have been used for deer population health (chronic wasting disease: Farnsworth et al., 2006). However these data are often prone to considerable biases due to the nature of hunting (choice of area, choice of animal, time of day, date of hunting) and it is not clear how representative these data are of the true population health. Faecal surveys may also be used.

Even with managed herds, the animals must be continually checked to find out if they are diseased, and this means that unless there is continual monitoring of uninfected animals over space and time then underascertainment is highly likely. Of course some modeling strategies are possible to deal with this issue as noted above (Section 11.4.2).

A

Basic R and WinBUGS

It is useful to be able to manipulate data, design models, and analyze output from posterior sampling with suitable tools. The package R, which is freely available (www.r-project.org), is a very useful tool for pre- and post-analysis of Bayesian models. Not only is R readily available, it also includes state-of-the-art procedures for manipulating/analyzing data and has very sophisticated graphics capabilities. It also has functionality for interacting with McMC programs and in particular has functions that can process McMC output (CODA, BOA).

A.1 Basic R Usage

R is an object-oriented language which is platform-independent and command-driven. This latter feature seems regressive given the common use of graphical user interfaces (such as in S-Plus). On the other hand, this allows wide availability across platforms. There is no doubt that this feature does frustrate the occasional user, particularly when data input must be command-based.

Review of basic R features is found in Maindonald and Braun (2003) and more extensive use in modeling is covered by Faraway (2006). We assume some basic familiarity with R.

A.1.1 Data

Most often data can be processed as vectors or matrices within R. For the South Carolina congenital anomaly mortality example, the data, consisting of county-based observed counts and expected rates, are read into a dataframe adat:

adat<-list(m=46,
y=c(0,7,1,5,1,1,5,16,0,17,4,0,0,1,1,7,1,3,0,0,8,2,13,7,0,8,0,3,2,4,1,
11,0,1,2,3,3,8,6,14,3,11,6,0,1,5),
e=c(1.129778827,6.667008775,0.650279674,6.988864371,0.95571406,
1.123210345,5.908349156,8.539026017,0.601016062,18.92051111,
2.272694617,1.73736337,2.019808077,1.688099759,1.747216093,
3.221840201,1.835890594,5.221942834,0.978703751,1.254579976,

6.407553754,2.676656232,16.57884744,3.077333607,1.087083697,
7.606301637,1.018114641,2.15774619,2.844152512,2.955816698,
0.985272233,9.22871658,0.38097193,1.855596038,1.579719813,
1.579719813,2.647098065,4.791707292,4.144711859,15.70852363,
0.765228101,11.32077795,6.256478678,1.500898035,2.085492893,7.297583004)).

The expected rates are computed from state-wide age-gender stratified rates and applied to county age-gender group populations and then summed.

Summarization of the basic data can be achieved via various commands, one of which is displayed here.

smr<-adat$y/adat$e
summary(smr)

This produces a numerical summary of the vector smr:

Min. 1st Qu. Median Mean 3rd Qu. Max.
0.0000 0.5404 0.8908 0.9282 1.2620 3.9200

A.1.2 Graphics

For non-spatial data a variety of summarization graphics are available. For example, histogram or density estimates are available. The commands

par(mfrow=c(1,2))
hist(adat$y,breaks=50,xlab="count",main = "")
plot(density(adat$y),main="")

produced the Figure A.1. We can also easily get crude relative risk estimates by computation of the standardized mortality ratio for the data which is just given by

plot(density(smr),main="")

Figure A.2 displays the density estimate for the SMR for this count data. Another display demonstrates some further range of measures available

par(mfrow=c(2,2))
hist(adat$y,xlab="count",main="")
hist(smr,xlab="standaridised incidence ratio",main="")
plot(adat$y,adat$e,xlab="expected",ylab="count",main="")
boxplot(smr,xlab="standaridised incidence ratio")

These commands lead to the display in Figure A.3.

A variety of functions are available for graphical representation of spatial data on R. A common practice with spatial data is to display a pixellated image of a smoothed (estimated) surface. In addition, contouring is useful and can be added easily to such a plot. Although perspective plotting is possible and visually attractive, these plots are not particularly informative and are not discussed here. Until recently, the AKIMA and SM packages provided the simplest functionality for such surface interpolation and plotting. In AKIMA,

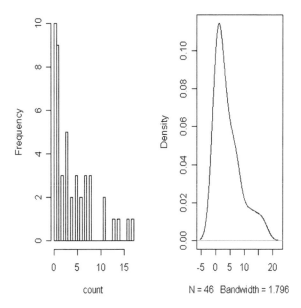

FIGURE A.1
Histogram and density estimate of the 46 county level counts for the South Carolina congenital anomaly mortality example.

the interp function provides a mathematical interpolator which allows the input of unordered x,y,z vectors:

library(akima)

asd1<-interp(xcen, ycen, smr).

The output is a list with 3 elements which are x,y grid points and associated interpolated values (z), These can be displayed via image or contour functions:

image(asd1,col=gray(20:0/20),xlab="x coordinate",ylab="y coordinate")

contour(asd1,add=T)

Figure A.4 displays the results of using interp for the smr data with county centroids (xcen,ycen).

The above commands yield a grey scale pixellated image map with a contour map overlain with default contour levels. An alternative package that can be used for such surface interpolation and plotting is the SM package. This package was developed to provide nonparametric density and regression estimation capability. The function sm.regression provides a nonparametric regression estimator for x,y,z data, based on the work described in Bowman and Azzalini (1997):

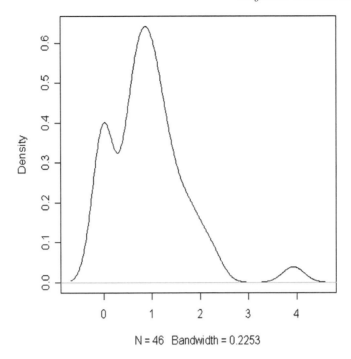

FIGURE A.2
Density estimate of the SMR for the SC anomaly mortality data.

```
xyMAT1<-cbind(xcen,ycen)
asd2<-sm.regression(xyMAT1,smr,h=c(3000,3000),display="image")
```
will display a contour surface for the interpolated smr with a fixed smoothing (specified by h). An evaluation grid can be specified also as well as different smoothing procedures.

More recently a more sophisticated smoothing and interpolation package has been developed. This package is called **MBA** and implements multilevel B spline surface interpolation. The main function is mba.surf and relies on a Multilevel B-spline Approximation (MBA) algorithm. The R implementation of the code was developed by Andrew Finley and Sudipto Banerjee at University of Minnesota. It is available in zip form from http://blue.fr.umn.edu/MBA/. The mba.surf function has considerable flexibility in the specification of the local smoothness and flexibility of the surface. An example of its application follows where we examine the smr surface for the SC congenital anomaly data:

```
xyMat2 <-cbind(xcen,ycen,smr)
asd2<-mba.surf(xyMat2,50,50)$xyz.est
 image(asd2,col=gray(20:0/20),xlab="x coordinate",ylab="y coordinate")
 contour(asd2,add=T)
 points(xcen,ycen)
```

Basic R and WinBUGS

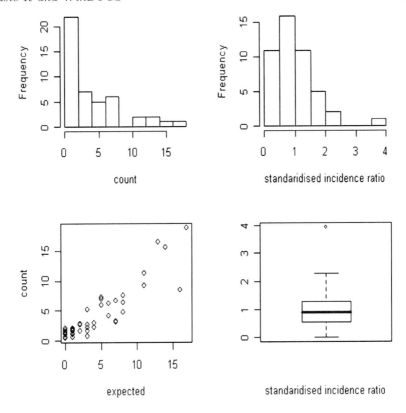

FIGURE A.3
Display of four basic plots of the observed data. Clockwise from top left: histogram of the observed counts; histogram of the SIR; boxplot of the SIR; scatterplot of observed versus expected count.

Figure A.5 displays the image and contour map for the smr example using the multilevel B spline code function mba.surf.

A.2 Use of R in Bayesian Modeling

R is a useful package for data manipulation and post McMC processing as it can a) pass information to WinBUGS and b) run WinBUGS. In addition to this specific purpose it can be useful in simpler scenarios also. R is used extensively in Bayesian course teaching (see for example www.mrc-bsu.cam.ac.uk/bugs/weblinks/webresource.shtml for extensive ex-

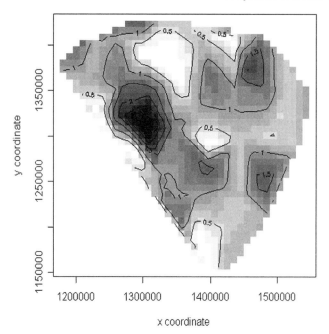

FIGURE A.4
Image and contour map of the interpolated SMR for the SC congenital anomaly mortality data example using interp.

amples). The simulation facilities in R allow a wide range of simple Bayesian modeling to take place. With simpler Bayesian models there could be two approaches to posterior inference. First, the exact form of the posterior distribution could be a known distribution and so summarization of the posterior information can made directly from the distribution. For instance, the beta posterior distribution that arises from the binomial likelihood and beta prior distribution can be viewed via the use of dbeta function (the distribution function for the beta distribution)

theta<-seq(0.3,0.6,0.001)
w=1;alpha=1;beta=1;
plot(theta,dbeta(theta,436,543),type="l",xlab="theta",
ylab="",xaxs="i",yaxs="i",yaxt="n",bty="n",cex=2)   # likelihood of θ
lines(theta,dbeta(theta,alpha,beta),lty=2)   # prior distribution
lines(theta,dbeta(theta,437+alpha,543+beta),lty=2)   # posterior distribution

This sequence of commands plots the likelihood, Beta (1,1) prior distribution (dotted line) and the posterior distribution, and results in the Figure A.6.

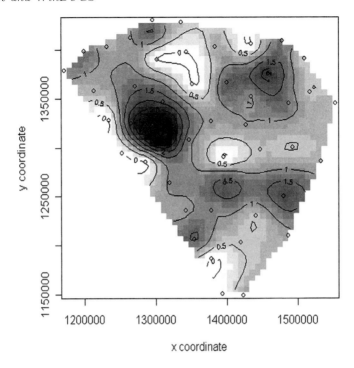

FIGURE A.5
The SMR image and contour map for thr SC congenital anomaly mortality example using the MBA package

Second, we may know the posterior distribution but may find it easier to *sample* from the posterior distribution. On R it is possible to simulate from a wide range of distributions and so if the form of the posterior distribution is known then we can take a posterior sample. For example a sample of 1000 values from a beta distribution with parameters 437+alpha,543+beta could be obtained in the vector samp by

samp<-rbeta(1000,437+alpha,543+beta)
summary(samp)

The final statement provides a summary of the sample values in samp.

Alternatively, sometimes within simulation studies, we would be interested in generating synthetic data from given parameter values for parametric distributions. For example, it is commonly assumed that incident counts of disease (y_i) within small areas (census tracts, counties, zip codes, etc.) follow a Poisson distribution with mean $e_i\theta_i$ where e_i is a known expected rate and θ_i is a relative risk. Under simulation it would be usual to assume a particular form for the risk, and then to simulate form the given Poisson distribution. Hence if for example $\theta_i = \exp(\alpha + v_i)$ where $\alpha = -.2$ and $v_i \sim N(0,1)$ then we could

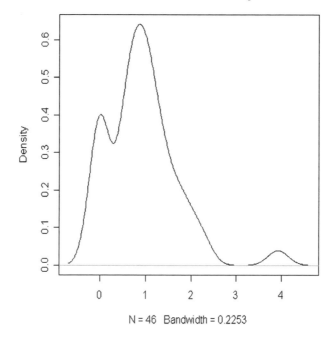

FIGURE A.6
The binomial likelihood, (flat) beta(1,1) prior distribution and the posterior distribution from R commands.

simulate this straightforwardly within R. For a sample size of m, the following code will leave a vector of m counts in ysim:

```
v<-rnorm(m,0,1)
theta<-exp(-.2+v)
mu<-e*theta
ysim<-rpois(m,mu).
```

A.3 WinBUGS

While it is not the intention to provide a detailed review of WinBUGS usage in disease mapping, some basic features of the package should be highlighted here to motivate further use. More advanced features will be introduced when they arise in later chapters. The reader is directed to Lawson et al. (2003) for a more detailed introduction to a selection of topics in this area. Example code for examples in that work are available from http://sph.sc.edu/alawson/default.htm. The WinBUGS online help facilities include a range of examples that are also very useful to follow. The examples can be executed within the package and so can be seen in operation. The MRC Biostatistics WinBUGS Web site (http://www.mrc-bsu.cam.ac.uk/bugs) is also a very useful resource.

A.3.1 Simulation

It is possible to use WinBUGS for simulation as it is a feature of the package that simulation-based methods are a kernel of the methodology. WinBUGS is based on the idea that Bayesian models can be represented as hierarchies with parameter nodes. Each parameter in the model can be represented as a stochastic node (in which case it has a distribution), a constant, in which case its fixed, or a logical node, in which case it is the result of an assignment operation based on an expression. Given below is an example of simple simulation code for a Poisson distributed variable y (y) with mean $\mu = e\theta$ (mu <- −exp ∗ theta) where a gamma distribution is assumed for θ at the next level of the hierarchy.

```
model;
{
y ~dpois(mu)
mu <- exp * theta
theta ~dgamma(0.05,0.05)
}
```

In this example, $\theta \sim Gamma(0.05, 0.05)$, the expected rate ($e = 1$) is loaded as data

list(exp=1)

while all stochastic nodes are initialized before the simulation starts:

list(theta=0.5,y=2).

Following the initialization, the number of values of y simulated is set in the sample monitor. Figure A.7 displays the series of counts obtained with 10000 iterations with the above model.

More sophisticated simulation is possible within WinBUGS of course, but it is often more convenient to exploit the interactive capabilities of R in simulation of data.

A.3.2 Model Code

The previous example demonstrated some features of WinBUGS model code, albeit in a simple simulation. When data is included within a model then a data likelihood must be specified. WinBUGS does not distinguish between a parameter node that has associated data as opposed to simply a stochastic node. The only difference is when data are loaded and a stochastic node with

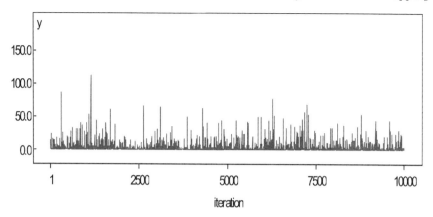

FIGURE A.7
Time series of simulated counts from WinBUGS under a simple Poisson–gamma model

data is not simulated. Hence, both data and parameters are treated alike. WinBUGS assumes that if data are not assigned to any stochastic node then the node must be initialized and updated. The model code below is an example where data are included. In this case, again we use the simple example of count data within small areas. Assume that there are m regions, the count in the i th region is y_i, the expected rate for the same region is e_i and the relative risk is θ_i. Assume also that $[y_i|e_i,\theta_i] \sim Poisson(\mu_i)$, and $\mu_i = e_i\theta_i$. At the next level of the hierarchy assume that $\theta_i \sim Gamma(a,b)$. Hence this model assumes that the relative risks are an exchangeable sample from a Gamma distribution. Finally, at the next level of the hierarchy, the gamma parameters: a, b, both have exponential distributions with fixed parameters $\lambda = 1/mean = 0.1$. This leads to a large variance and mean of 10. With these hyper-prior distributions the expected value of θ_i is 1 but with a relatively small variance. For this model the WinBUGS code is given below.

```
model{for (i in 1:m)
{     # Poisson likelihood for observed counts
    y[i]~dpois(mu[i])
    mu[i]<-e[i]*theta[i]
    # Relative Risk
    theta[i]~dgamma(a,b)}
# Prior distributions for "population" parameters
a~dexp(0.1)
b~dexp(0.1)
# Population mean and population variance
mean<-a/b
var<-a/pow(b,2)}
```

Basic R and WinBUGS

In the above specification the for loop extends over all subscripted variables (y[],mu[],e[],theta[]). Note that the for specification has the same loop format as that found in R or S-Plus. The prior distributions are specified outside the main loop for parameters which are not subscripted. Note that theta[i] is subscripted and so its distribution must be specified inside a for loop.

A variant of this model defines a log linear model for the relative risk:

```
model{for (i in 1:m)
{       # Poisson likelihood for observed counts
        y[i]~dpois(mu[i])
        mu[i]<-e[i]*theta[i]
        # Relative Risk
        log(theta[i])<-a+v[i]
        v[i]~dnorm(0,precv)}
# Prior distributions for "population" parameters
a~dnorm(0,preca)
siga<-1/preca
sigv<-1/precv
preca~dgamma(0.5,0.0005)
precv~dgamma(0.5,0.0005)}
```

This model can be run with the data from the adat list (m,y,e) and initialisation of the stochastic parameters (preca,precv,a,v) can be made using:

```
list(preca=0.1,precv=0.1,a=0.1,
v=c(0,0,0,0,0,0,0,0,0,0,0,0,0,0,0,
0,0,0,0,0,0,0,0,0,0,
0,0,0,0,0,0,0,0,0,0,0,
0,0,0,0,0,0,0,0,0,0,0,0)).
```

The Metropolis rate monitor from this program on WinBUGS for a single chain of 20,000 iterations is shown in Figure A.8. Metropolis–Hastings updating was chosen by WinBUGS in this case. This monitor checks for a reasonable acceptance rate. In general, the rate should lie between 0.25 and 0.5 (see Section 3.3.7.1). Hence, in this case the sampler seems to have reasonable acceptance rate.

The sample Monitor on WinBUGS also provides a wealth of information concerning parameters that have been monitored, the choice of monitored parameters being left to the user. In Figure A.9, the sample monitor has been set to record the parameters a,preca,precv,siga,sigv and the display shows part of the stats option on the Sample Monitor. Displayed are the posterior summaries for each parameter for the chosen sample size (mean, sd, MC error, 2.5% and 97.5% percentiles and the median). These are the default summaries, but can be altered to produced e.g., alternative percentiles. All parameters can be monitored and summarized in this way. For those that are indexed by region (such as theta[], mu[], v[]), then a complete list of summaries for each region is produced.

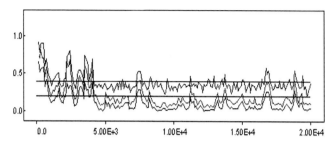

FIGURE A.8
Display of the acceptance rate monitor for the Metropolis-Hastings sampler adopted by WinBUGS, for the Poisson log linear frailty relative risk model for 20,000 iterations.

node	mean	sd	MC error	2.50%	median	97.50%
a	-0.00171	0.03814	0.001706	-0.08869	-3.20E-04	0.0733
preca	1258	1490	49.17	17.76	732.6	5429
precv	2183	1596	215.5	151.7	1947	6.00E+03
siga	0.01844	0.4321	0.009608	1.84E-04	0.001368	0.06029
sigv	0.001223	0.001814	2.47E-04	1.67E-04	5.14E-04	0.00663

FIGURE A.9
Display of summary of posterior parameter estimates as provided by WinBUGS for a sample of 2000 after a burn-in of 20,000 iterations for the log linear Poisson relative risk model with simple frailty random effect.

Figure A.10 displays the thematic map of the posterior expected spatial distribution of the theta[] parameter (relative risk) for the Poisson log linear model. This map is easily obtained from the GeoBUGS menu option within WinBUGS. The region geographies (in this case county boundaries) must be available as a GIS-format file (ArcView, Epimap, Splus, ArcInfo) and this can be imported to WinBUGS for use with GeoBUGS. Once the map geographies are input, any parameter indexed by the region index can be plotted via the Map Tool. The order of regions must be the same as the parameter vectors in the model of course. In the Poisson log linear model above, there are three parameter vectors that can be plotted: theta[], mu[], and v[]. Figure A.11 displays the map for the posterior mean uncorrelated frailty term v[].

Alternative models can be specified for different data formats of course. The above example assumed a Poisson likelihood for count data within counties. Another situation commonly found in spatial analysis of disease risk is that a finite population is measured within an area and within that population a binary variable is recorded on each unit. For example, an example would be total brain cancer cases recorded within counties and the number of male cases. Here we examine another South Carolina example where we have the

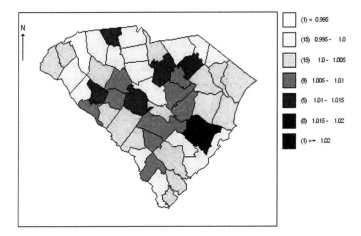

FIGURE A.10
Display of the GeoBUGS posterior expected relative risk map (theta[]) for the 46 counties of South Carolina for a 2000 sample size and the Poisson log linear frailty model.

total case count for 1996–2000 of brain cancer by county in SC. The male count is also known. The sex difference may have a spatial expression. The model code below can be applied to this case and is similar to the previous Poisson model except we have a binomial likelihood for the male count data (malc[i]) out of a total count (totc[i])

```
model{for (i in 1:m)
{     # binomial likelihood for observed counts
      malc[i]~dbin(p[i],totc[i])
      logit(p[i])<-a+v[i]
v[i]~dnorm(0,precv)}
# Prior distributions for "population" parameters
a~dnorm(0,preca)
preca<-1/siga
precv<-1/sigv
siga~dgamma(0.001,0.001)
sigv~dgamma(0.001,0.001)}
```

The mapped results of this analysis after 10,000 iterations are displayed for a sample size of 2000, in the Figures A.12 and A.13. It is clear that there is considerable spatial structure in this data, as the probability map has clear peaks of male over female incidence in the northern coastal regions of the state. The uncorrelated frailty term also shows this tendency which suggests that considerable spatial structure remains in the data.

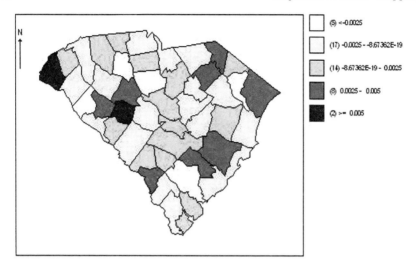

FIGURE A.11
Display of the posterior mean uncorrelated frailty term, v[], for a sample of size 2000, for the Poisson log linear relative risk model with uncorrelated frailty term.

A case event Bernoulli model can be specified, as a special case of the binomial model above, where y[i] is a binary vector and a Bernoulli likelihood is assumed. This is appropriate where you have individual level data, and is commonly found in putative pollution hazard studies but is by no means uncommon elsewhere. A logistic linear link can also be assumed here and the code below displays this as a WinBUGS model, similar to the binomial frailty model. The only difference lies in the outcome data and likelihood. For case-control event data, the binary outcome would be the case/control status and all other variables (such as individual covariates or functions of the x, y locational coordinates of the cases and controls) can be included as covariates. In the example code below, two covariates are read in (X1[],X2[]) and they have regression parameters B1 and B2 with zero-mean Gaussian prior distributions. The precision parameters have diffuse Gamma prior distributions. This a binary random effect logistic linear model.

```
model{for (i in 1:m)
{    # Bernoulli likelihood for observed counts
    y[i]~dbern(p[i])
    logit(p[i])<-a+v[i]+B1*X1+B2*X2
v[i]~dnorm(0,precv)}
# Prior distributions for "population" parameters
B1~dnorm(0,precB1)
B2~dnorm(0,precB2)
```

Basic R and WinBUGS 297

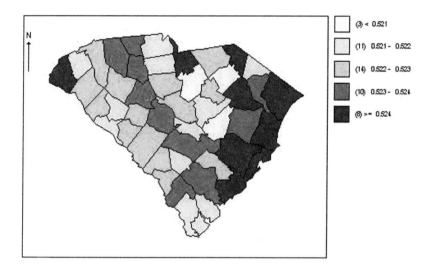

FIGURE A.12
Display of of the posterior mean probability of a male cancer by county in SC for the five year brain cancer case example (1996–2000).

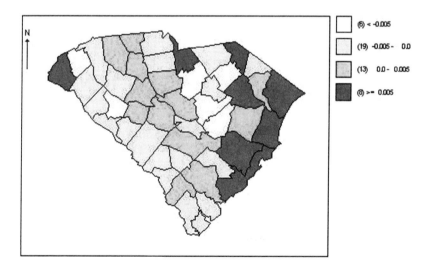

FIGURE A.13
Display of the posterior mean uncorrelated frailty (v[]) for the logistic linear model by county in SC for the five year brain cancer case example (1996–2000).

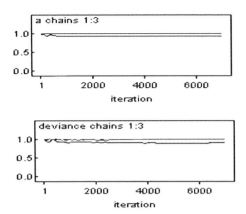

FIGURE A.14
Display of Brooks–Gelman–Rubin plots for the deviance and parameter a for the SC county-based brain cancer logistic linear frailty model for 1996–2000. The plots are closely linear and parallel which suggests close convergence by 2000 iterations.

```
a~dnorm(0,preca)
preca<-~dgamma(0.001,0.001)
precv~dgamma(0.001,0.001)
precB1~dgamma(0.001,0.001)
precB2~dgamma(0.001,0.001)
}
```

Convergence of McMC samplers is an important issue, as noted in Section 3.3.6. In all the examples above the samplers were monitored using a variety of diagnostics. The main diagnostic used on WinBUGS is the Brooks-Gelman-Rubin (BGR) diagnostic which examines the between and within sample variances for output streams. Figure A.14 displays the BGR diagnostic for two components of the logistic linear model for the SC brain cancer example: the deviance and parameter a. The display suggests that the models have converged early at least for these parameters.

A.4 R2WinBUGS Function

This function is very useful for executing WinBUGS programs from R. For an overview see Sturtz et al. (2005). The main advantage of this lies in the ability to further manipulate data within R before and after execution of

WinBUGS code. In addition, R2WinBUGS provides added facilities (such as a different DIC calculation, and graphical summaries). The main advantage of this functionality is found in simulation. When multiple datasets must be repeatedly analyzed then the ability to pass these to WinBUGS from R and to continue processing of further simulation datasets within a loop is a major advantage.

The main function call, once R2WinBUGS is loaded, is to a function called bugs which activates WinBUGS and passes the relevant information for processing:

bugs(data, inits, parameters.to.save, model.file = "model.txt",
n.chains = 3, n.iter = 2000, n.burnin = floor(n.iter/2),
n.thin = max(1, floor(n.chains * (n.iter - n.burnin)/1000)),
bin = (n.iter - n.burnin) / n.thin,
digits = 5, codaPkg = FALSE,
bugs.directory = "c:/Program Files/WinBUGS14/")

The main parameters for the function call are summarized below:

data: either a named list (corresponding to the names used in the model) OR a vector or list of the names of objects OR a name of a file containing the data.

inits: a list with n.chains elements, each element being a list of starting values, OR a function that creates these values.

parameters.to.save: a character vector of the names of parameters to be saved

model.file: file containing the WinBUGS code

n.chains: number of chains (defaults to 3)

n.iter: number of iterations (including burn-in)

n.burnin: number of burn-in iterations (default to half length of chain: n.iter/2).

Other parameters can be set to control e.g., thinning rate (n.thin), DIC calculation (DIC).

A simple example of a call is as follows. The Poisson–gamma model where the relative risk is assumed to have mean $e_i\theta_i$ is considered below for the South Carolina congenital anomaly mortality data. This data was first examined in Chapter 4. It consists of incident counts in 46 counties of South Carolina for one year. The expected rates associated with these counts are also available, computed from the statewide reference population. For purposes of data input to R2WinBUGS we need to set up a data structure which includes the outcome variable, the expected rate, and number of regions (y,e,m). The structure adat defined below is such a list structure:

adat<-list(m=46,
y=c(0,7,1,5,1,1,5,16,0,17,4,0,0,1,1,7,1,3,0,0,8,2,13,7,0,8,0,3,2,4,1,
11,0,1,2,3,3,8,6,14,3,11,6,0,1,5),

```
e=c(1.129778827,6.667008775,0.650279674,6.988864371,0.95571406,
1.123210345,5.908349156,8.539026017,0.601016062,18.92051111,
2.272694617,1.73736337,2.019808077,1.688099759,1.747216093,
3.221840201,1.835890594,5.221942834,0.978703751,1.254579976,
6.407553754,2.676656232,16.57884744,3.077333607,1.087083697,
7.606301637,1.018114641,2.15774619,2.844152512,2.955816698,
0.985272233,9.22871658,0.38097193,1.855596038,1.579719813,
1.579719813,2.647098065,4.791707292,4.144711859,15.70852363,
0.765228101,11.32077795,6.256478678,1.500898035,2.085492893,
7.297583004))
```

For initial values a list structure can be defined for this example that contains 3 initializations of the relevant parameters. The parameters to initialize are the predicted outcome values. To run R2WinBUGS the default is 3 chains. In this example, we can specify initial values for 3 chains as a list of lists with a,b,theta,ypred within each of 3 list elements (ab,ac,ad) which in turn, form the 3 component list ainits:

```
ab<-list(a=0.1,b=0.1,theta=c(1,1,1,1,1,1,1,1,1,1,1,1,1,1,1,1,1,1,
1,1,1,1,1,1,1,1,1,1,1,1,1,1,1,1,1,1,1,1,1,1,1,1,1,1,1,1,1),
ypred=c(0,7,1,5,1,1,5,16,0,15,4,0,0,1,1,7,1,3,0,0,8,2,13,7,0,
8,0,3,2,4,1,11,0,1,2,3,3,8,6,14,3,11,6,0,1,5))
ac<-list(a=0.12,b=0.12,theta=c(1.2,1,1,1,1,1,1,1,1,1.2,1,1,1,1,1,1,1,1,1,
1,1,1,1,1,1,1,1.2,1,1,1,1,1,1,1,1,1.5,1,1,1.6,1,1,1,1),
ypred=c(0,7,1,5,1,1,5,16,0,15,4,0,0,0,0,6,1,2,0,0,8,2,13,7,0,
8,0,3,2,4,1,11,0,1,2,3,3,8,6,14,3,11,6,0,1,5))
ad<-list(a=0.15,b=0.17,theta=c(1.4,1,1,1,1,1,1,1,1.8,1,1,1,1,1,1,1,1,1,1,
1,1,1,1,1,1,1,1.2,1,1,1,1,1,2.1,1,1,1,1,1,1,1,1,1,1.1,1,1),
ypred=c(0,7,1,5,0,0,0,0,0,0,4,0,0,1,1,7,1,3,0,0,8,2,13,7,0,8,0,
3,2,4,1,11,0,1,2,3,3,8,6,14,3,11,6,0,1,5))
ainits<-list(ab,ac,ad)
```

In the example, the parameters to monitor are the mean, residual, and predictive residual ("mu","r","rpred"), the data are in the object adat, the number of chains defaults to 3, the inits for the 3 chains are in ainits, the number of iterations is 4000 and the model is found in text file "map_modelWB_pred.txt":

```
parameters<-c("mu","r","rpred")
map.sim<-bugs(adat,inits = ainits,
parameters,n.iter=4000,
model.file ="map_modelWB_pred.txt")
```

The model file: 'map_modelWB_pred.txt' contains the following model code:

```
model
{for (i in 1:m)
{# Poisson likelihood for observed counts
    y[i]~dpois(mu[i])
```

Basic R and WinBUGS

```
        ypred[i]~dpois(mu[i])
        mu[i]<-e[i]*theta[i]
        probexc[i]<-step(theta[i]-1)
        # Relative Risk
        theta[i]~dgamma(a,b)
        r[i]<-y[i]-mu[i]
        rpred[i]<-y[i]-ypred[i]}
# Prior distributions for "population" parameters
a~dexp(10)
b~dexp(10)
# Population mean and population variance
mean<-a/b
var<-a/pow(b,2)
}
```

Submitting the command map.sim<-bugs(....) in R calls WinBUGS with the parameters specified, and stores the results of the call in object map.sim. Upon return from WinBUGS, map.sim contains a range of information. For example, some useful components and summaries are as follows:

map.sim $mean   yields the posterior mean estimate
map.sim$sd   yields the posterior standard deviations
map.sim$mean$rpred   yields the separate posterior mean for component rpred
map.sim$median   yields the posterior sample median
map.sim$summary[,3] yields the 2.5 posterior sample percentile
map.sim$summary[,7] yields the 97.5 posterior sample percentile
map.sim$sims.matrix   yields the complete output matrix

Diagnostic plots of the output are available via use of the plot command applied to the bugs object. For example, the caterpillar and box plots available on WinBUGS can be produced with the command plot(map.sim) (see Figure A.15). Alternatively simple plotting commands applied to elements of the output can be used. For example,

   plot(density(map.sim$summary[1:46,1]),main="")

will produce a density estimate of the mean posterior distribution of parameter mu (see Figure A.16), whereas the commands

   smat<-map.sim$sims.matrix
   dfsmat<-data.frame(smat[,47:92])
   boxplot(dfsmat)

will produce a multiple boxplot of the residuals in R for each unit (see Figure A.17).

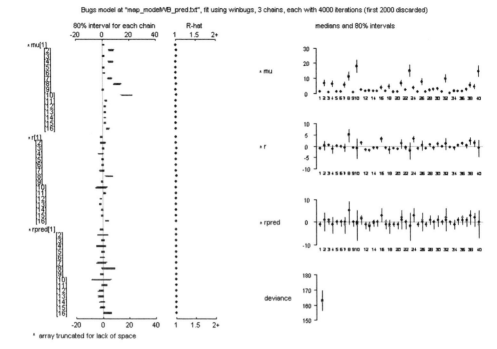

FIGURE A.15
Display of the plot command applied to the output object from the bugs(..) command. The output consists of caterpillar and box plots for the nominated output.

A.5 BRugs

The BRugs package is an implementation of OpenBUGS on R. This is a collection of R functions allowing users to analyze graphical models via McMC techniques. Most of the R functions provide a link to the BRugs dynamic link library (shared object file). Essentially this allows users considerable freedom to build graphical models from scratch, but it also requires that each step in the modeling process is a separate call to a BRugs function. Hence compilation, model checking, monitor setting, etc. are all separate calls to R functions. The advantage of BRugs over R2WinBUGS is that WinBUGS is not called and constraints of that windows-implemented package are not imposed. This means that relatively large datasets can be handled and there is no need to run under Windows.

Basic R and WinBUGS

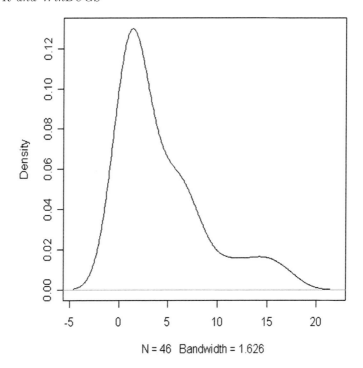

FIGURE A.16
Display of the density estimate for the posterior mean for parameter mu from the bugs output object map.sim.

For a BUGS model file stored in "log-linear-Poisson-model2.txt":

```
model{for (i in 1:m)
 { # Poisson likelihood for observed counts
 y[i]~dpois(mu[i])
 mu[i]<-e[i]*theta[i]
 # Relative Risk  log(theta[i])<-a+v[i]
 v[i]~dnorm(0,precv)}
 # Prior distributions for "population" parameters
 a~dnorm(0,preca)
 preca<-pow(sda,-2)
 precv<-pow(sdv,-2)
 sda~dunif(0,100)
 sdv~dunif(0,100)}
```

data stored in list format in "adat.txt":

```
list(m=46,
y=c(0,7,1,5,1,1,5,16,0,17,4,0,0,1,1,7,1,3,0,0,8,2,13,7,0,8,0,3,2,4,1,
11,0,1,2,3,3,8,6,14,3,11,6,0,1,5),
```

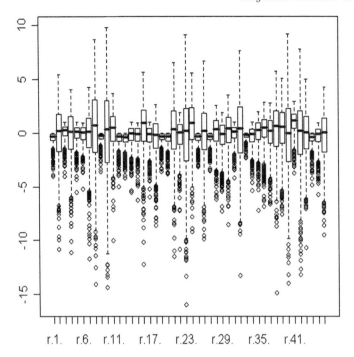

FIGURE A.17
Display of multiple boxplots of posterior mean residuals in r as output by R2WinBUGS.

```
e=c(1.129778827,6.667008775,0.650279674,6.988864371,0.95571406,
1.123210345,5.908349156,8.539026017,0.601016062,18.92051111,
2.272694617,1.73736337,2.019808077,1.688099759,1.747216093,
3.221840201,1.835890594,5.221942834,0.978703751,1.254579976,
6.407553754,2.676656232,16.57884744,3.077333607,1.087083697,
7.606301637,1.018114641,2.15774619,2.844152512,2.955816698,
0.985272233,9.22871658,0.38097193,1.855596038,1.579719813,
1.579719813,2.647098065,4.791707292,4.144711859,15.70852363,
0.765228101,11.32077795,6.256478678,1.500898035,2.085492893,
7.297583004))
```

and initialisation values in "inits.txt":

list(a=0.1,sda=0.1,sdv=0.1)

The following command sequence will fit the BUGS code model on R, with a burn-in of 10000 iterations with 2 chains monitored:

```
library(BRugs)
modelCheck("log-linear-Poisson-model2.txt")
modelData("adat.txt")
modelCompile(numChains=2)
modelInits(rep("inits.txt",2))
modelGenInits                              #......random effect initialisation
modelUpdate(10000)
samplesSet(c("theta","v","deviance"))
modelUpdate(2000)
samplesStats("*")
```

Note that output can be stored in a dataframe for later inspection:

asd<-samplesStats("*").

These functions can be built into a purpose built R function itself. Within the BRugs package there is a command already specified:

BRugsFit("log-linear-Poisson-model2.txt", data="adat.txt", numChains=2, parametersToSave=c("deviance","a"),nBurnin = 5000, nIter = 2000, nThin = 1)

This command will fit the model specified in "log-linear-Poisson-model2.txt" with a burn-in of 5000 and a sample of size 2000, and will display the deviance and estimate for parameter "a", as well as the DIC.

Note also that convergence checking can be done via the usual BGR statistic:

plotBgr("deviance",col=gray(0:3/3))

This command gives the BGR plot for the above model fit as in Figure A.18. The dark line is the ratio of between and within chain variance estimates.

A.6 Maps on R and GeoBUGS

Both on R and WinBUGS it is possible to handle mapped data.

On R the libraries maps and maptools provide facilities for the handling of polygon-based maps. In particular, maps allows the visualizing of polygonal data and with further programing allows the creation of thematic maps. The maptools package allows the importation of ArcView shape files to R (readShapePoly function) and also the export of maps to WinBUGS format via the function sp2WB. Both spatstat and splancs have functionality for mapping point process data.

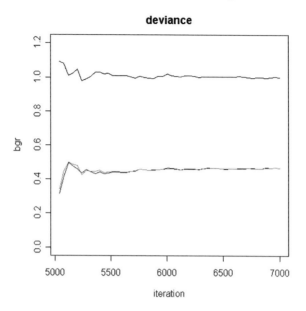

FIGURE A.18
BGR statistic plot for the Poisson log linear relative risk model applied to a 46 region dataset, from BRugs

A direct conversion process to GeoBUGS format is useful and the Windows program maps2WinBUGS is very useful in this regard. This program can be downloaded free from www.sourceforge.net. The program allows the import of both MapInfo MIF or TAB files as well as ArcView shapefiles and exports GeoBUGS format map files. Other R code facilties can be found on the Resource page of the WINBUGS site (http://193.60.86.19/bugs/weblinks/webresource.shtml). The GeoBUGS manual has conversion format recommendations for different GIS conversions to GeoBUGS format.

B
Selected WinBUGS Code

A selection of WinBUGS models are presented here to provide a sample of how code can be specified for the model discussed. ODC files for all the fitted models within the book are available for download from http://www.sph.sc.edu/alawson/default.htm.

B.1 Code for the Convolution Model (Chapter 5)

```
model
  {
  for( i in 1 : m ) {
  y[i] ~dpois(mu[i])
   mu[i] <- e[i] * rr[i]
  smr[i]<-(y[i]+eps2)/(e[i]+eps2)
   rr[i] <- exp(al0 +v[i]+u[i])
  v[i]~dnorm(0,1.0E-6)
  PP[i]<-step(rr[i]-1+eps)
  ypred[i] ~dpois(mu[i])
  PPL[i] <- pow(ypred[i]-y[i],2)
  PPL2[i] <- abs(ypred[i]-y[i])
   }
  mape <- mean(PPL2[])
  mspe <- mean(PPL[])
  # car normal
  u[1:m]~car.normal(adj[],weights[],num[],tau.u)
  for(k in 1:sumNumNeigh)
  {
       weights[k]<-1
  }
  al0 ~dflat()
  eps<-1.0E-6
  eps2~dnorm(0,1000)
```

```
tau.u~dgamma(0.005,0.005)
}
```

B.2   Code for Spatial Spline Model (Chapter 5)

This fits the model outlined in section 5.7.2 for count data:

```
model{
for (i in 1 : 100){
ME[i]<-0}
a
for (i in 1:100){
for (j in 1:100){
SOM[i,j]<-tauS*OM[i,j]
}}
for (i in 1 : m){
y[i]~dpois(mu[i])
mu[i]<-e[i]*theta[i]
ssum[i]<-inprod(gam[],CDZ[i,])
log(theta[i])<-a0+a1*xc[i]+a2*yc[i]+ssum[i]}
gam[1:100]~dmnorm(ME[],SOM[,])
a0~dflat()
a1~dnorm(0,0.0001)
a2~dnorm(0,0.0001)
tauS<-1/(sigmaS*sigmaS)
sigmaS~dunif(0,10)}
```

B.3   Code for the Spatial Autologistic Model (Chapter 6)

This model was specified as a simple regression model on the neighbor sums:

```
model
{
for (i in 1:m)
{
 yb[i]~dbern(p[i])
 logit(p[i])<-alpha0+alpha*x[i]+beta*ysum[i]}
alpha0~dflat()
alpha~dnorm(0.0,0.1)
beta~dnorm(0.0,0.1)}
```

B.4 Code for Logistic Spatial Case Control Model (Chapter 6)

Logistic spatial case control model with a fully specified spatial covariance model

```
model {
for (i in 1:N){
ind[i]~dbern(p[i])
 logit(p[i])<-f[i]
f[i]<-gam1*dis[i]+gam0+v[i]+W[i]
v[i]~dnorm(0,0.001)
res[i]<-(ind[i]-p[i])/sqrt(p[i]*(1-p[i]))
prexR2[i]<-step(res[i]-2)}
gam0~dflat()
gam1~dnorm(0,0.01)
 for(i in 1:N){ mu[i]<-0}
 W[1:N] ~spatial.exp(mu[], x[], y[], tau, phi,1)
tau~dexp(0.1)
phi~dexp(0.1)
}
```

B.5 Code for PP Residual Model (Chapter 6)

This fits a heterogeneous Poisson process Bayesian model using the Dirichlet tesselation weights. The background risk is estimated externally (e.g., in R) and appears evaluated at the case event locations in the variable den[]

```
#w[] Dirichlet weights
# d[] dist from putative source
#I[] indicator variable denoting 'real' data point or 'dummy' data point
# x,y,w,den,I and d must be read in
model{
C <- 10000
            for (i in 1:N) {
            f[i]<-1.+exp(bet1*d[i])
                zeros[i] <- 0
                    log(lam[i])<-bet0+log(f[i])
                    log(L[i])<-I[i]*log(lam[i])-w[i]*den[i]*lam[i]
                    phi[i] <- -log(L[i]) + C
```

```
                    zeros[i] ~dpois(phi[i])
                    res[i]<-((1/w[i])-den[i]*lam[i])
                    x1[i]<-x[i]
                    y1[i]<-y[i]
                    probexR[i]<-step(res[i]-0)}
  bet0~dnorm(0,0.001)
  bet1~dnorm(0,0.001)}
```

B.5.1 Same Model with Uncorrelated Random Effect

```
model{
  C <- 10000
                    for (i in 1:N) {
                f[i]<-1.+exp(bet1*d[i])
                    zeros[i] <- 0
                        log(lam[i])<-bet0+log(f[i])
                        log(L[i])<-I[i]*(log(lam[i])+v[i])-w[i]*den[i]*lam[i]*exp(v[i])
                        phi[i] <- -log(L[i]) + C
                        zeros[i] ~dpois(phi[i])
                        res[i]<-((1/w[i])-den[i]*lam[i])
                        x1[i]<-x[i]
                        y1[i]<-y[i]
                        probexR[i]<-step(res[i]-0)
                    v[i]~dnorm(0,tauv)
            }
  bet0~dnorm(0,0.001)
  bet1~dnorm(0,0.001)
  tauv<-0.1
        }
```

B.6 Code for the Logistic Spatial Case-Control Model (Chapter 6)

The specification is as for use with the R2winBUGS function

```
# x,y and ind as inputs.
# dis is distance
model
{
C <- 10000
for (i in 1:N){
```

Selected WinBUGS Code

```
f[i]<-(1+exp(-gam1*dis[i]))*exp(gam0+v[i])
zeros[i] <- 0
log(lam[i])<-log(f[i])
log(L[i])<-ind[i]*log(lam[i])-log(1.+lam[i])
phi[i] <- -log(L[i]) + C
zeros[i] ~dpois(phi[i])
mu[i]<-0
p[i]<-lam[i]/(1+lam[i])
x1[i]<-x[i]
y1[i]<-y[i]
v[i]~dnorm(0,0.1)
res[i]<-(ind[i]-p[i])/sqrt(p[i]*(1-p[i]))
prexR2[i]<-step(res[i]-2)
}
gam0~dnorm(0,0.1)
gam1~dnorm(0,0.1)
}
```

----

in file "logistic_case_control_P2.txt"

datalist elements: N, x, y, dis, ind
initial values: gam0, gam1, v
parameters: v, res, p, prexR2

B.6.0.1 R2WinBUGS commands:

The following commands set up the parameter vectors to report and the mapping tools. Use is made of the multivariate B spline package (MBA) for contouring:

```
>parameters<-c("res","v","prexR2","p")
>reslogist2<-bugs(data,inits,parameters,model.file="logistic_case_control_P2.txt",
n.iter=10000)
```

The following R code will plot the mapped surface of $\Pr(r_i > 2)$

```
>plot(data$x[data$ind==1],data$y[data$ind==1],xlab="easting",ylab="northing")
>zedprexR2<-cbind(data$x,data$y,reslogist2$mean$prexR2)
>ressurfprex<-mba.surf(zedprexR2,20,20)$xyz.est
>contour(ressurfprex,levels=c(0.05,0.1,0.2),add=T)
```

B.7 Code for Poisson Residual Clustering Example (Chapter 6)

This is a standard Poisson risk model for count data with a UH effect and residual and exceedences computed

```
model
{
  for( i in 1 : m ) {
  y[i] ~dpois(mu[i])
  mu[i] <- e[i] * rr[i]
  log(rr[i]) <- al0 +v[i]
  v[i]~dnorm(0,0.001)
  PP[i]<-step(rr[i]-1+eps)
  res[i]<-(y[i]-mu[i])/sqrt(mu[i])
  prexR1[i]<-step(res[i]-1)
  prexR2[i]<-step(res[i]-2)
  }
  al0~dflat()
  eps<-1.0E-6
}
```

B.8 Code for the **Proper CAR Model (Chapter 7)**

This is a proper CAR model for Ohio respiratory cancer 1988

```
model{
    for(i in 1 : 88) {
        m[i] <- 1/num[i]
    }
    cumsum[1] <- 0
    for(i in 2:(88+1)) {
    cumsum[i] <- sum(num[1:(i-1)])
    }
    for(k in 1 : sumNumNeigh) {
    for(i in 1:88) {
    pick[k,i] <- step(k - cumsum[i] - epsilon) * step(cumsum[i+1] - k) }
    C[k] <- 1 / inprod(num[], pick[k,])  # weight for each pair of neighbours
```

Selected WinBUGS Code

```
    }
    epsilon <- 0.0001
    for (i in 1 : 88) {
    y[i] ~dpois(mu[i])
    log(mu[i])<-log(e[i])+a0+a1*(pov[i]+ep1[i])+S[i]+v[i]
    RR[i] <- exp(a0+a1*(pov[i]+ep1[i])+S[i]+v[i]
    theta[i] <- a0+a1*(pov[i]+ep1[i])+S[i]+v[i]
    v[i]~dnorm(0,0.0001)
    ep1[i]~dnorm(0,0.0001)
    dis1[i]<-dist[i]}
    # Proper CAR prior distribution for spatial random effects:
    S[1:88] ~car.proper(theta[], C[], adj[], num[], m[], prec, gamma)
    # Other priors:
    prec <-1/pow(sdS,2)              # prior on precision
    sdS~dunif(0,100)                 # uniform sd prior
    sigma <- sqrt(1 / prec)          # standard deviation
    gamma.min <- min.bound(C[], adj[], num[], m[])
    gamma.max <- max.bound(C[], adj[], num[], m[])
    gamma ~dunif(0.5, gamma.max)
a0~dflat()
a1~dnorm(0,0.0001)
#tau.v<-1/pow(sdv,2)
#sdv~dunif(0,10)
#tau.p1<-1/pow(sdp1,2)
#sdp1~dunif(0,10)
}
```

B.9 Code for the Multiscale Model for PH and County Level Data (Chapter 8)

The multiscale model for PH districts and counties in Georgia where u and v for districts are included in county effects

```
model{
for( i in 1:18){
yph[i]~dpois(muph[i])
log(muph[i])<-log(eph[i])+log(thph[i])
thph[i]<-exp(aph0+uph[i]+vph[i])
vph[i]~dnorm(0,0.0001)
}
for( j in 1:159){
yc[j]~dpois(muc[j])
```

```
log(muc[j])<-log(ec[j])+log(thc[j])
thc[j]<-exp(ac0+uc[j]+vc[j]+uph[phc[j]]+vph[phc[j]])
vc[j]~dnorm(0,0.0001)
}
for (k in 1 :nsumph){weiph[k]<-1}
uph[1:18]~car.normal(adj[],weiph[],num[],tauph)
#tau.vph<-1/pow(sdvph,2)
tauph<-1/pow(sdph,2)
aph0~dflat()
#sdvph~dunif(0,100)
sdph~dunif(0,100)
for (o in 1 :nsumc){weic[o]<-1}
uc[1:159]~car.normal(adj2[],weic[],num2[],tauc)
#tau.vc<-1/pow(sdvc,2)
tauc<-1/pow(sdc,2)
ac0~dflat()
#sdvc~dunif(0,100)
sdc~dunif(0,100) }
```

B.10 Code for the Shared Component Model for Georgia Asthma and COPD (Chapter 9)

The joint shared component model for asthma and COPD in Georgia counties

```
MODEL{
for(i in 1:N){
asthma[i]~dpois(Muas[i])
log(Muas[i])<-log(Easthma[i])+log(thetAS[i])
log(thetAS[i])<-b0+Was[i]+Wcom[i]*delta
COPD[i]~dpois(MuC[i])
log(MuC[i])<-log(ECOPD[i])+log(thetCOPD[i])
log(thetCOPD[i])<-b1+Wcom[i]/delta
angina1[i]<-angina[i]
Eangina1[i]<-Eangina[i]
}
delta<-exp(sdf)
sdf~dnorm(0,5.88)
for( i in 1:sumNumNeigh){weights[i]<-1}
Wcom[1:N]~car.normal(adj[],weights[],num[],tWcom)
Was[1:N]~car.normal(adj[],weights[],num[],tWas)
b0~dflat()
b1~dflat()
```

```
tWcom<-pow(sdWcom,-2)
sdWcom~dunif(0,100)
tWas<-pow(sdWas,-2)
sdWas~dunif(0,100) }
```

B.11 Code for the Seizure Example with Spatial Effect (Chapter 10)

The seizure longitudinal example with a spatial contextual effect w

```
model{
for(i in 1:59){
 y[i,1]<-t1[i]
 y[i,2]<-t2[i]
 y[i,3]<-t3[i]
 y[i,4]<-t4[i]
}
ep[1]~dnorm(0,0.0001)
for (i in 1:59){
for (k in 1:4){
y[i,k]~dpois(mu[i,k])
log(mu[i,k])<-log(baseC[i])+alp0+alp1*group[i]+alp2*age[i]+ep[k]+w[ind[i]]
}}
for (k in 2:4){
ep[k]~dnorm(ep[k-1],alp3)}
alp0~dflat()
alp1~dnorm(0,0.0001)
alp2~dnorm(0,0.0001)
alp3~dgamma(0.001,0.001)
w[1:46]~car.normal(adj[],weights[],num[],tauW)
for (i in 1:sumNumNeigh){weights[i]<-1}
tauW<-pow(sdW,-2)
sdW~dunif(0,100)}}
```

B.12 Code for the Knorr-Held Model for Space–Time Relative Risk Estimation (Chapter 11)

This is a space-time model for the VLBW example from Georgia with a time dependent temporal effect and time dependent interaction term.

```
model{
for (i in 1:159){
for (j in 1: 11){
vlbw[i,j]~dbin(p[i,j],birth[i,j])
logit(p[i,j])<-a0+v[i]+u[i]+g[j]+psi[i,j]
}
psi[i,1]~dnorm(0,taupsi)
for (j in 2:11){
psi[i,j]~dnorm(psi[i,j-1],taupsi)}
v[i]~dnorm(0,tauv)      }
u[1:159]~car.normal(adj[], weights[], num[], tauu)
    for(k in 1:sumNumNeigh) {
        weights[k] <- 1}
g[1]~dnorm(0,0.0001)
time1[1]<-time[1]-1993
for (j in 2: 11){
g[j]~dnorm(g[j-1],tau.g)
time1[j]<-time[j]-1993}
taupsi<-1/pow(sdpsi,2)
sdpsi~dunif(0,100)
tau.g<-1/pow(sdg,2)
sdg~dunif(0,100)
tauu<-1/pow(sdu,2)
sdu~dunif(0,100)
tauv<-1/pow(sdv,2)
sdv~dunif(0,100)
a0~dflat()}
```

B.13 Code for the Space–Time Autologistic Model (Chapter 11)

This is the ST autologistic model with temporally dependent intercept and slope parameters for the neighbor sum terms

Selected WinBUGS Code

```
model
{
for (i in 1:m)
{
 logit(p[i,1])<-a[1]+b[1]*ys1[i,1]
 ypred[i,1]~dbern(p[i,1])
 diff[i,1]<-pow(y[i,1]-ypred[i,1],2)
      for (k in 2:T)
      {
           y[i,k]~dbern(p[i,k])
 logit(p[i,k])<-a[k]+b[k]*ys1[i,k]+c[k]*ys1[i,(k-1)]
      ypred[i,k]~dbern(p[i,k])
      diff[i,k]<-pow(y[i,k]-ypred[i,k],2)
      }
}
for (k in 1:T){
 a[k] ~dnorm(0,0.01)
 b[k] ~dnorm(0,0.01)
 c[k] ~dnorm(0,0.01)
           }
           mspe<-mean(diff[,])
}
```

C

R Code for Thematic Mapping

The following R code was developed by Bo Cai and can be used to produce thematic maps from ArcView shapefiles on R. The function is called fillmap and must be called with a complete set of ArcView .shp, .sbn, .dbf, .shx files resident in the active work directory.

```
fillmap<-function(map, figtitle, y , n.col, bk="e", cuts,legendtxt="", legendpos=c(466000,4330000), titlepos=c(350000,4650000)){
polylist<- Map2poly(map)
if(bk=="q"){p <- seq(0,1, length=n.col+1)
 br <- round(quantile(y, probs=p),2)}
if(bk=="e"){br <- round(seq(min(y), max(y), length=n.col+1),2)}
if(bk=="c"){if (length(cuts)!= (n.col+1)) {cat("Cut off and color categories do not match. ", "\n")
 break} else {br <- cuts} }
# 0: dark 1: light light Current shading ranges from darkest to light gray white (to distinguish with lakes).
shading<-gray((n.col-1):0/(n.col-1))
y.grp<-findInterval(y, vec=br, rightmost.closed = TRUE, all.inside = TRUE)
y.shad<-shading[y.grp]
plot.polylist(polylist,col=y.shad,axes=F)
br<-round(br, 2)
if (legendtxt=="")
{
cn<-length(y[y>=br[n.col]]) # number of regions in this intervals
 leg.txt<-paste(" [",br[n.col],",",br[n.col+1],"]","(",cn,")",sep="")
for(i in (n.col-1):1){
cn<-length(y[(y>=br[i])&(y<=br[i+1])])
leg.txt<-append(leg.txt,paste(" [",br[i],",",br[i+1],")","(",cn,")",sep="")) }
leg.txt<-rev(leg.txt)
} else {leg.txt<-legendtxt}
legend(legendpos[1],legendpos[2],legend=leg.txt,fill=shading,cex=0.7,ncol=1,bty="n")
text(titlepos[1],titlepos[2], figtitle, cex=1.2)}
```

Calling the function

1) Load libraries: foreign, maptools, and sp
>library(foreign)
>library(sp)
>library(maptools)
2) To call the routine for the Georgia shapefile with 159 counties with values generated from a Gaussian distribution with mean 1 and standard deviation 0.1:
>x<-read.shape("Georgia.shp")
>y<-rnorm(159,1,.1)
>maintitle<-paste("")
>fillmap(x,maintitle,y,n.col=5)

This produced the figure below:

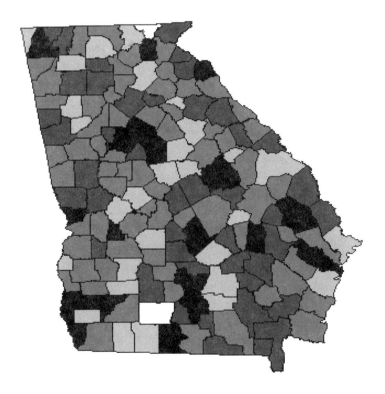

References

Agarwal, D., A. Gelfand, and S. Citron-Pousty (2002). Zero-inflated models with application to spatial count data. *Environmental and Ecological Statistics 9*, 341–355.

Anto, J. and J. Sunyer (1990). Epidemiological studies of asthma epidemics in Barcelona. *Chest 98*, 185–190.

Arya, S. P. (1998). *Air Pollution Metereology and Dispersion*. New York: Oxford University Press.

Baddeley, A. and R. Turner (2000). Practical maximum pseudolikelihood for spatial point patterns. *Australia and New Zealand Journal of Statistics 42*, 283–322.

Baddeley, A., R. Turner, J. Møller, and M. Hazelton (2005). Residual analysis for spatial point processes. *Journal of the Royal Statistical Society B 67*, 617–666, with discussion.

Balakrishnan, S. and D. Madigan (2006). Algorithms for sparse linear classifiers in the massive data setting. *Journal of Machine Learning Research 1*.

Banerjee, S. and B. P. Carlin (2003). Semiparametric spatio-temporal frailty modeling. *Environmetrics 14*, 523–535.

Banerjee, S., B. P. Carlin, and A. E. Gelfand (2004). *Hierarchical Modeling and Analysis for Spatial Data*. London: Chapman and Hall/CRC Press.

Banerjee, S., M. M. Wall, and B. P. Carlin (2003). Frailty modeling for spatially correlated survival data, with application to infant mortality in Minnesota. *Biostatistics 4*, 123–142.

Bankman, I. (Ed.) (2007). *Handbook of Medical Image Processing and Analysis* (2nd ed.), Volume 1. New York: Academic Press.

Barndorff-Nielsen, O. E., W. S. Kendall, and M. N. M. van Lieshout (1999). *Stochastic Geometry: Likelihood and Computation*. Boca Raton, FL: Chapman & Hall.

Bastos, L. and D. Gamerman (2006). Dynamical survival models with spatial frailty. *Lifetime Data Analysis 12*, 441–460.

Berman, M. and T. R. Turner (1992). Approximating point process likelihoods with GLIM. *Applied Statistics 41*, 31–38.

Bernardinelli, L., D. G. Clayton, C. Pascutto, C. Montomoli, M. Ghislandi, and M. Songini (1995). Bayesian analysis of space-time variation in disease risk. *Statistics in Medicine 14*, 2433–2443.

Bernardo, J. M. and A. F. M. Smith (1994). *Bayesian Theory*. New York: Wiley.

Besag, J. (1975). Statistical analysis of non-lattice data. *The Statistician 24*, 179–195.

Besag, J. and P. J. Green (1993). Spatial statistics and Bayesian computation. *Journal of the Royal Statistical Society, Series B 55*, 25–37.

Besag, J. and J. Newell (1991). The detection of clusters in rare diseases. *Journal of the Royal Statistical Society, Series A 154*, 143–155.

Besag, J. and J. Tantrum (2003). Likelihood analysis of binary data in space and time. In P. Green, N. Hjort, and S. Richardson (Eds.), *Highly Structured Stochastic Systems*, Chapter 9A, 289–295. New York: Oxford University Press.

Besag, J., J. York, and A. Mollié (1991). Bayesian image restoration with two applications in spatial statistics. *Annals of the Institute of Statistical Mathematics 43*, 1–59.

Best, N., S. Richardson, and A. Thomson (2005). A comparison of Bayesian spatial models for disease mapping. *Statistical Methods in Medical Research 14*, 35–59.

Best, N. G. and J. C. Wakefield (1999). Accounting for inaccuracies in population counts and case registration in cancer mapping studies. *Journal of the Royal Statistical Society A 162*, 363–382.

Bishop, C. (2006). *Pattern Recognition and Machine Learning*. New York: Springer.

Boehning, D., E. Dietz, and P. Schlattmann (2000). Space-time mixture modelling of public health data. *Statistics in Medicine 19*, 2333–2344.

Boehning, D., E. Dietz, P. Schlattmann, L. Mendonca, and U. Kirchner (1999). The zero-inflated poisson model and the decayed, missing and filled teeth index in dental epidemiology. *Journal of the Royal Statistical Society A 162*, 195–209.

Bottai, M., M. Geraci, and A. Lawson (2007). Testing for unusual aggregation of health risk in semiparametric models. *Statistics in Medicine*. DoI: 10/1002/sim.3126.

References

Bowman, A. and A. Azzalini (1997). *Applied Smoothing Techniques for Data Analysis: the Kernel Approach with S-Plus Illustrations.* New York: Oxford University Press.

Breslow, N. and D. G. Clayton (1993). Approximate inference in generalised linear mixed models. *Journal of the American Statistical Association 88*, 9–25.

Brix, A. and P. Diggle (2001). Spatio-temporal prediction for log-gaussian cox processes. *Journal of the Royal Statistical Society B 63*, 823–841.

Brooks, S. and A. E. Gelman (1998). General methods for monitoring convergence of iterative simulations. *Journal of Computational and Graphical Statistics 7*, 434–455.

Brooks, S., P. Guidici, and A. Philippe (2003). Nonparamteric convergence assessment for mcmc model selection. *Journal of Computational and Graphical Statistics 12*, 1–22.

Brooks, S. P. (1998). Quantitative convergence assessment for MCMC via Cusums. *Statistics and Computing 8*, 267–274.

Buckland, S. T., D. R. Anderson, K. P. Burnham, J. L. Laake, D. L. Borchers, and L. Thomas (2001). *Introduction to Distance Sampling.* New York: Oxford University Press.

Byers, S. and A. Raftery (2002). Bayesian estimation and segmentation of spatial point processes using voronoi tilings. In A. B. Lawson and D. Denison (Eds.), *Spatial Cluster Modelling*, Chapter 6. New York: CRC Press.

Cai, B. and D. Dunson (2008). Bayesian variable selection in nonparametric random effects models. *submitted*.

Cai, B. and A. B. Lawson (2008). Space-time latent structure models with gam-like covariate linkage. *submitted*.

Caragea, P. and M. Kaiser (2006). Covariates and time in the autologistic model. Technical Report 2006-19, Iowa State University, Department of Statistics.

Carlin, B. P. and J. Hodges (1999). Hierarchical proportional hazards regression models for highly stratified data. *Biometrics 55*, 1162–1170.

Carlin, B. P. and T. Louis (2000). *Bayes and Empirical Bayes Methods for Data Analysis* (2nd ed.). New York: Chapman & Hall/CRC Press.

Carroll, R., D. Ruppert, L. Stefanski, and C. Crainiceanu (2006). *Measurement Error in Nonlinear Models* (2nd ed.). New York: Chapman & Hall.

Cassella, G. and E. I. George (1992). Explaining the Gibbs Sampler. *The American Statistician 46*, 167–174.

Chaix, B., M. Rosvall, J. Lynch, and J. Merlo (2006). Disentangling contextual effects on cause specific mortality in a longitudinal 23-year follow-up study: impact of population density or socio-economic environment? *International Journal of Epidemiology 35*, 633–643.

Clark, A. B. and A. B. Lawson (2002). Spatio-temporal cluster modelling of small area health data. In A. B. Lawson and D. Denison (Eds.), *Spatial Cluster Modelling*, Chapter 14, 235–258. New York: CRC Press.

Clayton, D. G. and L. Bernardinelli (1992). Bayesian methods for mapping disease risk. In P. Elliott, J. Cuzick, D. English, and R. Stern (Eds.), *Geographical and Environmental Epidemiology: Methods for Small-Area Studies*. Oxford: Oxford University Press.

Clayton, D. G., L. Bernardinelli, and C. Montomoli (1993). Spatial correlation in ecological analysis. *International Journal of Epidemiology 22*, 1193–1202.

Clayton, D. G. and J. Kaldor (1987). Empirical Bayes estimates of age-standardised relative risks for use in disease mapping. *Biometrics 43*, 671–691.

Cliff, A. D. and J. K. Ord (1981). *Spatial Processes: Models and Applications*. London: Pion.

Congdon, P. (2003). *Applied Bayesian Modelling*. New York: Wiley.

Congdon, P. (2005). *Bayesian Models for Categorical Data*. New York: Wiley.

Congdon, P. (2007). *Bayesian Statistical Modelling* (2nd ed.). New York: Wiley.

Congdon, P., M. Almog, S. Curtis, and R. Ellerman (2007). A spatial structural equation modelling framework for health count responses. *Statistics in Medicine 26*, 5267–5284.

Cook, R. and J. Lawless (2007). *The Statistical Analysis of Recurrent Events*. New York: Springer.

Cooner, F., S. Banerjee, and M. McBean (2006). Modelling geographically referenced survival data with a cure fraction. *Statistical Methods in Medical Research 15*(4), 307–324.

Cowles, M. K. (2004). Review of WinBUGS 1.4. *The American Statistician 58*, 330–336.

Cressie, N., S. Richardson, and I. Jaussent (2004). Ecological bias: use of maximum-entropy approximations. *Australia New Zealand Journal of Statistics 46*, 233–255.

Cressie, N. A. C. (1993). *Statistics for Spatial Data* (revised ed.). New York: Wiley.

References

Cressie, N. A. C. (1996). Change of support and the modifiable areal unit problem. *Geographical Systems 2*, 83–101.

Cressie, N. A. C. and A. Mugglin (2000). Spatio-temporal hierarchical modelling of an infectious disease from (simulated) count data. In J. Bethlehem and P. van der Heijden (Eds.), *Compstat 2000*. Heidelberg: Physica verlag.

Cressie, N. C. and A. Lawson (2000). Hierarchical probability models and Bayesian analysis of mine locations. *Advances in Applied Probability 32*(2), 315–330.

Dabney, A. and J. Wakefield (2005). Issues in the mapping of two diseases. *Statistical Methods in Medical Research 14*, 83–112.

Davison, A. C. and D. V. Hinkley (1997). *Bootstrap Methods and their Application*. London: Cambridge University Press.

Dellaportas, P., J. Forster, and I. Ntzoufras (2002). On Bayesian model and variable selection using mcmc. *Statistics and Computing 12*, 27–36.

Denison, D., N. Adams, C. Holmes, and D. Hand (2002). Bayesian partition modelling. *Computational Statistics and Data Analysis 38*, 475–485.

Denison, D. and C. Holmes (2001). Bayesian partitioning for estimating disease risk. *Biometrics 57*, 143–149.

Denison, D., C. Holmes, B. Mallick, and A. Smith (2002). *Bayesian Methods for nonlinear Classification and Regression*. New York: Wiley.

Diggle, P. (2007). Spatio-temporal point processes: Methods and applications. In B. Finkenstädt, L. Held, and V. Isham (Eds.), *Statistical Methods for Spatio-Temporal Systems*, Chapter 1, 1–45. London: CRC Press.

Diggle, P. and P. Elliott (1995). Statistical issues in the analysis of disease risk near point sources using individual or spatially aggregated data. *Journal of Epidemiology and Community Health 49*, s20–s27.

Diggle, P., T. Fiksel, Y. Ogata, D. Stoyan, and M. Tanemura (1994). On parameter estimation for pairwise interaction processes. *International Statistical Review 62*, 99–117.

Diggle, P., P. Heagerty, K.-Y. Liang, and S. Zeger (2002). *Analysis of Longitudinal Data* (2 ed.). New York: Oxford University Press.

Diggle, P., S. Morris, P. Elliott, and G. Shaddick (1997). Regression modelling of disease risk in relation to point sources. *Journal of the Royal Statistical Society, A 160*, 491–505.

Diggle, P. and P. Ribeiro Jr. (2007). *Model-based Geostatistics*. New York: Springer.

Diggle, P. and B. Rowlingson (1994). A conditional approach to point process modelling of elevated risk. *Journal of the Royal Statistical Society A 157*, 433–440.

Diggle, P., P. Zheng, and P. Durr (2005). Nonparametric estimation of spatial segregation in a multivariate point process: bovine tuberculosis in cornwall, UK. *Journal of the Royal Statistical Society C 54*, 645–658.

Diggle, P. J. (1985). A kernel method for smoothing point process data. *Journal of the Royal Statistical Society C 34*, 138–147.

Diggle, P. J. (1990). A point process modelling approach to raised incidence of a rare phenomenon in the vicinity of a prespecified point. *Jour. Royal Statist. Soc. A 153*, 349–362.

Diggle, P. J. (2003). *Statistical Analysis of Spatial Point Patterns* (2nd ed.). London: Arnold.

Diggle, P. J. (2005). Spatio-temporal point processes, partial likelihood, foot and mouth disease. *Statistical Methods in Medical Research 15*, 325–336.

Diggle, P. J., S. Morris, and J. C. Wakefield (2000). Point-source modelling using matched case-control data. *Biostatistics 1*, 1–17.

Diggle, P. J., J. Tawn, and R. Moyeed (1998). Model-based Geostatistics. *Journal of the Royal Statistical Society C 47*, 299–350.

Duan, J., M. Guidani, and A. Gelfand (2007). Generalized spatial dirichlet process models. *Biometrika 94*, 809–825.

Durr, P. and A. Gatrell (Eds.) (2004). *GIS and Spatial Analysis in Veterinary Science*. London: Cabi.

Durr, P., N. Tait, and A. B. Lawson (2005). Bayesian hierarchical modelling to enhance the epidemiological value of abattoir surveys for bovine fasciolosis. *Preventive Veterinary Medicine 71*, 157–172.

Eberley, L. and B. P. Carlin (2000). Identifiability and convergence issues for Markov chain Monte Carlo fitting of spatial models. *Statistics in Medicine 19*, 2279–2294.

Elliott, P., J. Cuzick, D. English, and R. Stern (Eds.) (1992). *Geographical and Environmental Epidemiology:Methods for Small-Area Studies*. Oxford University Press.

Elliott, P., J. C. Wakefield, N. G. Best, and D. J. Briggs (Eds.) (2000). *Spatial Epidemiology: Methods and Applications*. London: Oxford University Press.

Esman, N. A. and G. M. Marsh (1996). Applications and limitations of air dispersion modeling in environmental epidemiology. *Journal of Exposure Analysis and Environmental Epidemiology 6*, 339–353.

References

Faraway, J. (2006). *Extending the Linear Model with R*. New York: Chapman & Hall/CRC.

Farnsworth, M., J. Hoeting, N. Hobbs, and M. Miller (2006). Linking chronic wasting disease to mule deer movement: a hiererchical Bayesian approach. *Ecological Applications 16*, 1026–1036.

Fernandez, C. and P. Green (2002). Modelling spatially correlated data via mixtures: a Bayesian approach. *Journal of the Royal Statitsical Society B 64*, 805–826.

Ferreira, J., D. Denison, and C. Holmes (2002). Partition modelling. In A. B. Lawson and D. Denison (Eds.), *Spatial Cluster Modelling*, Chapter 7, 125–145. New York: CRC Press.

Fotheringham, A. S., C. Brunsdon, and M. Charlton (2002). *Geographically Weighted Regression : The Analysis of Spatially Varying Relationships*. New York: Wiley.

French, J. and M. Wand (2004). Generalized additive models for cancer mapping with incomplete covariates. *Biostatistics 5*, 177–191.

Gamerman, D. and H. Lopes (2006). *Markov Chain Monte Carlo: Stochastic Simulation for Bayesian Inference*. New York: CRC Press.

Gangnon, R. (2006). Impact of prior choice on local Bayes factors for cluster detection. *Statistics in Medicine 25*, 883–895.

Gangnon, R. and M. Clayton (2000). Bayesian detection and modeling of spatial disease clustering. *Biometrics 56*, 922–935.

Gardner, M. J. (1989). Review of reported increases of childhood cancer rates in the vicinity of nuclear installations in the UK. *Journal of the Royal Statistical Society 152*, 307–325.

Gelfand, A. and S. Ghosh (1998). Model choice: A minimum posterior predictive loss approach. *Biometrika 85*, 1–11.

Gelfand, A., A. Kottas, and S. MacEachern (2005). Bayesian nonparametric spatial modeling with dirichlet process mixing. *Journal of the American Statistical Association 100*, 1021–1035.

Gelfand, A. and P. Vounatsou (2003). Proper multivariate conditional autoregressive models for spatial data. *Biostatistics 4*, 11–25.

Gelman, A. (2006). Prior distributions for variance parameters in hierarchical models. *Bayesian Analysis 1*, 515–533.

Gelman, A., J. B. Carlin, H. S. Stern, and D. Rubin (2004). *Bayesian Data Analysis*. London: Chapman & Hall/CRC press.

Gelman, A. and J. Hill (2007). *Data Analysis Using Regression and Multilevel-Hierarchical Models*. New York: Cambridge University Press.

Gelman, A. E. and D. Rubin (1992). Inference from iterative simulation using multiple sequences (with discussion). *Statistical Science 7*, 457–511.

Geweke, J. (1992). Evaluating the accuracy of sampling -based approaches to the calculation of posterior moments. In J. Bernardo, J. Berger, A. Dawid, and A. Smith (Eds.), *Bayesian Statistics 4*. New York: Oxford University Press.

Ghosh, M. and J. N. K. Rao (1994). Small area estimation: an appraisal. *Statistical Science 9*, 55–93.

Ghosh, S., P. Mukhopadhyay, and J.-C. Lu (2006). Bayesian analysis of zero-inflated regression models. *Journal of Statistical Planning and Inference 136*, 1360–1375.

Gilks, W. R., D. G. Clayton, D. J. Spiegelhalter, N. G. Best, A. J. McNeil, L. D. Sharples, and A. J. Kirby (1993). Modelling complexity: Applications of Gibbs Sampling in medicine. *Journal of the Royal Statistical Society B 55*, 39–52.

Gilks, W. R., S. Richardson, and D. J. Spiegelhalter (Eds.) (1996). *Markov Chain Monte Carlo in Practice*. London: Chapman & Hall.

Gneiting, T., M. Genton, and P. Guttorp (2007). Geostatistical space-time models, stationarity, separability, and full symmetry. In B. Finkenstädt, L. Held, and V. Isham (Eds.), *Statistical Methods for Spatio-Temporal Systems*, Chapter 4, pp. 151–175. London: CRC Press.

Godsill, S. (2001). On the relationship between Markov chain Monte Carlo methods for model uncertainty. *Journal of Computational and Graphical Statistics 10*, 230–248.

Goldstein, H. and A. Leyland (2001). Further topics in multilevel modelling: Context and composition. In A. Leyland and H. Goldstein (Eds.), *Multilevel Modelling of Health Statistics*, Chapter 12.4, 181–186. New York: Wiley.

Greco, F., A. B. Lawson, D. Cocchi, and T. Temples (2005). Voronoi tesselation and other methods for misaligned small area health data. *Environmental and Ecological Statistics 12*, 379–395.

Green, P. J. (1995). Reversible jump MCMC computation and Bayesian model determination. *Biometrika 82*, 711–732.

Green, P. J. and S. Richardson (2002). Hidden Markov models and disease mapping. *Journal of the American Statitsical Association 97*, 1055–1070.

Greenland, S. (1992). Divergent biases in ecologic and individual-level studies. *Statistics in Medicine 11*, 1209–1223.

Greenland, S. and J. Robins (1994). Ecologic studies—biases, misconceptions and counterexamples. *American Journal of Sociology 139*, 747–759.

References

Griffin, J. and M. Steel (2006). Order-based dependent Dirichlet processes. *Journal of the American Statistical Asociation 101*, 179–194.

Gustafson, P. (1998). Flexible Bayesian modelling for survival data. *Lifetime Data Analysis 4*, 281–299.

Gustafson, P. (2004). *Measurement Error and Misclassification in Statistics and Epidemiology*. London: Chapman & Hall.

Heagerty, P. and S. Lele (1998). A composite likelihood approach to binary spatial data. *Journal of the American Statistical Association 93*, 1099–1111.

Hegarty, A. and D. Barry (2008). Bayesian disease mapping using product partition models. *Statistics in Medicine (to appear)*.

Heisterkamp, S., G. Doornbos, and N. Nagelkerke (2000). Assessing health impact of environmental pollution sources using space-time models. *Statistics in Medicine 19*, 2569–2578.

Held, L., I. Natario, S. Fenton, H. Rue, and N. Becker (2005). Towards joint disease mapping. *Statistical Methods in Medical Research 14*, 61–82.

Henderson, R., S. Shimakura, and D. Gorst (2002). Modeling spatial variation in leukemia survival data. *Journal of the American Statistical Association 97*, 965–972.

Höhle, M., E. Jorgensen, and P. O'Neill (2005). Inference in disease transmission experiments by using stochastic epidemic models. *Journal of the Royal Statistical Society C 54*, 349–366.

Hossain, M. and A. Lawson (2008). Approximate methods in Bayesian point process spatial models. *submitted*.

Hossain, M. and A. B. Lawson (2005). Local likelihood disease clustering: Development and evaluation. *Environmental and Ecological Statistics 12*, 259–273.

Hossain, M. and A. B. Lawson (2006). Cluster detection diagnostics for small area health data: with reference to evaluation of local likelihood models. *Statistics in Medicine 25*, 771–786.

Ibrahaim, J., M. Chen, and D. Sinha (2000). *Bayesian Survival Analysis*. New York: Springer.

Inskip, H., V. Beral, P. Fraser, and P. Haskey (1983). Methods for age-adjustment of rates. *Statistics in Medicine 2*, 483–493.

Ishwaran, H. and L. James (2001). Gibbs sampling methods for stick-breaking priors. *Journal of the American Statistical Association 96*, 161–173.

Ishwaran, H. and L. James (2002). Approximate Dirichlet process computing in finite normal mixtures: Smoothing and prior information. *Journal of Computational and Graphical Statistics 11*, 1–26.

Jin, X., S. Banerjee, and B. Carlin (2008). Order-free coregionalized areal data models with application to multiple disease mapping. *Journal of the Royal Statistical Society B.* (to appear).

Jin, X., B. Carlin, and S. Banerjee (2005). Generalized hierarchical multivariate car models for areal data. *Biometrics 61*, 950–961.

Kauermann, G. and J. D. Opsomer (2003). Local likelihood estimation in generalized additive models. *Scandinavian Journal of Statistics 30*, 317–337.

Kelsall, J. and P. Diggle (1998). Spatial variation in risk of disease: a nonparametric binary regression approach. *Applied Statistics 47*, 559–573.

Kelsall, J. and J. Wakefield (2002). Modelling spatial variation in disease risk: A geostatistical approach. *Journal of the American Statitsical Association 97*, 692–701.

Kim, H., D. Sun, and T. R (2001). A bivariate bayes method for improving the estimates of mortality rates with a twofold autoregressive model. *Journal of the American Statistical Association 96*, 1506–1521.

Kim, S., M. Tadesse, and M. Vanucci (2006). Variable selection in clustering via Dirichlet process mixture models. *Biometrika 93*, 877–893.

Knorr-Held, L. (2000). Bayesian modelling of inseparable space-time variation in disease risk. *Statistics in Medicine 19*, 2555–2567.

Knorr-Held, L. and J. Besag (1998). Modelling risk from a disease in time and space. *Statistics in Medicine 17*, 2045–2060.

Knorr-Held, L. and N. G. Best (2001). A shared component model for detecting joint and selective clustering of two diseases. *Journal of the Royal Statistical Society 164*, 73–85.

Knorr-Held, L. and G. Rasser (2000). Bayesian detection of clusters and discontinuities in disease maps. *Biometrics 56*, 13–21.

Knorr-Held, L. and S. Richardson (2003). A hierarchical model for space-time surveillance data on meningococcal disease incidence. *Journal of the Royal Statistical Society 52*, 169–183.

Kottas, A., J. Duan, and A. Gelfand (2007). Modelling disease incidence data with spatial and spatio-temporal dirichlet process mixtures. *Biometrical Journal 49*(5), 1–14.

Kulldorff, M. and N. Nagarwalla (1995). Spatial disease clusters: detection and inference. *Statistics in Medicine 14*, 799–810.

Kunsch, H. (1987). Intrinsic autoregressions and related models on the two-dimensional lattice. *Biometrika 74*, 517–524.

References

Kuo, L. and B. Mallick (1998). Variable selection for regression models. *Sankhya 60*, 65–81.

Lachin, J. L. (2000). Statistical considerations in the intent-to-treat principle. *Controlled Clinical Trials 21*, 167–189.

Lambert, D. (1992). Zero-inflated poisson regression, with application to defects in manufacturing. *Technometrics 34*, 1–14.

Lambert, P., A. Sutton, P. Burton, K. Abrams, and P. Jones (2005). How vague is vague? A simulation study of the impact of the use of vague prior distributions in mcmc using winbugs. *Statistics in Medicine 24*, 2401–2428.

Lawson, A. B. (1992a). GLIM and normalising constant models in spatial and directional data analysis. *Computational Statistics and Data Analysis 13*, 331–348.

Lawson, A. B. (1992b). On fitting non-stationary markov point processes on glim. In Y. Dodge and J. Whittacker (Eds.), *Computational Statistics I*. Physica Verlag.

Lawson, A. B. (1993a). A deviance residual for heterogeneous spatial Poisson processes. *Biometrics 49*, 889–897.

Lawson, A. B. (1993b). On the analysis of mortality events around a prespecified fixed point. *Journal of the Royal Statistical Society A 156*, 363–377.

Lawson, A. B. (1995). Markov chain Monte Carlo methods for putative pollution source problems in environmental epidemiology. *Statistics in Medicine 14*, 2473–2486.

Lawson, A. B. (1996). Use of deprivation indices in small area studies: letter. *Journal of Epidemiology and Community Health 50*, 689–690.

Lawson, A. B. (2000). Cluster modelling of disease incidence via rjmcmc methods: a comparative evaluation. *Statistics in Medicine 19*, 2361–2376.

Lawson, A. B. (2001). *Statistical Methods in Spatial Epidemiology*. New York: Wiley.

Lawson, A. B. (2002). The analysis of putative sources of health hazard. In D. Zimmerman (Ed.), *Encyclopedia of Life Support Systems*. UNESCO.

Lawson, A. B. (2004). Some issues in the spatio-temporal analysis of public health surveillance data. In R. Brookmeyer and D. Stroup (Eds.), *Monitoring the Health of Populations: Statistical Principles and Methods for Public Health Surveillance*, Chapter 11, pp. 289–314. New York: Oxford University Press.

Lawson, A. B. (2006a). Disease cluster detection: a critique and a Bayesian proposal. *Statistics in Medicine 25*, 897–916.

Lawson, A. B. (2006b). *Statistical Methods in Spatial Epidemiology* (2nd ed.). New York: Wiley.

Lawson, A. B. and S. Banerjee (2008). Bayesian spatial analysis. In S. Fotheringham and P. Rogerson (Eds.), *Handbook of Spatial Analysis*, Chapter 9. New York: Sage.

Lawson, A. B., A. Biggeri, D. Boehning, E. Lesaffre, J.-F. Viel, A. Clark, P. Schlattmann, and F. Divino (2000). Disease mapping models: An empirical evaluation. *Statistics in Medicine 19*, 2217–2242. special issue: Disease Mapping with emphasis on evaluation of methods.

Lawson, A. B., A. Biggeri, and E. Dreassi (1999). Edge effects in disease mapping. In A. B. Lawson, D. B̈ohning, E. Lesaffre, A. Biggeri, J.-F. Viel, and R. Bertollini (Eds.), *Disease Mapping and Risk assessment for Public Health*, Chapter 6, 85–97. New York: Wiley.

Lawson, A. B., D. Böhning, E. Lessafre, A. Biggeri, J. F. Viel, and R. Bertollini (Eds.) (1999). *Disease Mapping and Risk Assessment for Public Health*. Chichester: Wiley.

Lawson, A. B., W. J. Browne, and C. L. Vidal-Rodiero (2003). *Disease Mapping with WinBUGS and MLwiN*. New York: Wiley.

Lawson, A. B. and A. Clark (1999a). Markov chain Monte Carlo methods for clustering in case event and count data in spatial epidemiology. In M. E. Halloran and D. Berry (Eds.), *Statistics and Epidemiology: Environment and Clinical Trials*, pp. 193–218. New York: Springer Verlag.

Lawson, A. B. and A. Clark (1999b). Markov chain Monte Carlo methods for putative sources of hazard and general clustering. In A. B. Lawson, D. Böhning, E. Lesaffre, A. Biggeri, J.-F. Viel, and R. Bertollini (Eds.), *Disease Mapping and Risk Assessment for Public Health*, Chapter 9, 119–141. Wiley.

Lawson, A. B. and A. Clark (2002). Spatial mixture relative risk models applied to disease mapping. *Statitsics in Medicine 21*, 359–370.

Lawson, A. B. and D. Denison (2002). Spatial cluster modelling: an overview. In A. B. Lawson and D. Denison (Eds.), *Spatial Cluster Modelling*, Chapter 1, pp. 1–19. New York: CRC press.

Lawson, A. B. and P. Leimich (2000). Approaches to space-time modelling of infectious disease behaviour. *IMA journal of Mathematics applied to Medicine and Biology 17*, 1–13.

Lawson, A. B. and J.-F. Viel (1995). Tests for directional space-time interaction in epidemiological data. *Statistics in Medicine 14*, 2383–2392.

Lawson, A. B. and F. Williams (2000). Spatial competing risk models in disease mapping. *Statistics in Medicine 19*, 2451–2468.

Lawson, A. B. and F. L. R. Williams (2001). *An Introductory Guide to Disease Mapping.* New York: Wiley.

Lawson, A. B. and H. Zhou (2005). Spatial statistical modeling of disease outbreaks: with particular reference to the UK FMD epidemic of 2001. *Preventive Veterinary Medicine 71*, 141–156.

Leonard, T. and H. Hsu (1999). *Bayesian Methods.* London: Cambridge University Press.

Leyland, A. and H. Goldstein (Eds.) (2001). *Multilevel Modelling in Health Statistics.* London: Wiley.

Lindsay, B. G. (1988). Composite likelihood methods. *Contemporary Mathematics 80*, 221–238.

Little, R. and D. Rubin (2002). *Statistical Analysis with Missing Data.* New York: Wiley.

Liu, J. S. (2001). *Monte Carlo Strategies in Scientific Computing.* New York: Springer.

Liu, X., M. Wall, and J. Hodges (2005). Generalised spatial structural equation models. *Biostatistics 6*, 539–557.

Louie, M. M. and E. D. Kolaczyk (2006). Multiscale detction of localised anomalous structure in aggregate disease incidence data. *Statistics in Medicine 25*, 787–810.

Lu, H. and B. Carlin (2005). Bayesian areal wombling for geographical boundary analysis. *Geographical Analysis 37*, 265–285.

Ma., B., A. B. Lawson, and Y. Liu (2007). Evaluation of Bayesian models for focused clustering in health data. *Environmetrics 18*, 1–16.

Ma, H., B. Carlin, and S. Banerjee (2006). Hierarchical and joint site edge methods for medicare hospice service region boudary analysis. *submitted*.

Macnab, Y. (2007). Spline smoothing in Bayesian disease mapping. *Environmetrics 18*, 727–744.

Maheswaran, R. and M. Craglia (Eds.) (2004). *GIS in Public Health Practice.* New York: CRC press.

Maindonald, J. and J. Braun (2003). *Data Analysis and Graphics Using R.* New York: Cambridge University Press.

Marin, J.-M. and C. Robert (2007). *Bayesian Core: A Practical Approach to Computational Bayesian Statistics.* New York: Springer.

McCullagh, P. and J. Nelder (1989). *Generalised Linear Models* (2nd ed.). London: Chapman & Hall.

McKeague, I. and M. Loiseaux (2002). Perfect sampling for point process cluster modelling. In A. B. Lawson and D. Denison (Eds.), *Spatial Cluster Modelling*, Chapter 5. New York: CRC Press.

Mira, A., J. Moller, and G. Roberts (2001). Perfect slice samplers. *Journal of the Royal Statistical Society B 63*, 583–606.

Molenberghs, G. and G. Verbeke (2005). *Models for Discrete Longitudinal Data*. New York: Springer.

Móller, J. and O. Skare (2001). Coloured voronoi tessselations for Bayesian image analysis and reservoir modelling. *Statistical Modelling 1*, 213–232.

Møller, J., A. Syversveen, and R. P. Waagepetersen (1998). Log Gaussian Cox processes. *Scandinavian Journal of Statistics 25*, 451–482.

Móller, J. and R. Waagepetersen (1998). Markov connected component fields. *Advances in Applied Probability 30*, 1–35.

Móller, J. and R. Waagpetersen (2004). *Statistical Inference and Simulation for Spatial Point Processes*. New York: CRC Press.

Morgan, O., M. Vreiheid, and H. Dolk (2004). Risk of low birth weight near eurohazcon hazardous waste landfill sites in england. *Archives of Environmental Health 59*, 149–151.

Morton, A. and B. Finkenstädt (2005). Discrete time modelling of disease incidence time series by using Markov chain Monte Carlo methods. *Journal of the Royal Statistical Society C 54*, 575–594.

Mugglin, A., N. Cressie, and I. Gemmel (2002). Hierarchical statistical modelling of influenza epidemic dynamics in space and time. *Statitsics in Medicine 21*, 2703–2721.

Neal, P. and G. Roberts (2004). Statistical inference and model selection for the 1861 hagelloch measles epidemic. *Biostatistics 5*, 249–261.

Neal, R. M. (2003). Slice sampling. *Annals of Statistics 31*, 1–34.

Nott, D. J. and T. Rydén (1999). Pairwise likelihood methods for inference in image models. *Biometrika 86*, 661–676.

Oden, N., R. Sokal, M. Fortin, and H. Goebl (1993). Categorical wombling: Detecting regions of significant change in spatially located categorical variables. *Geographical Analysis 25*(4).

O'Loughlin, J. (2002). The Electoral Geography of Weimar Germany: Exploratory Spatial Data Analyses (ESDA) of Protestant Support for the Nazi Party. *Political Analysis 10*(3), 217–243.

Pfeiffer, D. U., T. P. Robinson, M. Stevenson, K. B. Stevens, D. J. Rogers, and A. C. A. Clements (Eds.) (2008). *Spatial Analysis in Epidemiology*. New York: Springer.

References

Plummer, M. and D. G. Clayton (1996). Estimation of population exposure in ecological studies. *Journal of the Royal Statistical Society B 58*, 113–126.

Propp, J. and D. Wilson (1996). Exact sampling with coupled markov chains. *Random Structures and Algortithms 9*, 223–252.

Rao, J. K. N. (2003). *Small Area Estimation.* New York: Wiley.

Richardson, S., A. Thomson, N. Best, and P. Elliott (2004). Interpreting posterior relative risk estimates in disease mapping studies. *Environmental Health Perspectives 112*, 1016–1025.

Ripley, B. D. (1981). *Spatial Statistics.* New York: Wiley.

Ripley, B. D. (1987). *Stochastic Simulation.* New York: Wiley.

Ripley, B. D. (1988). *Statistical Inference for Spatial Processes.* New York: Cambridge University Press.

Robert, C. and G. Casella (2005). *Monte Carlo Statistical Methods* (2nd ed.). New York: Springer.

Royle, A. and M. Berliner (1999). A hierarchical approach to multivariate spatial modeling and prediction. *Journal of Agricultural, Biological and Environmental Statistics 4*, 29–56.

Rue, H. and L. Held (2005). *Gaussian Markov Random Fields: Theory and Applications.* New York: Chapman & Hall/CRC.

Salway, R. and J. Wakefield (2005). Sources of bias in ecological studies of non-rare events. *Environmental and Ecological Statistics 12*, 321–347.

Schlattman, P. and D. Böhning (1993). Mixture models and disease mapping. *Statistics in Medicine 12*, 1943–1950.

Smith, A. F. M. and A. E. Gelfand (1992). Bayesian statistics without tears: A sampling-resampling perspective. *American Statistician 46*, 84–88.

Smith, A. F. M. and G. Roberts (1993). Bayesian computation via the Gibbs Sampler and related Markov chain Monte Carlo methods. *Journal of the Royal Statistical Society B 55*, 3–23.

Snow, J. (1854). *On the Mode of Communication of Cholera* (2nd ed.). London: Churchill Livingstone.

Spiegelhalter, D., A. Thomas, N. Best, and D. Lunn (2007). *WinBUGS manual.* Cambridge,UK: MRC Biostatistics Unit. version 1.4.3.

Spiegelhalter, D. J., N. G. Best, B. P. Carlin, and A. van der Linde (2002). Bayesian deviance, the effective number of parameters and the comparison of arbitrarily complex models. *Journal of the Royal Statistical Society B 64*, 583–640.

Spiegelhalter, D. J., N. G. Best, W. R. Gilks, and H. Inskip (1996). Hepatitis B: a case study in MCMC methods. In W. R. Gilks, S. Richardson, and D. J. Spiegelhalter (Eds.), *Markov Chain Monte Carlo in Practice*. London: Chapman & Hall.

Steele, F., H. Goldstein, and W. Browne (2004). A general multilevel multistate competing risks model for event history data, with an application to a study of contraceptive use dynamics. *Statistical Modelling 4*, 145–159.

Stern, H. and Y. Jeon (2004). Applying structural equation models with incomplete data. In A. Gelman and X.-L. Meng (Eds.), *Applied Bayesain Modeling and Causal Inference from Incomplete-Data Perspectives*, Chapter 30, 331–342. London: Wiley.

Stern, H. S. and N. A. C. Cressie (1999). Inference for extremes in disease mapping. In A. B. Lawson, A. Biggeri, D. Boehning, E. Lesaffre, J. F. Viel, and R. Bertollini (Eds.), *Disease Mapping and Risk Assessment for Public Health*, Chapter 5. New York: Wiley.

Stern, H. S. and N. A. C. Cressie (2000). Posterior predictive model checks for disease mapping models. *Statistics in Medicine 19*, 2377–2397.

Stevenson, M., R. Morris, A. B. Lawson, J. Wilesmith, J. M. Ryan, and R. Jackson (2005). Area level risks for BSE in British cattle before and after the July 1988 meat and bone meal feed ban. *Preventive Veterinary Medicine 69*, 129–144.

Stevenson, M. A., R. S. Morris, A. B. Lawson, J. W. Wilesmith, J. B. M. Ryan, and R. A. Jackson (2001). Spatial regression analysis of the bovine spongiform encephalopathy epidemic in Great Britain. In A. C. Gatrell and P. Durr (Eds.), *The GISVET conference- Application of Geographical Information Systems and Spatil Analysis to Veterinary Science*, Lancaster.

Stevenson, M. A., J. W. Wilesmith, J. B. M. Ryan, R. S. Morris, A. B. Lawson, D. U. Pfeiffer, and D. Lin (2000). Descriptive spatial analysis of the epidemic of bovine spongiform encephalopathy in Great Britain to June 1997. *Veterinary Record 147*, 379–384.

Sturtz, S., U. Ligges, and A. Gelman (2005). R2WinBUGS: A package for running WinBUGS from R. *Journal of Statistical Software 12*, 1–16.

Sun, D., R. Tsutakawa, H. Kim, and Z. He (2000). Spatio-temporal interaction with disease mapping. *Statistics in Medicine 19*, 2015–2035.

Sutton, S. (2005). *The modeling of spatially-referenced recurrent event data in a South Carolina Population*. PhD thesis, University of South Carolina.

Tanner, M. A. (1996). *Tools for Statistical Inference* (3rd ed.). New York: Springer Verlag.

References

Thomas, A., N. Best, D. Lunn, R. Arnold, and D. Spiegelhalter (2004). Geobugs user manual v 1.2. www.mrc-bsu.cam.ac.uk/bugs.

Tibshirani, R. and T. Hastie (1987). Local likelihood estimation. *Journal of the American Statistical Association 82*, 559–568.

Tierney, L. and J. Kadane (1986). Accurate approximations for posterior moments and marginal densities. *Journal of the American Statistical Association 81*, 82–86.

van Lieshout, M. and A. Baddeley (2002). Extrapolating and interpolating spatial patterns. In A. B. Lawson and D. Denison (Eds.), *Spatial Cluster Modelling*, Chapter 4, 61–86. New York: CRC Press.

Varin, C., G. Høst, and Ø. Skare (2005). Pairwise likelihood infernce in spatial generalized linear mixed models. *Computational Statistics and Data Analysis 49*, 1173–1191.

Verbeke, G. and G. Molenberghs (2000). *Linear Mixed Models for Longitudinal Data*. New York: Springer.

Vidal-Rodiero, C. L. and A. B. Lawson (2005). An evaluation of edge effects in disease mapping. *Computational Statistics and Data Analysis 49*, 45–62.

Voss, D. (2004). Using ecological inference for contextual research. In G. King, O. Rosen, and M. Tanner (Eds.), *Ecological Inference: New Methodological Strategies*, Chapter 3. New York: Cambridge University Press.

Wackernagel, H. (2003). *Multivariate Geostatistics* (3rd ed.). New York: Springer.

Wakefield, J. and G. Shaddick (2006). Health-exposure modeling and the ecological fallacy. *Biostatistics 7*, 438–455.

Wakefield, J. C. and S. Morris (2001). The Bayesian modeling of disease risk in relation to a point source. *Journal of the American Statistical Association 96*, 77–91.

Waller, L. and C. Gotway (2004). *Applied Spatial Statistics for Public Health Data*. New York: Wiley.

Waller, L. A., B. P. Carlin, H. Xia, and A. E. Gelfand (1997). Hierarchical spatio-temporal mapping of disease rates. *Journal of the American Statistical Association 92*, 607–617.

Wang, F. and M. Wall (2003). Generalized common spatial factor model. *Biostatitsics 4*, 569–582.

Wikle, C. K. (2002). Spatial modelling of count data: A case study in modelling breeding bird survey data on large spatial domains. In A. B. Lawson and D. G. T. Denison (Eds.), *Spatial Cluster Modelling*, Chapter 11. London: CRC Press.

Williams, F., A. Lawson, and O. Lloyd (1992). Low sex ratios of births in areas at risk from air pollution from incinerators, as shown by geographical analysis and 3-dimensional mapping. *International Journal of Epidemiology 21*, 311–319.

Wolpert, R. L. and K. Ickstadt (1998). Poisson/gamma random field models for spatial statistics. *Biometrika 85*, 251–267.

Xia, H. and B. P. Carlin (1998). Spatio-temporal models with errors in covariates: Mapping Ohio lung cancer mortality. *Statistics in Medicine 17*, 2025–2043.

Yu, B. and P. Mykland (1998). Looking at Markov samplers through cusum path plots: A simple diagnostic idea. *Statistics and Computing 8*, 275–286.

Zhang, S., D. Sun, C. He, and M. Schootman (2006). A Bayesian semiparametric model for colorectal cancer incidences. *Statistics in Medicine 25*, 285–309.

Zhou, H., A. B. Lawson, J. Hebert, E. Slate, and E. Hill (2007). A Bayesian hierarchical modeling approach for studying the factors affecting the stage at diagnosis of prostate cancer. *Statistics in Medicine 27*(9). DOI 10.1002/sim:3024.

Zhou, H., A. B. Lawson, J. Hebert, E. Slate, and E. Hill (2008). Joint spatial survival modelling for the date of diagnosis and the vital outcome for prostate cancer. *Statistics in Medicine*. to appear.

Zhu, L., B. Carlin, and A. Gelfand (2003). Hierarchical regression with misaligned spatial data: relating ambient ozone and pediatric asthma er visits in Atlanta. *Environmetrics 14*, 537–557.

Zhu, L. and B. P. Carlin (2000). Comparing hierarchical models for spatio-temporally misaligned data using the DIC criterion. *Statistics in Medicine 19*, 2265–2278.

Zia, H., B. P. Carlin, and L. A. Waller (1997). Hierarchical models for mapping Ohio lung cancer rates. *Environmetrics 8*, 107–120.

Index

accelerated failure time models
    AFT, 229
adaptive rejection sampling
    ARS, 40
adjacent category logits model, 107
air pollution source, 165
Akaike information criterion
    AIC, 55, 56
Auto models
    binomial, 100
    Logistic
        space–time, 266
    logistic, 21, 96, 126
    Poisson, 21
autocorrelation
    residual, 60
    spatial, 20, 94
avian influenza, 276

baseline category logits model, 107
Bayes estimate, 29
Bayes Factor, 56
Bayesian information criterion
    BIC, 56
Bayesian linear model, 79
Bayesian structural equation model, 163
Berkson error, 162, 174, 195–197
Berman-Turner integration scheme, 74, 89
Bernoulli model, 266
Beta mixtures, 232
bias
    confounding, 159
    contextual effect, 159
    effect modification, 159
    measurement error, 159

binomial distribution, 274, 279
binomial model, 128, 260, 262
block effects, 194
block updating, 40
Brooks-Gelman-Rubin statistic
    BGR, 45, 298, 305
BRugs, 302
BSE epidemic, 276

CAR-Voronoi model, 195
censoring, 229
    right, 230
center rate
    parameter, 135
    posterior marginal, 138
cholesky decomposition, 220
cluster centers, 133
cluster detection, 119, 130
cluster models, 133
    Bayesian, 134
cluster process
    Cox, 133
    Neyman-Scott, 133
    Poisson, 133
clustering
    general, 133
    hot spot, 120, 130
    residual, 121
CODA package, 41, 283
comorbidity, 251
competing risk
    model, 202, 248, 278
Conditional autoregressive models
    improper
        ICAR, 86, 108, 200
    proper
        PCAR, 88

339

conditional logistic model, 75, 125, 177
conditional mark models, 245
conditional predictive ordinates, 63, 68
confounder effect, 171, 181
conjugacy, 25, 82
    conditional, 25
    gamma-gamma, 34
contextual effects, 162–164, 181, 186, 231, 248, 249
control disease, 167, 168, 211
convergence, 40
    monitoring Metropolis-like samplers, 46
    multi-chain methods, 44
    single chain methods, 41
convex hull, 143
convolution models, 91, 96, 146, 156, 157, 206, 239, 267, 276
correlation
    cross-, 204, 216, 219
        multivariate, 219
    spatial, 20, 80
correlation function
    exponential, 192
    Matérn, 90, 231
    power exponential, 90, 196, 231
    separable, 256
count accumulation, 247
Covariance models
    fully specified, 90
covariate acceleration, 229
credible interval, 36
cross-covariances, 219, 221
cusum plot, 42

data augmentation, 106
decoupling approximation, 172
Delauney triangulation, 49
dependence
    nonseparable, 178, 180
deprivation indices, 163, 171

deviance, 41, 55
Deviance information criterion DIC, 56, 57, 67, 92, 98
directed acyclic graph DAG, 28, 33, 34
directional covariates, 78
directional effect, 166, 170, 178, 180
Dirichlet distribution, 110
Dirichlet process, 244
Dirichlet process mixture models DP, 111, 139, 271
Dirichlet tesselation, 49, 60, 89, 123, 139, 171, 195, 230, 257
dispersal models, 170
distance covariates, 78
distance effect, 166, 180
Distance sampling, 281
dropout, 241

ecological analysis, 151
ecological bias, 158, 181, 185, 187, 194
edge detection, 149
    derivative-based methods, 150
edge effects
    guard areas, 114
    misalignment, 200
    partition models, 145
    space-time, 115
    spatial, 111, 124, 171
edge weighting, 113
empirical Bayes method, 117
endpoint distribution
    extreme value, 228
    gamma, 228
    log-normal, 228
    Weibull, 228, 231, 279
epidemic curve, 279
epidemic period, 274
event history analysis, 243
exceedence probability, 55, 64, 69, 123, 130, 181, 264
expected predictive deviance

Index

EPD, 57
exposure model, 169
    putative hazard, 169

Fernald example, 173
foot-and-mouth epidemic, 276
    FMD, 273
frailty
    correlated, 230–232
    individual, 241, 248, 251
    uncorrelated, 230, 236

Gamma process, 232, 244
Gaussian convolution, 178
Gaussian process
    Model, 101, 192
    multivariate, 219, 220
    Spatial, 20, 85, 89, 133, 157, 168, 178, 192, 196, 204, 220, 230
    spatio-temporal, 256
    zero mean, 90
general linear mixed model, 237
generalised linear spatial modelling, 90
GeoBUGS, 294
geoR, 196
geoRglm, 46, 90
Geweke statistic, 41
Gibbs Sampler, 37, 136, 259
Gibbs sampler, 87
Gibbs updates, 38
goodness-of-fit
    diagnostics, 55, 92
    measures, 55, 62

hazard
    baseline, 232
    cumulative, 232
hazard function, 228
hidden process models, 143
hierarchical model, 27
hull peeling, 112
hybrid-additive model, 170
hyperprior distribution, 28, 82, 83

identifiability, 23, 180, 207, 268, 269
independence
    conditional, 20
infectious disease models, 261
influenza models, 273, 275
intention-to-treat
    ITT, 242
interaction effect
    nonseparable, 261
    random walk Type II, 261, 264
    space-time, 178, 180, 255, 259
    Type III, 261
    Type IV, 261
interpolation, 191
intervention, 281
intraclass correlation, 91

Kaplan-Meyer curves, 233
Kriging
    Bayesian, 192, 194, 196
    block, 193, 194

L1 norm model, 108
Laplace approximation, 50, 52
larynx cancer, 165
Latent structure models, 108, 268
likelihood
    approximation, 48
    function, 19
    local, 141, 146
    marginal, 30
    maximum, 52
    models, 19
    pairwise, 49
    partial, 272
linear model of coregionalisation, 219, 221
linked component model, 204
log Gaussian Cox process, 133, 157, 160
    LGCP, 89
log-normal model, 80, 84
logistic-normal model, 84

marked process, 244

Markov chain Monte Carlo, 56
    McMC, 36
        spatial birth-death, 136
        subsampling and thinning, 45
Markov connected component models, 148
Markov process, 135
Markov random field model, 220
    Gaussian, 273
    MRF, 87
maximum entropy approach, 161
$MCAR(B, \Sigma)$, 221
mean square predictive error
    MSPE, 58, 268
measurement error, 161, 171, 174, 181, 196
Metropolis algorithm, 37
Metropolis updates, 38
Metropolis–Hastings algorithm, 37, 39, 43, 46
Metropolis–Hastings updates, 38
Metropolis–Langevin sampler, 46, 90
Metropolized Carlin–Chib algorithm, 136, 140
misaligned data problem
    MIDP, 190
misalignment
    outcome, 198
    predictor, 191
    space-time, 198, 261
misclassification error, 158, 161, 171
missing data, 241
mixture intensity, 134
mixture models, 108, 110, 146, 269
MLwiN, 37
modifiable areal unit problem
    MAUP, 159, 185
Monte Carlo estimator, 36
Monte Carlo integration, 193
Moran's I statistic, 60, 94
Multilevel modeling, 186
multinomial distribution, 216, 217, 249

multinomial logit model, 211, 212
multiple category data
    nominal, 107
    ordinal, 107
multiple event types, 244
Multiscale analysis, 185, 187
multivariate CAR models
    $MCAR(B, \Sigma)$, 220
    MVCAR, 220
Multivariate normal distribution, 21, 79, 85, 173, 249
Multivariate spatial correlation, 219
mv.car, 221

negative binomial distribution, 26, 82
noninteger Poisson data, 199
normal approximation, 52, 53

overdispersion, 80

parametric bootstrap, 59
partition models, 139, 144
Pearson correlation coefficient, 43
perfect sampling, 47
point process
    marked, 278
Poisson process model
    heterogeneous, 73, 75, 123, 176, 202, 243, 245
Poisson–gamma model, 27, 35, 117
Poisson-gamma model, 26, 80
Poisson-log-normal model, 84
posterior
    appproximation, 48
    density
        marginal, 36
        functional, 36
        mean, 35
posterior
    variance, 35
posterior distribution, 24
    proper, 25
posterior expected mixing field, 108

Index

posterior predicted survival, 233
Posterior predictive loss, 57, 92
posterior sampling, 35
predictive capability, 281
predictive distribution, 26, 62, 82
    posterior, 26, 34
    prior, 26, 27
principal component decomposition, 269
Prior distribution
    auto-regressive, 180
    Bernoulli, 141
    beta, 33
    Gamma, 80
    gamma, 23
    improper, 22
    Inverse gamma, 80
    inverse gamma, 23
    inverse Wishart, 222
    Jeffrey's, 22
    Laplace, 23
    multivariate normal
        MVN, 230, 232, 260
    multivariate normal
        MVN, 90
    noninformative, 22
    Potts, 146
    proper, 22
    random walk, 236, 238, 248, 251, 260–262, 269, 279
    vague, 22
    Wishart, 222, 270
    zero-mean Gaussian, 23, 31, 34, 79, 87, 132, 154, 169, 170, 196, 216, 232, 236, 238, 248, 251, 260
prior distribution
    zero-mean Gaussian, 80
process mixture, 105
proportional hazard models, 228
    Cox, 273
proportional mortality model, 205
proportional odds model, 107
proportionality of hazard, 229
propriety, 22

proprotional hazard models
    Cox, 232
pseudolikelihood, 48, 96, 100, 266
putative hazard models, 165, 210
putative health hazards, 78, 123, 136
    spatio-temporal, 176
putative source analysis, 165

quantile-quantile plots, 43

R, 37, 283
R2WinBUGS, 57
R2WinBUGS, 126, 298
random effect
    correlated
        CH, 84, 164, 171, 200, 260
    uncorrelated
        UH, 80, 97, 154, 171, 260
random field
    Gaussian, 195, 256
recursive Bayesian updating, 281
relative risk, 77
relative risk map, 82
repeated event analysis, 243
residual
    Bayesian, 61, 126, 128, 181
    deviance, 60, 69
    predictive, 62, 128, 181
residuals, 55, 59
reversible jump samplers, 47, 136, 140
rgamma, 35, 82

SatScan, 138
scale labeling, 186
sex ratios of births, 76
shared component model, 205, 207
singular information methods, 102
space–time infection model, 275
spatial factor analysis, 110
spatial trend, 78
spatially-contextual, 231
spatiotemporal interaction, 178
Spline models, 101, 171, 198

low rank, 154
low rank Kriging, 102
standardisation, 77
stickieness, 47
sticky sampler, 43
Strauss distribution, 135, 136
structural equation models , 111
survival function, 228
susceptible-infected-removed models
    SIR, 274
swine fever, 279

temporally weighted regression, 279
time to endpoint, 228
transmission distribution, 275
trend component, 279

underascertainment, 274, 281
unobserved confounding, 163

variable selection, 269
vector autoregressive model, 273
Von Mises distribution, 178

WinBUGS, 37, 290
within area exposure distribution, 159
Wombling, 149

Zip regression model, 102, 107
zoonosis, 276